Real World Multicore Embedded Systems

Expert Guide

Real World Multicore Embedded Systems

Embedded Systems

A Practical Approach

Expert Guide

Bryon Moyer

AMSTERDAM • BOSTON • HEIDELBERG • LONDON
NEW YORK • OXFORD • PARIS • SAN DIEGO
SAN FRANCISCO • SINGAPORE • SYDNEY • TOKYO
Newnes is an imprint of Elsevier

Newnes is an imprint of Elsevier
The Boulevard, Langford Lane, Kidlington, Oxford, OX5 1GB, UK
225 Wyman Street, Waltham, MA 02451, USA

First published 2013

British Library Cataloguing-in-Publication Data
A catalogue record for this book is available from the British Library

Library of Congress Cataloging-in-Publication Data
A catalog record for this book is available from the Library of Congress

ISBN: 978-0-12-416018-7

For information on all Newnes publications visit our
web site at books.elsevier.com

Typeset by MPS Limited, Chennai, India
www.adi-mps.com

Contents

About the Editor

Bryon Moyer

Technology Writer and Editor, EE Journal
Email: bryon@moyerwriting.com

Bryon Moyer is a technology writer and an editor/writer for EE Journal. He has over 30 years' experience as an engineer and marketer in Silicon Valley, having worked for MMI, AMD, Cypress, Altera, Actel, Teja Technologies, and Vector Fabrics. He has focused on PLDs/FPGAs, EDA, embedded systems, multicore processing, networking protocols, and software analysis. He has a BSEE from UC Berkeley and an MSEE from Santa Clara University.

About the Authors

Jim Holt

Systems Architect for Freescale's Networking and Multimedia group, Austin, TX, USA; E-mail: RWBL70@freescale.com

Dr. Jim Holt is a Systems Architect for Freescale's Networking and Multimedia group, and is also a Research Affiliate at the Massachusetts Institute of Technology where he is investigating self-aware 1000-core chips of the future. He has 29 years of industry experience focused on microprocessors, multicore systems, software engineering, distributed systems, design verification, and design optimization. He earned a Ph.D. in Electrical and Computer Engineering from the University of Texas at Austin, and an MS in Computer Science from Texas State University-San Marcos.

John Carbone

E-mail: jcarbone@expresslogic.com

John A. Carbone, vice president of marketing for Express Logic, has 35 years experience in real-time computer systems and software, ranging from embedded system developer and FAE to vice president of sales and marketing.
Mr. Carbone's experience includes embedded computers, array processors, attached processors, development tools, and both commercial and proprietary real-time operating systems. Mr. Carbone has a BA degree in mathematics from Boston College.

William Lamie

E-mail: blamie@expresslogic.com

William E. Lamie is co-founder and CEO of Express Logic, Inc., and is the author of the ThreadX® RTOS. Prior to founding Express Logic, Mr. Lamie was the author of the Nucleus™ RTOS and co-founder of Accelerated Technology, Inc. Mr. Lamie has over 25 years' experience in embedded systems development, over 20 of which are in the development of real-time operating systems for embedded applications. He has a BS in Computer Science from San Diego State University.

Max Domeika

Technical Project Manager, Intel Corporation, Intel Corporation, Hillsboro, OR, USA; E-mail: max.j.domeika@intel.com

Max Domeika is a technical product manager at Intel Corporation working on HTML5. Over the past 16 years, Max has held several positions in compiler and software tools development. Max earned a BS in Computer Science from the University of Puget Sound, an MS in Computer Science from Clemson University, and a MS in Management in Science & Technology from Oregon Graduate Institute. Max is the author of Software Development for Embedded Multi-core Systems from Elsevier and Break Away with Intel® AtomTM Processors from Intel Press.

Kenn Luecke

E-mail: kenn.r.luecke@boeing.com

Kenn Luecke currently works in Boeing Test & Evaluation in the Vehicle Management Subsystem Labs. He has 23 years of experience in network systems, avionics system, multicore and parallel embedded software development, mission-critical computing, and security. He has a Bachelor's degree in Computer Science from the University of Missouri-St. Louis and a Masters of Business Administration from Washington University in St. Louis.

Neal Stollon

Principal Engineer with HDL Dynamics, Dallas, TX, USA; E-mail: neals@hdldynamics.com

Dr. Neal Stollon is Principal Engineer with HDL Dynamics, which provides on-chip digital IP consulting, integration, and analysis services for SoC design. Dr. Stollon is chairman of the Nexus 5001 Forum, a consortium developing standards based on-chip instrument solutions, and is also co-chair of the OCP-IP Debug Working Group, which also addresses SoC debug requirements. Dr. Stollon has a Ph.D in EE from Southern Methodist University and is author of the book "On-Chip Instrumentation".

Yosinori Watanabe

Senior Architect, Cadence Design Systems, Berkeley, CA, USA;
E-mail: watanabe@cadence.com

Dr. Yosinori Watanabe is a Senior Architect at Cadence Design Systems. He has 21 years of experience in microprocessor design, embedded system design, logic and behavioral synthesis, and the design and verification of embedded systems and software. He has a Ph.D. in Electrical Engineering and Computer Sciences from the University of California at Berkeley.

Paul Stravers

Chief Architect and Co-founder of Vector Fabrics

Dr. Paul Stravers is chief architect and co-founder of Vector Fabrics. He has 24 years of experience in academics and industry in the fields of computer architecture, design automation and embedded system design. He holds a Ph.D. in electrical engineering from Delft University of Technology.

Tom Dickens

Boeing Associate Technical Fellow at Boeing, Renton, WA, USA; E-mail: thomas.p.dickens@boeing.com

Tom Dickens is a Boeing Associate Technical Fellow at Boeing, with 28 years of experience building software systems for engineers (C, C++, Java) and designing, building, and programing embedded systems (assembly code). He also has over 20 years teaching at the university level. He has a BS in Electronics Engineering from Henry Cogswell College and an MS in Computer Engineering from NTU.

David Kleidermacher

Chief Technology Officer, Green Hills Software, Inc., Santa Barbara, CA, USA;
E-mail: davek@ghs.com

David Kleidermacher is Chief Technology Officer at Green Hills Software, where he is responsible for technology strategy, platform planning, and solutions design. Kleidermacher has spent 21 years working in the areas of systems software and security, including secure operating systems, virtualization technology, and high robustness infrastructure. Kleidermacher earned his bachelor of science in Computer Science from Cornell University.

Gitu Jain

Software Engineer, Synopsys; E-mail: gitujain@yahoo.com

Dr. Gitu Jain is a Software Engineer at Synopsys and also teaches at the UC Santa Cruz Extension in Silicon Valley. She has 20 years of experience in software R&D at semiconductor companies, with expertise in parallel computing and EDA algorithm design. She has a Ph.D. in Electrical and Computer Engineering from the University of Iowa.

Sanjay Lal

Cofounder of Kyma Systems, Danville, CA, USA; E-mail: sanjayl@kymasys.com

Sanjay Lal is cofounder of Kyma Systems, an engineering consulting firm. He has 15 years of experience, with specific expertise in the areas of operating systems, hypervisors, bare-metal data-path programming, computer networks, and processor architectures. He has a Bachelors and Masters of Applied Science from the University of Toronto.

Sanjay R. Deshpande

Design Manager, Freescale Semiconductor, Austin, TX, USA;
E-mail: R58977@freescale.com

Dr. Sanjay R. Deshpande is a Distinguished Member of the Technical Staff and a Design Manager at Freescale Semiconductor. He has over 30 years of experience in the areas of multicore interconnect technology, cache coherency, and I/O virtualization. He has a Bachelor of Technology degree from Indian Institute of Technology, Bombay, India, and an MS in Computer Science and a Ph.D. in Electrical and Computer Engineering, both from the University of Texas, Austin.

Frank Schirrmeister

Senior Director, System Development Suite at Cadence Design Systems, Santa Clara, CA, USA; E-mail: frank@schirrmeister.com

Frank Schirrmeister is Senior Director for Product Management of the System Development Suite at Cadence Design Systems. He has over 20 years of experience in IP and semiconductor design, embedded software development, hardware/software co-development and electronic design automation. He holds a MSEE (Dipl.-Ing.) from the Technical University of Berlin, Germany.

CHAPTER 1

Introduction and Roadmap

Bryon Moyer
Technology Writer and Editor, EE Journal

Chapter Outline

Multicore is here

Actually, multicore has been around for many years in the desktop and supercomputing arenas. But it has lagged in the mainstream embedded world; it is now here for embedded as well.

Real World Multicore Embedded Systems.
DOI: http://dx.doi.org/10.1016/B978-0-12-416018-7.00001-8

Up until recently, multicore within embedded has been restricted primarily to two fields: mobile (assuming it qualifies as an embedded system, a categorization that not everyone agrees with) and networking. There have been multiple computing cores in phones for some time now. However, each processor typically owned a particular job — baseband processing, graphics processing, applications processing, etc. — and did that job independently, with little or no interaction with other processing cores. So multicore really wasn't an issue then. That's changed now that application processors in smartphones have multiple cores: it's time to start treating them as full-on multicore systems.

Meanwhile, networking (or, more specifically, packet-processing) systems have used multicore for a long time, well before any tools were available to ease the multicore job. This has been a highly specialized niche, with rockstar programmers deeply imbued with the skills needed to extract the highest possible performance from their code. This has meant handcrafting for specific platforms and manual programming from scratch. This world is likely to retain its specialty designation because, even as multicore matures, the performance requirements of these systems require manual care.

For the rest of embedded, multiple cores have become an unavoidable reality. And multicore has not been enthusiastically embraced for one simple reason: it's hard. Or it feels hard. There's been a huge energy barrier to cross to feel competent in the multicore world.

Some parts of multicore truly are hard, but as it reaches the mainstream, many of the issues that you might have had to resolve yourself have already been taken care of. There are now tools and libraries and even changes to language standards that make embedded multicore programming less of a walk in the wild.

And that's where this book comes in. There have been people quietly working for years on solving and simplifying multicore issues for embedded systems. And let's be clear: what works for desktops is not at all acceptable for embedded systems, with their limitations on size, resources and power. It has taken extra work to make some of the multicore infrastructure relevant to embedded systems.

Some of the people involved in those processes or simply with experience in using multicore have contributed from their vast knowledge to help you understand multicore for embedded. Most importantly, the intent is that, by taking in the various topics we've covered, you'll cross over that energy barrier and be able to start doing the multicore work you need to do.

Scope

The term "embedded system" is broad and ill-defined. You probably know it when you see it, although community standards may vary. We won't try to define what is included; it's probably easier to say what isn't included:

— desktop-style general computing (although desktop computers are sometimes harnessed for use in embedded applications)
— high-performance computing (HPC), the realm of supercomputers and massively parallel scientific and financial computing.

Many of the concepts discussed actually apply to both of those realms, but we will restrict examples and specifics to the embedded space, and there will be topics (like MPI, for example) that we won't touch on.

Who should read this book?

This book is for anyone that will need to work on embedded multicore systems. That casts a wide net. It includes:

• Systems architects that are transitioning from single-core to multicore systems.
• Chip architects that have to implement sophisticated systems-on-chips (SoCs).
• Software programmers designing infrastructure and tools to support embedded multicore.
• Software programmers writing multicore applications.
• Software programmers taking sequential programs and rewriting them for multicore.
• Systems engineers trying to debug and optimize a multicore system.

This means that we deal with hardware, firmware/middleware, software, and tools. The one area we don't deal with is actual hardware circuit design. We may talk about the benefits of hardware queuing, for example, but we won't talk about how to design a hardware queue on a chip.

We have assumed that you are an accomplished engineer with respect to single-core embedded systems. So we're not going to go too far into the realm of the basic (although some review is helpful for context). For example, we won't describe how operating systems work in general — we assume you know that. We talk about those elements of operating systems that are specific to multicore.

Organization and roadmap

In order to give you the broad range of information that underpins multicore technology, we've divided the book into several sections. These have been ordered in a way that allows a linear reading from start to finish, but you can also dive into areas of particular interest directly if you have experience in the other areas. Each chapter is an independent "essay", although we've tried to avoid too much duplication, so you will find some references from chapter to chapter.

Concurrency

We start with a review of the concept that underlies everything that matters in this book: concurrency. The reason everything changes with multicore is that our old assumption that one thing happens before another no longer holds true. More than one thing can happen at the same time, meaning we get more work done more quickly, but we also open up a number of significant challenges that we haven't had to deal with before. Understanding concurrency in detail is important to making sense out of everything else that follows.

Architecture

The next section concedes the fact that hardware is expensive to design, therefore hardware platforms will be created upon which software will be written. It's nice to think that embedded systems are a perfect marriage between purpose-built hardware and software that have been co-designed, but, in reality, especially when designing an expensive SoC, the hardware must serve more than just one application.

High-level architecture

This chapter takes a broad view of multicore hardware architecture as it applies to embedded systems. Author Frank Schirrmeister focuses on the arrangement of processing cores, but he must necessarily include some discussion of memory and interconnect as well. The intent of this chapter is to help you understand either how to build a better architecture or how to select an architecture.

Memory architecture

Memory can be arranged in numerous different ways, each of which presents benefits and challenges. In this chapter, author Gitu Jain picks up on the more general description in the high-level architecture chapter and dives deeper into the implications not only of main memory, but also of cache memory and the various means by which multiple caches can be kept coherent.

Interconnect

Processors and memory are important, but the means by which they intercommunicate can have a dramatic impact on performance. It can also impact the cost of a chip to the extent that simply over-provisioning interconnect is not an option. Sanjay Deshpande provides a broad treatment of the considerations and trade-offs that apply when designing or choosing an interconnect scheme.

Infrastructure

Once the hardware is in place, a layer of services and abstraction is needed in order to shield applications from low-level details. This starts with something as basic and obvious as the operating system, but includes specialized libraries supporting things like synchronization.

Operating systems

Operating systems provide critical services and access to resources on computing platforms. But with embedded systems, the number of options for operating systems is far larger than it is for other domains. Some of those options can't deal with multicore; others can. Some are big and feature-rich; others are small or practically non-existent. And they each have different ways of controlling how they operate. So in this chapter, with the assistance of Bill Lamie and John Carbone, I discuss those elements of operating systems that impact their performance in multicore systems, including the ability of associated tools to assist with software-level debugging.

Virtualization

The increased need for security and robustness has made it necessary to implement varying levels of virtualization in embedded systems. Author David Kleidermacher describes the many different ways in which virtualization can be implemented, along with the benefits and drawbacks of each option.

Multicore-related libraries

The details of multicore implementation can be tedious and error-prone to put into place. A layer of libraries and middleware can abstract application programs away from those details, making them not only more robust and easier to write, but also more portable between systems that might have very different low-level features. Author Max Domeika takes us on a tour of those multicore-related resources that are available to help ease some of the burden.

Application software

The goal of a good multicore system is to make the underlying configuration details as irrelevant as possible to the application programmer. That's less possible for embedded systems, where programmers work harder to optimize their software to a specific platform. But multicore raises specific new issues that cannot be ignored by programmers.

Languages and tools

Some languages are better than others at handling the parallelism that multicore systems make possible. That said, some languages are more popular than others without regard to their ability to handle parallelism. Author Gitu Jain takes us through a variety of languages, covering their appropriateness for multicore as well as their prevalence.

Meanwhile, tailoring an application for a multicore platform suggests analysis and tools that don't apply for single-core systems. Author Kenn Luecke surveys a range of tools from a number of sources that are applicable to multicore design.

Partitioning applications

Since multicore systems can do more than one thing at a time, application programs that were once sequential in nature can be teased apart to keep multiple cores busy. But that process of splitting up a program can be very difficult. I'm assisted by Paul Stravers in this chapter that describes the issues surrounding partitioning and then shows you how to do it both manually and with new specialized tools designed to help.

Synchronization

The success of various pieces of a program running in parallel to yield a correct result depends strongly on good synchronization between those different pieces. This concept lies at the heart of what can make or break a multicore application. Author Tom Dickens runs through a litany of dos and don'ts for keeping an application program on track as it executes.

Hardware assistance

While there may appear to be a bright line defining what is expected in hardware and what should be in software, there are some gray areas. In some cases, hardware can assist with specific functions to improve performance.

Hardware accelerators

There are times when a hardware block can increase the performance of some compute-intensive function dramatically over what software can manage. Those accelerators can be designed to operate in parallel with the cores, adding a new concurrency dimension. In this chapter, Yosinori Watanabe and I discuss hardware accelerators and their considerations for integration into embedded multicore systems.

Synchronization hardware

Much of the bookkeeping and other details required for correct functioning of a multicore system is handled by low-level software services and libraries such as those described in the earlier chapter on multicore libraries. But hardware infrastructure can help out here as well — to the point of the processor instruction set having an impact. So author Jim Holt takes a look at some important low-level hardware considerations for improving embedded system performance.

System-level considerations

We close out with system-level optimization concepts: bare-metal systems and debugging. These chapters complement the preceding material, leveraging all of the concepts presented in one way or another.

Bare-metal systems

The performance of some systems — notably, the packet-processing systems mentioned at the outset of this introduction — is so critical that the overhead of an operating system cannot be tolerated. This suggests a completely different way of doing multicore design. Sanjay Lal and I describe both hardware and software considerations in such a system.

Debug

Finally, debug can be much more difficult in a multicore system than in a single-core one. It's no longer enough to stop and start a processor and look at registers and memory. You have multiple cores, multiple clocks, multiple sets of interrupts and handlers, and multiple memories and caches, each of which may be marching to the beat of a different drummer. Successful debug relies heavily on the existence of robust control and observability features, and yet these features entail costs that must be balanced. Author Neal Stollon discusses those trade-offs and trends for multicore debug.

A roadmap of this book

Figure 1.1 illustrates the relationships between the various chapters and their relationship to system design. The particular depiction of the design process may feel oversimplified, and, in fact it is. But, as with everything embedded, each project and process is different, so this should be viewed as a rough abstraction for the purposes of depicting chapter relevance.

Based on this model:

- Everyone should read the Concurrency chapter.
- System architects and designers can jump into the various platform-design-related chapters, with Architecture, Memory, and Interconnect being the most fundamental. Hardware synchronization is important if designing accelerated infrastructure.

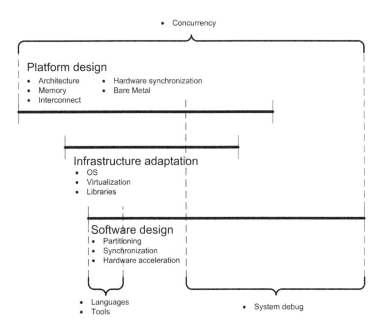

Figure 1.1
A rough depiction of embedded multicore system development.

- For those engineers adapting an OS or virtualization platform or writing drivers or adapting libraries, there are chapters specifically introducing those topics.
- For application writers, Partitioning and Synchronization are important chapters. Hardware accelerators are also included here because our focus is not on designing the hardware for the accelerator, but rather on using it within the system: invoking it in software, writing drivers for it, and handling the concurrency that it can bring. There is also an overview of languages and tools as they pertain to multicore to help make choices at the start of a project.
- Integration and verification engineers can tackle the Debug chapter.
- System designers and programmers trying to extract maximal performance can take on the Bare-metal chapter.

That said, everyone can benefit from the topics they're not specifically involved in. A hardware designer can design a better platform if he or she understands the real-world issues that software programmers face either when writing an application or building firmware. A programmer writing an application can write better code if he or she understands the nuances and trade-offs involved in the platform that will run the code.

Our goal is to bring to you the experiences of those that have been tackling these issues before it was mainstream to do so. Each of their chapters encapsulates years of learning what works and what doesn't work so that you don't have to repeat those lessons. We hope that this serves you well.

Welcome to the age of embedded multicore.

CHAPTER 2

The Promise and Challenges of Concurrency

Bryon Moyer
Technology Writer and Editor, EE Journal

Chapter Outline

The opportunities and challenges that arise from multicore technology – or any kind of multiple processor arrangement – are rooted in the concept of concurrency. You can loosely conceive of this as "more than one thing happening at a time". But when things happen simultaneously, it's very easy for chaos to ensue. If you create an "assembly line" to make burgers quickly in a fast food joint, with one guy putting the patty on the bun and the next guy adding a dab of mustard, things will get messy if the mustard guy doesn't wait for a burger to be in place before applying the mustard. Coordination is key, and yet, as obvious as this may sound, it can be extremely challenging in a complex piece of software.

The purpose of this chapter is to address concurrency and its associated challenges at a high level. Specific solutions to the problems will be covered in later chapters.

Real World Multicore Embedded Systems.
DOI: http://dx.doi.org/10.1016/B978-0-12-416018-7.00002-X

Concurrency fundamentals

It is first important to separate the notion of inherent concurrency and implemented parallelization. A given algorithm or process may be full of opportunities for things to run independently from each other. An actual implementation will typically select from these opportunities a specific parallel implementation and go forward with that.

For example, in our burger-making example, you could make burgers more quickly if you had multiple assembly lines going at the same time. In theory, given an infinite supply of materials, you could make infinitely many burgers concurrently. However, in reality, you only have a limited number of employees and countertops on which to do the work. So you may actually implement, say, two lines even though the process inherently could allow more. In a similar fashion, the number of processors and other resources drives the decision on how much parallelism to implement.

It's critical to note, however, that a chosen implementation relies on the inherent opportunities afforded by the algorithm itself. No amount of parallelization will help an algorithm that has little inherent concurrency, as we'll explore later in this chapter.

So what you end up with is a series of program sections that can be run independently punctuated by places where they need to "check in" with each other to exchange data — an event referred to as "synchronization."

For example, one fast food employee can lay a patty on a bun completely independently from someone else squirting mustard on a different burger. During the laying and squirting processes, the two can be completely independent. However, after they're done, each has to pass his or her burger to the next guy, and neither can restart with a new burger until a new one is in place. So if the mustard guy is a lot faster than the patty-laying guy, he'll have to wait idly until the new burger shows up. That is a synchronization point (as shown in Figure 2.1).

A key characteristic here is the fact that the two independent processes may operate at completely different speeds, and that speed may not be predictable. Different employees on different shifts, for example, may go

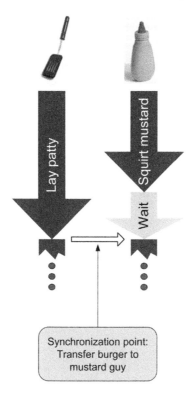

Figure 2.1
Where the two independent processes interact is a synchronization point.

at different speeds. This is a fundamental issue for parallel execution of programs. While there are steps that can be taken to make the relative speeds more predictable, in the abstract, they need to be considered unpredictable. This concept of a program spawning a set of independent processes with occasional check-in points is shown in Figure 2.2.

Depending on the specific implementation, the independent portions of the program might be threads or processes (Figure 2.3). At this stage, we're really not interested in those specifics, so to avoid getting caught up in that detail, they are often generically referred to as "tasks". In this chapter, we will focus on tasks; how those tasks are realized, including the definitions of SMP and AMP shown in the figure, will be discussed in later chapters.

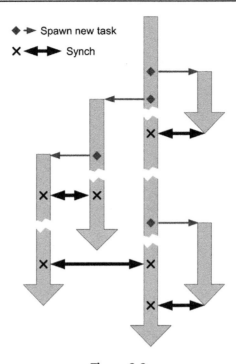

Figure 2.2
A series of tasks run mutually asynchronously with occasional synchronization points.

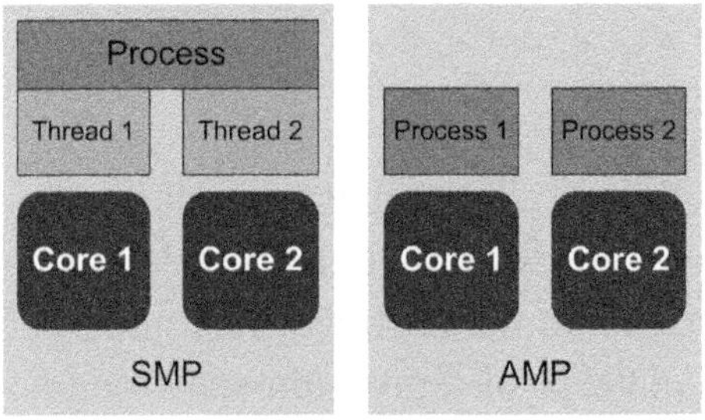

Figure 2.3
Tasks can be different threads within a process or different processes.

Two kinds of concurrency

There are fundamentally two different ways to do more than one thing at a time: bulk up so that you have multiple processors doing the same thing, or use division of labor, where different processors do different things at the same time.

Data parallelism

The first of those is the easiest to explain. Let's say you've got a four-bit vector that you want to operate on. Let's make it really simple for the sake of example and say that you need to increment the value of every entry in the vector. In a standard program, you would do this with a loop:

```
for i=1 to 4 {
    Increment the ith value
}
```

This problem is exceedingly easy to parallelize. In fact, it belongs to a general category of problems whimsically called "embarrassingly parallel" (Figure 2.4) Each vector entry is completely independent and can be incremented completely independently. Given four processors, you could easily have each processor work on one of the entries and do the entire vector in ¼ the time it takes to do it on a single processor.

In fact, in this case, it would probably be even less than ¼ because you no longer have the need for an iterator − the i in the pseudocode above; you no longer have to increment i each time and compare it to 4 to see if you're done (Figure 2.5).

This is referred to as data parallelism; multiple instances of data can be operated on at the same time. The inherent concurrency allows a four-fold speed-up, although a given implementation might choose less if fewer than four processors are available.

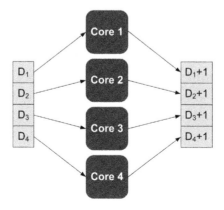

Figure 2.4
Embarrassingly parallel computation.

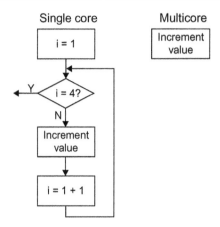

Figure 2.5
Looping in a single core takes more cycles than multicore.

Two key attributes of this problem make it so easy to parallelize:

— the operation being performed on one entry doesn't depend on any other entry
— the number of entries is known and fixed.

That second one is important. If you're trying to figure out how to exploit concurrency in a way that's static — in other words, you know exactly how the problem will be parallelized at compile time — then the number of loop iterations must be known at compile time. A "while" loop or a "for" loop where the endpoint is calculated instead of constant cannot be so neatly parallelized because, for any given run, you don't know how many parallel instances there might be.

Functional parallelism

The other way of splitting things up involves giving different processors different things to do. Let's take a simple example where we have a number of text files and we want to cycle through them to count the number of characters in each one. We could do this with the following pseudo-program:

```
for each file {
    Open the file
    Count the characters
    Close the file
}
```

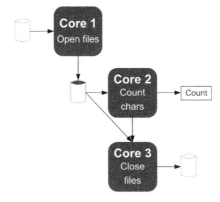

Figure 2.6
Different cores performing different operations.

We can take three processors and give each of them a different task. The first processor opens files; the second counts characters; and the third closes files (Figure 2.6).

There is a fundamental difference between this and the prior example of data parallelism. In the vector-increment example, we took a problem that had been solved by a loop and completely eliminated the loop. In this new example, because of the serial nature of the three tasks, if you only had one loop iteration, then there would be no savings at all. It only works if you have a workload involving repeated iterations of this loop.

As illustrated in Figure 2.7, when the first file is opened, the second and third processors sit idle. After one file is open, then the second processor can count the characters, while the third processor is still idle. Only when the third file is opened do all processors finally kick in as the third processor closes the first file. This leads to the descriptive term "pipeline" for this kind of arrangement, and, when executing, it doesn't really hit its stride until the pipeline fills up. This is also referred to as "loop distribution" because the duties of one loop are distributed into multiple loops, one on each processor.

This figure also illustrates the fact that using this algorithm on only one file provides no benefit whatsoever.

Real-world programs and algorithms typically have both inherent data and functional concurrency. In some situations, you can use both. For

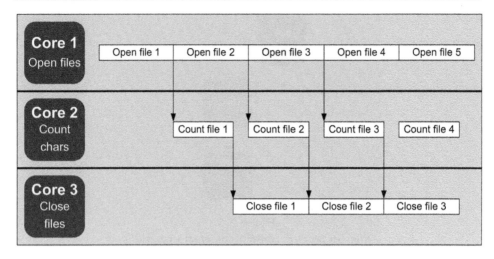

Figure 2.7
The pipeline isn't full until all cores are busy.

example, if you had six processors, you could double the three-processor pipeline to work through the files twice as fast. In other situations, you may have to decide whether to exploit one or the other in your implementation.

One of the challenges of a pipeline lies in what's called balancing the pipeline. Execution can only go as fast as the slowest stage. In Figure 2.7, opening files is shown as taking longer than counting the characters. In that situation, counting faster will not improve performance; it will simply increase the idle time between files.

The ideal situation is to balance the tasks so that every pipeline stage takes the same amount of time; in practice, this is so difficult as to be more or less impossible. It becomes even harder when different iterations take more or less time. For instance, it will presumably take longer to count the characters in a bigger file, so really the times for counting characters above should vary from file to file. Now it's completely impossible to balance the pipeline perfectly.

Dependencies

One of the keys to the simple examples we've shown is the independence of operations. Things get more complicated when one calculation depends on the results of another. And there are a number of ways in which these dependencies crop up. We'll describe some basic cases here, but a complete theory of dependencies can be quite intricate.

It bears noting here that this discussion is intended to motivate some of the key challenges in parallelizing software for multicore. In general, one should not be expected to manually analyze all of the dependencies in a program in order to parallelize it; tools become important for this. For this reason, the discussion won't be exhaustive, and will show concept examples rather than focusing on practical ways of dealing with dependencies, which will be covered in the chapter on parallelizing software.

Producers and consumers of data

Dependencies are easier to understand if you think of a program of consisting of producers and consumers of data (Figure 2.8). Some part of the program does a calculation that some other part will use: the first part is the producer and the second is the consumer. This happens at very fine-grained instruction levels and at higher levels, especially if you are taking an object-oriented approach — objects are also producers and consumers of data.

At its most basic, a dependency means that a consumer of data must wait to consume its data until the producer has produced the data (Figure 2.9). The concept is straightforward, but the implications vary depending on the language and approach taken. At the instruction level, many compilers have been designed to exploit low-level concurrency, doing things like instruction reordering to make execution more efficient while making sure that no dependencies are violated.

It gets more complicated with languages like C that allow pointers. The concept is the same, but compilers have no way of understanding how various pointers relate, and so can't do any optimization. There are two reasons why this is so: pointer aliasing and pointer arithmetic.

Pointer aliasing is an extremely common occurrence in a C program. If you have a function that takes a pointer to, say, an image as a parameter, that function may name the pointer `imagePtr`. If a program needs to call

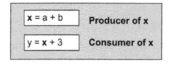

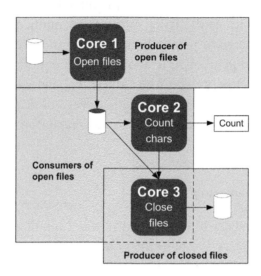

Figure 2.8
Producers and consumers at the fine- and coarse-grained level. Entities are often both producers and consumers.

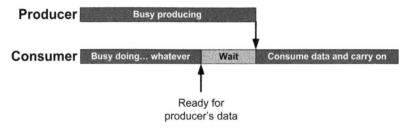

Figure 2.9
A consumer cannot proceed until it gets its data from the producer.

that function on behalf of two different images — say, leftImage and rightImage, then when the function is called with leftImage as the parameter, then leftImage and imagePtr will refer to the same data. When called for rightImage, then rightImage and imagePtr will point to the same data (Figure 2.10).

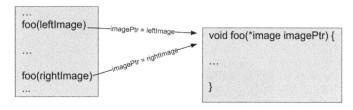

Figure 2.10
Different pointers may point to the same locations at different times.

This is referred to as aliasing because a given piece of data may be accessed by variables of different names at different times. There's no way to know statically what the dependencies are, not only because the names look completely different, but also because they may change as the program progresses. Thorough dynamic analysis is required to understand the relationships between pointers.

Pointer arithmetic can also be an obvious problem because, even if you know where a pointer starts out, manipulating the actual address being pointed to can result in the pointer pointing pretty much anywhere (including address 0, which any C programmer has done at least once in his or her life). Where it ends up pointing may or may not correlate to a memory location associated with some other pointer (Figure 2.11).

For example, when scanning through an array with one pointer to make changes, it may be very hard to understand that some subsequent operation, where a different pointer scans through the same array (possibly using different pointer arithmetic), will read that data (Figure 2.12). If the second scan consumes data that the first scan was supposed to put into place, then parallelizing those as independent will cause the program to function incorrectly. In many cases, this dependency cannot be identified by static inspection; the only way to tell is to notice at run time that the pointers address the same space.

These dependencies are based on a consumer needing to wait until the producer has created the data: writing before reading. The opposite situation also exists: if a producer is about to rewrite a memory location, you want to be sure that all consumers of the old data are done before you overwrite the old data with new data (Figure 2.13). This is called an

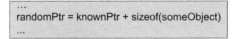

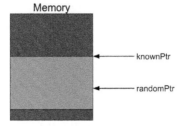

Figure 2.11
Pointer arithmetic can cause a pointer to refer to some location in memory that may or may not be pointed to by some other pointer.

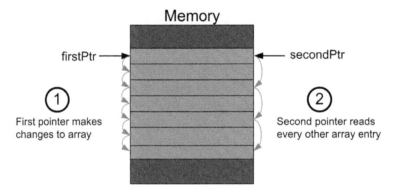

Figure 2.12
Two pointers operating on the same array create a dependency that isn't evident by static inspection.

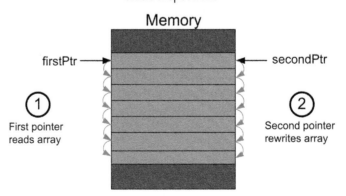

Figure 2.13
The second pointer must wait before overwriting data until the first pointer has completed its read, creating an anti-dependency.

"anti-dependency". Everything we've discussed about dependencies also holds for anti-dependencies except that this is about waiting to write until all the reads are done: reading before writing.

This has been an overview of dependencies; they will be developed in more detail in the Partitioning chapter.

Loops and dependencies

Dependencies become more complex when loops are involved – and in programs being targeted for parallelization – loops are almost always involved. We saw above how an embarrassingly parallel loop can be parallelized so that each processor gets one iteration of the loop. Let's look at an example that's slightly different from that example.

```
for i=1 to 4 {
    add the (i-1)ᵗʰ value to the iᵗʰ value
}
```

Note that in this and all examples like this, I'm ignoring what happens for the first iteration, since that detail isn't critical for the discussion.

This creates a subtle change because each loop iteration produces a result that will be consumed in the next loop iteration. So the second loop iteration can't start until the first iteration has produced its data. This means that the loop iterations can no longer run exactly in parallel: each of these parallel iterations is offset from its predecessor (Figure 2.14). While the total computation time is still less than required to execute the loop on a single processor, it's not as fast as if there were no dependencies between the loop iterations. Such dependencies are referred to as "loop-carry" (or "loop-carried") dependencies.

It gets even more complicated when you have nested loops iterating across multiple iterators. Let's say you're traversing a two-dimensional matrix using i to scan along a row and using j to scan down the rows (Figure 2.15).

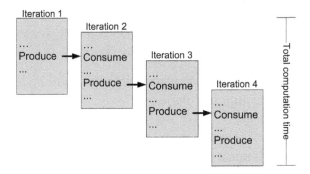

Figure 2.14
Even though iterations are parallelized, each must wait until its needed data is
produced by the prior iteration, causing offsets that increase overall computation
time above what would be required for independent iterations.

Memory

1,1	2,1	3,1	4,1
1,2	2,2	3,2	4,2
1,3	2,3	3,3	4,3
1,4	2,4	3,4	4,4

Figure 2.15
4×4 array with *i* iterating along a row (inner loop) and *j* iterating down the rows
(outer loop).

And let's assume further that a given cell depends on the *new* value of
the cell directly above it (Figure 2.16):

```
for j=1 to 4 {
    for i=1 to 4 {
        add the (i,j-1)ᵗʰ value to the (i,j)ᵗʰ value
    }
}
```

First of all, there are lots of ways to parallelize this code, depending on
how many cores we have. If we were to go as far as possible, we would

Memory

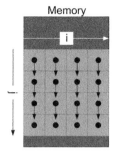

Figure 2.16
Each cell gets a new value that depends on the new value in the cell in the prior row.

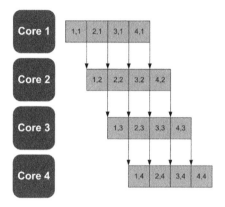

Figure 2.17
If each row gets its own core, then each row must wait until the first cell in the prior row is done before starting.

need 16 cores since there are 16 cells. Or, with four cores, we could assign one row to each core.

If we did the latter, then we couldn't start the second row until the first cell of the first row was calculated (Figure 2.17).

If we completely parallelized it, then we could start all of the first-row entries at the same time, but the second-row entries would have to wait until their respective first-row entries were done (Figure 2.18).

Note that using so many cores really doesn't speed anything up: using only four cores would do just as well since only four cores would be executing at any given time (Figure 2.19). This implementation assigns one column to each core, instead of one row, as is done in Figure 2.17. As a result, the loop can be processed faster because no core has to wait for any other core. There is no way to parallelize this set of nested loops any further because of the dependencies.

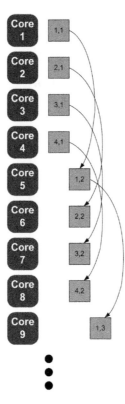

Figure 2.18
An implementation that assigns each cell to its own core.

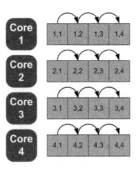

Figure 2.19
Four cores can implement this loop in the same time as 16.

Such nested loops give rise to the concept of "loop distance". Each iterator gets a loop distance. So in the above example, in particular as shown in Figure 2.16, where the arrows show the dependency, the loop distance for i is 0 since there is no dependency; the loop distance for j is 1, since the data consumed in one cell depends on the cell directly above it, which is the prior j iteration. As a "vector", the loop distance for i and j is [0,1].

If we changed the code slightly to make the dependency on $j-2$ instead of $j-1$:

```
for j=1 to 4 {
    for i=1 to 4 {
        add the (i,j-2)ᵗʰ value to the (i,j)ᵗʰ value
    }
}
```

then the loop distance for j is 2, as shown in Figure 2.20.

This means that the second row doesn't have to wait for the first row, since it no longer depends on the first row. The third row, however, does have to wait for the first row (Figure 2.21). Thus we can parallelize further with more cores, if we wish, completing the task in half the time required for the prior example.

While it may seem obscure, the loop distance is an important measure for synchronizing data. It's not a matter of one core producing data and

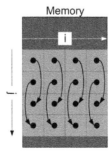

Figure 2.20
Example showing *j* loop distance of 2.

the other immediately consuming it; the consuming core may have to wait a number of iterations before consuming the data, depending on how things are parallelized. While it's waiting, the producer continues with its iterations, writing more data. Such data can be, for example, written into some kind of first-in/first-out (FIFO) memory, and the loop distance determines how long that FIFO has to be. This will be discussed more fully in the Partitioning chapter.

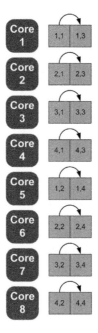

Figure 2.21
With loop distance of 2, two rows can be started in parallel.

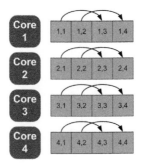

Figure 2.22
A four-core implementation with loop distance [0,2].

Let's take the prior example and implement it with only four cores instead of eight, as shown in Figure 2.22.

Let's look at Core 1. When it's done with cell [1,1], it must move on to cell [1,2]. But cell [1,3] needs the result from [1,1]. Strictly speaking, this is an anti-dependency: the [1,1] result must be kept around until [1,3] reads it. Depending on how we implement things, cell [1,2] might destroy the result.

Now, as shown above, we can really just implement this as an array in each core, keeping all the results separate. But in some multicore systems, the operating system will determine which cores get which threads, and if each cell is spawned as a thread, then things could be assigned differently. For example, the first two cores might exchange the last two cells (Figure 2.23).

Now Core 1 has to hand its results to Core 2 (and vice versa, not illustrated to avoid clutter). The solution is for Core 1 to put the result of [1,1] somewhere for safekeeping until [1,3] is ready for it. Then [1,2] can proceed, and Core 2 can pick up the result it needs when it's ready. But the [1,2] result will also be ready before Core 2 is ready for the [1,1]

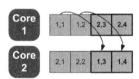

Figure 2.23
The first two of four cores, with a different assignment of cells.

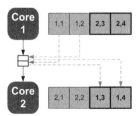

Figure 2.24
FIFO used to communicate results between cores. The minimum FIFO size is related to the loop distance.

result. So the [1,2] result can't just be put in the same place as the [1,1] result or it will overwrite it.

One solution, at the risk of getting into implementation details, is to use some kind of FIFO structure between Core 1 and Core 2 (Figure 2.24). Because the loop distance for j is 2, the FIFO needs to be at least 2 deep to avoid stalling things. Additionally, by using a FIFO instead of trying to hard-code an array implementation, the solution is robust against any arbitrary thread assignments that the operating system may make.

FIFOs are sometimes thought to be expensive, depending on how they are implemented. The intent here isn't to focus on the details of the FIFO, but rather to illustrate its relationship to the loop distance. Specific synchronization mechanisms will be discussed in future chapters. More concrete examples of dependencies and synchronization are presented in the Partitioning chapter.

Manual determination of loop distance can, frankly, be quite confusing. In fact, the body of a loop may have numerous variables, each with different loop distances. Branches further complicate things. The existence of tools to handle this will be covered in a subsequent chapter. Because of these tools, we will not delve further into the intricacies, but rather leave the discussion here as a motivation of the concept of loop distance as it shows up in tools.

Shared resources

The second major challenge that concurrent tasks present is the fact that different tasks may need to access the same resources at the same time. For the most part, the challenges are exactly the same as those presented by a multi-threaded program on a single-core system. The use of critical sections and locks and their ilk proceeds exactly as before.

However, the implementations of solutions that work for single-core systems may not work for multicore systems. For example, one simple brute-force way to block any other thread from interrupting a critical section of code is to suspend interrupts while within that critical section. While that might work for a single core, it doesn't work if there is another core that could be accessing (or corrupting) a shared memory location.

The other new concept that multicore adds is the fact that each core has its own cache, and global data replicated in the cache may at times be out of sync with the latest version. And this gets complex because cache coherency strategies can themselves be complex, and different platforms will have different schemes.

So, while a programmer can ignore the cache on a single-core system, that's no longer possible for multicore, and, as we'll see in subsequent chapters, the handling of synchronization may depend on the caching strategy of the platform.

Summary

All of the challenges of multicore computing arise from concurrency, the fact that different things may happen at the same time. If we're used to events occurring in a prescribed order, then it can require a bit of mental gymnastics to get used to the idea that two operations in two different parallel threads may happen in any order with respect to each other.

Concurrency is a good thing — it lets us do things in parallel so that we achieve a goal more quickly. But it can also make things go haywire, so most of this book is dedicated to managing the challenges of concurrency in order to realize the promise of concurrency.

CHAPTER 3

Multicore Architectures

Frank Schirrmeister

Senior Director, System Development Suite at Cadence Design Systems, Santa Clara, CA, USA

Chapter Outline

This chapter will introduce the concepts of multicore related issues, while the subsequent chapters will go into further details. We will start with a general analysis of how electronic design trends lead to multicore

Real World Multicore Embedded Systems.
DOI: http://dx.doi.org/10.1016/B978-0-12-416018-7.00003-1

hardware-software architectures as the only viable solution addressing consumer requirements on cost, performance and power. We will then categorize multicore architecture concepts by processing and communication requirements and show how different processing techniques combine to form multicore architectures that address the specific needs of different application domains. Special attention will be given to the programmability of the different hardware architectures and the impact that hardware has on software. We will close the chapter with a brief review of existing hardware architectures available on the market, as well as a brief discussion about programming models capable of expressing parallel functionality, which can then be mapped into multiple processor cores.

The need for multicore architectures

The answer to increasing complexity, which has resulted in more silicon real estate in the last several decades, has historically been abstraction, combined with automation, translation, and verification against the next lowest level of abstraction. This approach has moved the hardware side of the electronics industry forward from layout to transistors to gates and then to the register-transfer level (RTL), enabling verification and automated synthesis to gates. Figure 3.1 illustrates these trends together with the increase in complexity of software from the early days of languages like Fortran and Cobol to sequential C and C++.

Since then, designs have grown even more complex, moving the industry forward (while also compounding the power problem) to block-level designs, whereby the design of the sub-system blocks is separated from the design of the systems-on-chips (SoC) platform. While this transition is still very much a hardware-based approach, with ever-increasing complexity, the next logical step is to move to parallel hardware and software via a sea of programmable processors, combined with programming models that express parallel software running on multiprocessor systems-on-chips (MPSoCs).

The center of Figure 3.1 shows the principles as outlined in Herb Sutter's article [1] published in 2005. While the race for processor speed has pushed technology over the past couple of decades, processor speed

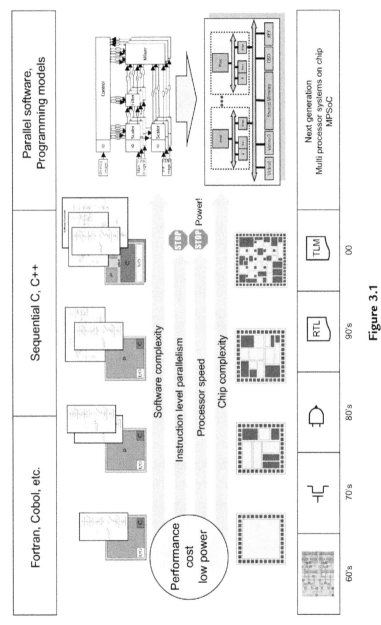

Figure 3.1
Design methodology evolution.

growth itself has now come to a screeching halt and leveled out at around 3 GHz. The fundamental reason that the growth of processor speed has stopped is that increasing clock speeds have made energy consumption unacceptably high, and power budgets can no longer be met. In addition, the relative performance per cycle achieved by way of instruction-level parallelism (ILP) is leveling out. Simply issuing more instructions in parallel has little effect on most applications.

As a result, the serial microprocessor processing speed is reaching a physical limit, which leads to two main consequences: processor manufacturers will focus on products that better support multithreading (such as multicore processors), and software developers will be forced to develop massively multithreaded programs as a way to better use such processors. The hardware design of multiple processors on a single die is reasonably well understood, but software designers are now facing the more daunting questions around MPSoC programming, debugging, simulation, and optimization.

But before we look into the details of the programming challenges, let's dig into the issues that motivate multicore architectures.

Multicore architecture drivers

Three main requirements are driving the trend to multicore architectures: low power, performance/throughput and memory bandwidth.

Arguably, low power requirements have been the single most significant cause for the shift to multicore design. During the 2006 Hot Chips conference at Stanford University, IBM fellow Bernie Meyerson asked whether there's such a thing as "too hot". Meyerson joked that it was foolish for the semiconductor industry to keep pursuing the highest performance for chips at the expense of power efficiency. Meyerson also noted that chips at the time gave off more heat per square inch than a steam iron.

Chip designers were at risk of going off the power cliff. Charting processor properties over the previous three decades showed that the performance of a single processor has leveled off at the turn of the century. Moreover, the efficiency of built-in instruction-level parallelism, which was effectively shielding software designers from having to think about parallelism proactively themselves, was no longer growing because issuing more than four instructions in parallel had little or no effect on most applications. The attempted increases in performance have led dangerously close to the energy consumption ceiling, essentially stopping performance progress in conventional processor development. For a while, dedicated hardware accelerators seemed to be a valid alternative for some applications, but they do not offer the flexibility and programmability required to allow an expensive chip development project to meet market window targets or to optimize the chip's time-in-market.

The second driver for multicore architectures is throughput and the resulting required processing performance. According to the Cisco Visual Networking Index 2011, global mobile data traffic grew 2.3-fold in 2011 — more than doubling for the fourth year in a row. The 2011 mobile data traffic was eight times the size of the entire global Internet in 2000 and mobile video traffic exceeded 50 percent for the first time in 2011. Looking forward, the prediction is that global mobile data traffic will increase 18-fold between 2011 and 2016. Mobile data traffic will grow at a compound annual growth rate (CAGR) of 78 percent from 2011 to 2016, reaching 10.8 exabytes per month by 2016. All this data needs to be collected from users, processed in devices and presented to users. Given the power limitations of scaling individual processors, multicore architectures are the only viable answer. The sheer amount of data also has fundamental impact on the communication structures on and off chip.

The rapidly increasing amount of data and its processing needs to be supported by storage on and off chip, leading to memory architecture and memory bandwidth as the third driver impacting multicore architectures. For instance, a detailed analysis of the Apple iPhone shows that, over the course of five generations from the original iPhone to the iPhone 4 S, the memory architecture and its access has changed fundamentally. As in most SoCs deployed in smartphone designs, the memory is implemented as a package-on-package (PoP) stack of DRAM on top of the SoC package. Memory has grown in size from 128 MB to 512 MB, and memory bandwidth has grown from 532 MB/s to 6400 MB/s. The number of CPU cores has been doubled for the A5 processor in the iPhone 4 S, with each one even more bandwidth-hungry than the single A4 core. In addition, a 4-fold increase in graphics processing combined with an increase in clock speeds makes the A5 another big consumer of bandwidth, explaining the increased need for memory bandwidth.

So what did these observations on low power, performance/throughput and memory bandwidth mean in the first decade of the 21st century? They meant that the electronics industry had to look at alternate solutions, one of the most promising being MPSoCs.

The move to MPSoC design elegantly addressed the power issues faced on the hardware side by creating multiple processors that execute at lower frequency, resulting in comparable overall MIPS performance while allowing designers to slow down the clock speed, a major constraint for low-power design. However, such a switch meant that the challenges have effectively been moved from the hardware to the software domain.

Traditional sequential software paradigms break

The traditional SoC was dominated by hardware as illustrated in Figure 3.2a. The functionality of an application was mapped into individual hardware blocks. The demand for more features in every new product generation has driven a trend to keep flexibility during product design and even during product lifetime by using programmable processors on chip as shown in Figure 3.2b.

The availability of configurable processors and co-processor synthesis allows users to tailor processors specifically to the functionally they are supposed to execute. As a result even more functions which traditionally can be optimized in specific hardware blocks are now run on processors, leading to MPSoCs organized in "silos" as indicated in Figure 3.2c.

The silo organization lends itself very well to situations in which the different portions of a sequential algorithm can be mapped to individual, specific processors. However, processor-based design using functional silos with only some limited communication between the different functions and processors does not scale well with the demands for even

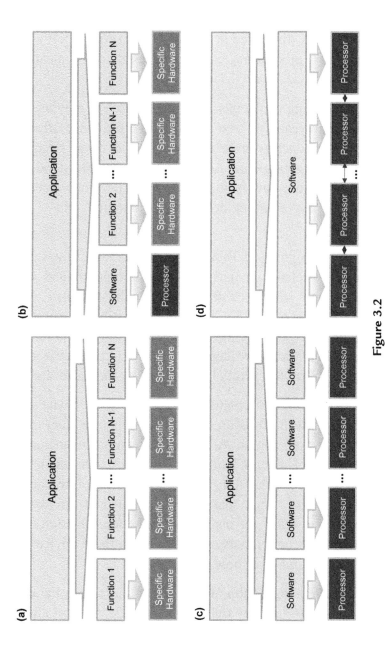

Figure 3.2

Evolution from traditional systems-on-chips (SoCs) to MPSoCs. (a) Traditional SoC. (b) Single-processor SoC. (c) MPSoC using "silos". (d) MPSoC with distributed software.

more performance in consumer designs because it does not address the fundamental issues of power consumption at higher clock speeds and limited additional improvements using ILP.

The next natural step is illustrated in Figure 3.2d, in which software is now distributed across processors, causing increased communication between them.

With that step, the design challenges are now effectively moving into the software domain. The years of "free" increased performance offered by processors — the free lunch — are over. While the hardware design of multiple processors on a single die is reasonably well understood, software programmers are now facing challenging questions around software migration, programming, debugging, simulation, and optimization.

Analysis of today's designs (see Table 3.1 further below) shows that the majority of multicore-related designs (excluding computing and networking) are using a silo-based design style as illustrated in Figure 3.2c. Approaches for MPSoCs with fully distributed software as illustrated in Figure 3.2d in the embedded space (Ambric, Tilera, XMOS etc.) require users to learn a new language or APIs to be inserted into the C code in order to utilize the parallelism offered by the hardware.

While the providers of these second-generation multicore devices are certainly more aware of the software challenges, it is important to recall the lessons of the first generation. A long list of companies including, but not limited to, Quicksilver, Chameleon, Morphics, Chromatic Research, Triscend, BOPS, Equator, PACT, Systolix and Intrinsity delivered very innovative, functional silicon but failed to provide programming solutions ready for mainstream adoption. This left only a very small number of programmers able to program these devices and eventually led to the demise of these companies.

The biggest challenge is migration of software to these new, parallel architectures and is unfortunately today often not addressed at all. Designers are asked to learn a new language or APIs to express parallelism when their first challenge is to understand and identify the potential for parallelism in the first place.

Scope of multicore hardware architectures

For multicore architectures, both hardware and software need to be considered as related and dependent on each other. The best hardware multicore architecture in the world will not be successful if only a limited number of people can program it with software to execute functionality. Conversely, the most inventive algorithm will not execute as intended if the underlying hardware architecture does not provide sufficient compute, storage and communication resources to support it.

As outlined in Figure 3.3, the effects of multiple cores in a design are visible at the system level, within SoCs as well as sub-systems within SoCs.

At the PCB system-level depicted on the right side, several of the different components may contain processors executing software. The application processor in the center serves as the main processor, but typically several other components including the GPS receiver, cellular modem, Bluetooth receiver, motion sensor and multimedia processor will contain processors and will execute software too. At this level of scope, the software and the processors are well separated from each other. Only minor interactions between the components will have to be considered from a multicore perspective. In this system, for instance, one would not expect any software function to be distributed between the actual different processor cores spanning multiple chips. In some application domains, like compute and networking, such partitioning of software may be more likely at the chip-level of scope.

As depicted on the left side of Figure 3.3, multiple functions like graphics, multimedia and audio will be integrated on the SoC. Similarly to the scope of the PCB system-level, some of the functions will be well separated and not span across the different processors. However, at this level, considerations as to whether to execute functions most efficiently on dedicated hardware accelerators or specialized multicore sub-systems will impact the architecture partitioning decisions when architecting the system.

Finally, at the level of SoC sub-system, multicore architecture decisions will be very visible. For example, the CPU sub-system will have processor cores of different performance and power consumption. In this

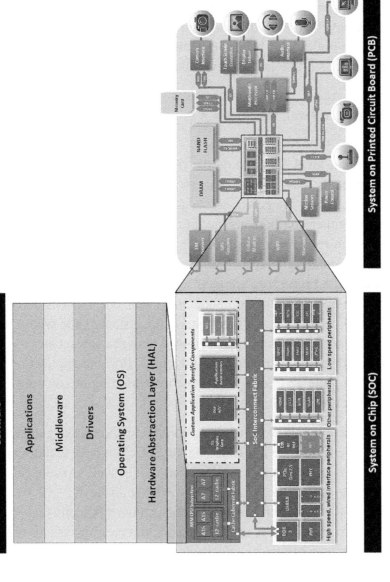

Figure 3.3
Example of a mobile multicore system.

case, switching between cores as well as partitioning of software between cores will become an important task to consider when programming multicore devices. Similarly to the compute sub-system, the graphics engine and multimedia processors can be specialized multicore architectures with cores optimized for the target application.

While the effects of multicore architectures can be seen at all three levels of scope — PCB, SoC and SoC sub-system — this chapter will focus mostly on the effects with SoCs and their sub-systems.

Basic multicore hardware architecture overview

Regardless of their scope, multicore architectures will exhibit some main general characteristics which we will further outline below and then dive into deeper in the following chapters.

From a processing perspective, the basic distinction between architectures — besides the performance of individual hardware processing elements — is the arrangement into either heterogeneous or homogeneous structures. Heterogeneous hardware architectures feature processing elements of different type and/or performance, and, often, the tasks executing on the different domains may not even interact with each other much. Modern mobile application processors, as structurally shown in Figure 3.3, may have such characteristics, with domains for application, graphics and multimedia processing. The software architecture within a heterogeneous hardware architecture may be either homogeneous, i.e. each CPU runs the same type and version of the operating system or software, or heterogeneous, i.e. each CPU runs either a different operating system/software or a different version of the same operating system/software.

Homogeneous hardware architectures feature the same processing element several times. A good example is the Intel quad-core processor or the MIT RAW architecture using the same cores several times in an array structure. The use of heterogeneous vs. homogeneous networks directly impacts programmability significantly, the latter often requiring run-time scheduling vs. the former requiring compile-time scheduling. From a software perspective, in homogeneous architectures, identical cores are typically assigned portions of the combined workload under the control of a single symmetric multiprocessing (SMP) operating system. This is common in many computer-oriented applications, like medical imaging, big-data analysis and scientific computing, running on a general purpose OS like Linux. Users can also have asymmetric multiprocessing (AMP) on a homogeneous hardware architecture by running different OS instances on different cores. These are described in further detail in the OS chapter.

A variation of SMP architectures is so-called bound multi processing architectures (BMP), which preserve the scheduling control of an AMP architecture with the unified hardware abstraction and process management of an SMP architecture. Instead of scheduling processes and tasks on any of the available processing elements, in BMP architectures the tasks can be assigned to specific processors. A mix is also possible, in which "bound" tasks are combined with tasks that can migrate between processors. These different configurations are presented in more detail in the OS chapter.

Communication between the different hardware processing elements can be done either with specific point-to-point connections, shared busses, on-chip crossbars or programmable routing networks like networks on chip (NoC). Communication structures directly impact data latencies and buffering requirements, and have to be considered carefully as they have a fundamental impact on programmability.

Memory access is the third component characterizing multicore architectures, in which asymmetric multi processing (AMP) and symmetric multi processing (SMP) arrangements directly influence both the performance of tasks running on processors and the communication between tasks and processors. SMP systems must present memory as a single shared block that looks the same to every core. Cores in an AMP system, by contrast, can have a mixture of shared and private memory since each core has its own OS. Either one can have either shared external memory or local memories per processor (typically referred to as a non-uniform-memory architecture, or NUMA), but in the case of SMP they must connected in a way that makes them appear to be a single shared block.

Often SMP architecture has more inherent code dependencies than AMP architecture because the cores are more likely to contend for the same data as they execute similar code. In AMP architectures sharing information between cores happens using inter-processor communication (IPC) and requires semaphores. Sharing data in memories can easily lead to data corruption.

Specific multicore architecture characteristics

In attempting a clear and crisp classification of multicore processors and MPSoCs, there are three main issues and the application itself to consider.

- The type of processing elements used doing the actual computation determines the choice of compilers and how specific tools need to be customized to support a specific architecture.
- The communication on chip and between chips determines how long processors have to wait before data can be accessed.
- The types of memory architectures used on and off chip have a profound impact on latencies in the processes accessing data.
- Optimization of the hardware architecture for the application running on them very much impacts whether the programming is straightforward, i.e. whether or not it is obvious which portions of a multicore system are running which tasks.

Processing architectures

The complexity of the processing elements has a very high impact on the programmability of devices and poses fundamentally different challenges:

- Lightweight processing elements — typically focused on arithmetic operations and often not having their own independent program counter — are more likely to see processor-specific code, assembler and macros. They are very sensitive to how well the code exploits architectural features of the processors.
- Mediumweight processing elements are likely to be programmed at a higher level using C, libraries and real-time operating systems, but

are still programmed directly, i.e. the programmer needs some understanding of the hardware architecture.

- Heavyweight processing elements are processors offering enough performance to run Linux, Android or Java, their key characteristic being that the programming is largely separated from the hardware architecture and can be done without intimate knowledge of the underlying hardware.

There is a wide spectrum of architectures for processing elements ranging from very fine-grain architectures like arithmetic logic units (ALUs), to lightweight processing elements that are typically programmed in assembler, to middleweight processing elements with their own C compiler, and to full heavyweight processors like the Intel® Core™ architecture supported by sophisticated software development environments and executing high-level operating systems like Linux or Android on them.

ALU processing architectures

Figure 3.4 shows the structure of a very lightweight processing element as used in the PACT XPP architecture. It contains three objects and routing busses. All objects have input registers which store the data or event packets. After the input register, a one-stage FIFO stores an additional packet if required. The ALU object in the center of the figure provides logical operators, basic arithmetic operators (i.e. adders) and special arithmetic operators including comparators and multipliers.

To allow efficient programming of arrays containing such an element, specific tools exploiting the specific functionality are required. If automated compilers for such a specific architecture are not available, then users often have to do manual low-level code optimization.

Lightweight processing architectures

A still lightweight but more complex processing element is shown in Figure 3.5. The Ambric "Bric" was composed of a compute unit and a RAM unit. Each compute unit contains two 32-bit streaming RISC processors and two 32-bit streaming RISC processors with DSP extensions.

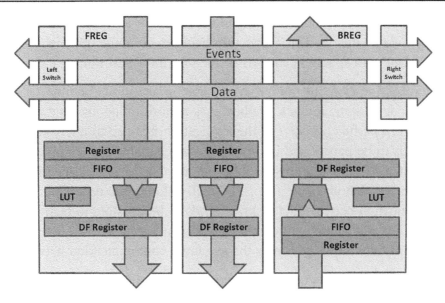

Figure 3.4
Arithmetic logic unit (Pact XPP III).

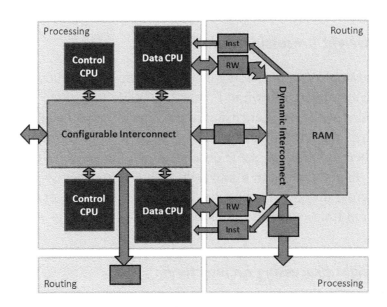

Figure 3.5
Lightweight processing element (Ambric Bric). *Source: http://i.cmpnet.com/eet/news/06/08/AMBRICS1437_PG_6.gif, FPF 2006.*

The "bric" is the physical building block of a core array in which "brics" connect via abutment. The connection between compute units is established with specific channels formed by the RAM units.

The processing elements are programmable in assembler and the software development challenge becomes a partitioning and routing challenge. The functionality of the application mapped to a device like this needs to be partitioned into tasks in sizes suitable for the processing units. There is typically no scheduler running on the processing element, which means that tasks are mapped directly at compile time. For communication between the tasks, it needs to be possible to map them into the communication structures available in hardware, and the connections are "routed" across the available communication network.

Modern FPGA architectures as provided by Altera and Xilinx contain similar repeatable structures of varying complexity, but would strictly speaking not be considered multicore architectures in a classic sense (except for those devices that have multiple hard processor cores integrated alongside the FPGA fabric or when the fabric is used to implement multiple soft processor cores).

Mediumweight processing architectures

Another example of a processing element which is of slightly more heavyweight nature is shown in Figure 3.6. While they often allow programming with compilers using higher-level languages like C, programmers are not fully shielded from the effects of the hardware because there is still quite some dependency of the software on the hardware. In comparison to lightweight architectures they typically can be used across different application domains. Often a mix of compile-time and run-time scheduling is used here.

Heavyweight processing architectures

Heavyweight processing elements are typically programmable in a high-level language and run simple schedulers to allow the execution of more than one task.

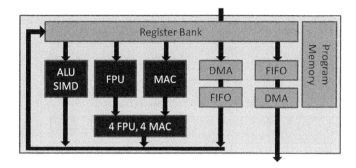

Figure 3.6
Heavyweight processing element (Cradle DSP) in the Cradle CT3616. *Source: http://www.cradle.com/downloads/Cradle_TOL_71305_new.pdf.*

The software programming challenge for heavyweight processing elements often lies in mapping tasks to the right processors, again a placement and routing problem, for which solutions are determined by processing performance and the capabilities of the available communication structures to allow communication between the tasks. Scheduling of tasks has become mostly a run-time issue and is not done at compile time.

The most complex heavyweight processing elements are processors like the Intel® Core™ Architecture as indicated in Figure 3.7.

Today processors are often arranged in dual-core, quad-core or more complex architectures in which each of the processors is programmable in high-level languages, and complex real-time operating systems (RTOS) guard the access to task scheduling and communication between cores. In systems like this, the programming challenge often focuses on the ability to partition an application appropriately into tasks and threads (a topic covered in detail in the Partitioning chapter of this book). A significant amount of scheduling and mapping of tasks to processors is done at run time, with schedulers determining which processor has the most compute bandwidth available.

Communication architectures

The second fundamental aspect of multicore processors and systems is the way processors communicate with each other and with memory, on and off chip.

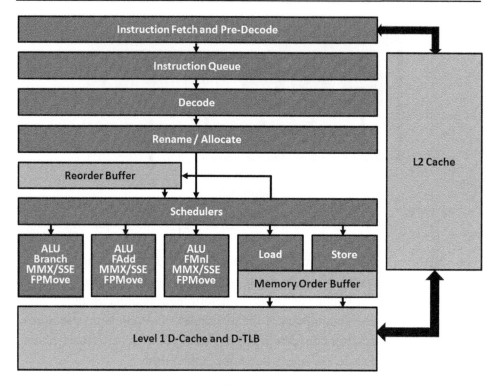

Figure 3.7
Example processor (Intel® Core™ Architecture).

In the compute space, different approaches for off-chip communication exist, by using either dedicated links to connect chips together or a central memory-controller architecture.

HyperTransport is a generalized point-to-point I/O interface that's been enhanced to support cache coherency. As an example, it is used for the I/O with AMD's Athlon chips, but also for the nVidia nForce 500, 600 and 700 series. It also finds its use for I/O and non-uniform memory access (NUMA) support in AMD's Opteron chips, which lets designers create a multiple-chip system without any intervening chips like the memory host controllers found in other multi-chip solutions. Programming of those systems is influenced by the number of hops between chips and the frequency of multi-hop memory accesses, which is application- and software-environment-specific.

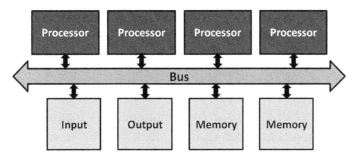

Figure 3.8
Classic SoC with shared bus.

In contrast, Intel keeps the processor chips simpler by leaving the memory and I/O interface management up to a central memory host controller.

As with off-chip interconnect, there are various options for interconnect between processors on a single die. Figure 3.8 shows the most basic approach of a central bus system, which may be hierarchical. It connects several masters with peripherals and memory.

Software programmability highly depends on the bus availability and locality of data for tasks running on the individual processors. If memory accesses are not guarded correctly, the locks and data races (as addressed in more detail in the software synchronization chapter) can lead to deadlocks and potentially functionally incorrect systems. The most common issues are as follows:

- Data races occur when two or more threads or processors are trying to access the same resource at the same time, where at least one of them is changing its state. If the threads or processors are not synchronized effectively, it is impossible to know which one will access the resource first. This leads to inconsistent results in the running program.
- Stalls happen when users have one thread or processor that has locked a certain resource and then moves on to other work in the program without first releasing the lock. When a second thread or processor tries to access that resource it is forced to wait for a possibly infinite amount of time, causing a stall. Even if a resource does not get locked for infinite amounts of time, they can cause

severe performance issues if a processor is not well used because a thread running on it is stalled.

- Deadlocks are similar to a stall, but occur when using a locking hierarchy. If, for example, Thread 1 or Processor 1 locks variable or memory region A and then wants to lock variable or memory region B while Thread 2 or Processor 2 is simultaneously locking variable or memory region B and then trying to lock variable or memory region A, the threads or processors are going to deadlock.

- False sharing is not necessarily an error in the program, but an issue affecting performance. It occurs when two threads or processors are manipulating different data values that lie on the same cache line. On an SMP system, the memory system must ensure cache coherency and will therefore swap the values every time an access occurs.

- Memory corruption occurs when a program writes to an incorrect memory region. This can happen in serial programs but it is even more difficult to find in parallel ones.

Figure 3.9 shows two very specific communication structures: an on-chip routing network and a crossbar for accessing peripherals and memory.

Software programming on SoCs with communication structures like these becomes a layout and routing issue. In a routing network, the distance between processors running tasks that are communicating with each other may influence system performance. It is non-trivial to assign tasks to processors.

Similar effects can appear in SoCs with crossbars, but mostly the arbitration between tasks accessing the same memory region will have impact on system performance. Again, in these SoCs placement of software tasks and routing of communication among them are non-trivial issues.

The last category of on-chip SoC communication is shown in Figure 3.10. Application-specific communication can be implemented when the target hardware architecture is very dependent on the application itself. As an example the data flow in video, audio and imaging applications can often be determined in advance because of fixed peak throughput requirements.

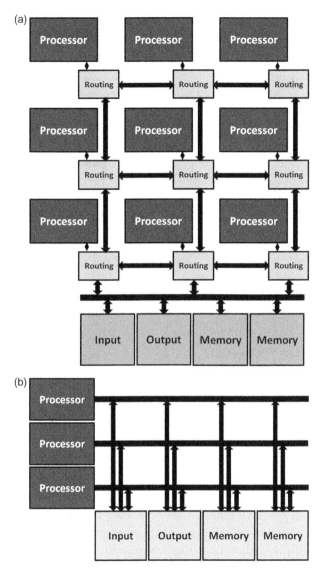

Figure 3.9
(a) Specific routing networks and (b) crossbar communication.

Especially in cases in which both application-specific communication and general communication via shared memory are available, software programmability becomes challenging as it is not clear at compile time which communication method will produce better system performance. A new class of system design tools may be required to help software designers to make these assessments in advance.

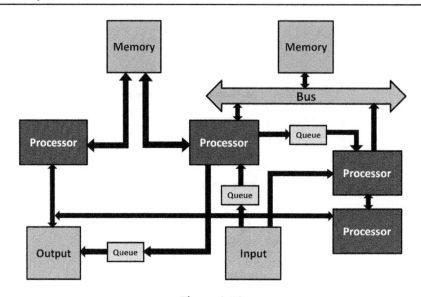

Figure 3.10
Application-specific communication. *Source: Tensilica, MCExpo Germany 2006.*

Memory architectures

The third fundamental aspect of multicore processors and systems is the way the memory is organized. Memory architecture and performance influence both the performance of tasks running on processors and the communication between tasks and processors.

Especially when task performance depends on locality of data in caches, a smart memory architecture and appropriate cache size will have a profound impact on performance. When tasks are communicating with each other via shared memory or are accessing the same memory regions for data processing, it becomes crucial that memory regions are locked appropriately to avoid data corruption with data races and that locking is done appropriately to avoid stalls and deadlocks.

Figure 3.11 shows simplified diagrams of symmetric multi processing (SMP) and non-uniform memory architecture (NUMA) systems as they can be found in the Intel Quad Xeon and AMD Quad Opteron processors.

An SMP architecture offers all processors a common memory bus. This architecture provides almost identical memory access latencies for any

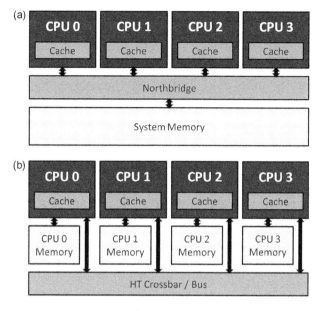

Figure 3.11
Simplified (a) SMP and (b) NUMA systems.

processor. However, the common system bus is a potential bottleneck of the entire memory. If a multi threaded application has a critical dependency on memory bandwidth, its performance will be limited by this memory organization. Especially cache coherency is often an issue affecting performance. If the processors in an SMP system have separate caches and share a common memory, it is necessary to keep the caches in a state of coherence by ensuring that any shared operand that is changed in any cache is changed throughout the entire system.

In the case of a NUMA system, each processor has its own local memory and therefore processors experience low latencies for accessing local memory. At the same time, remote memory is accessed at higher latencies. That's where the non-uniform memory organization notion originates from. If memory access is organized correctly, i.e. each processor operates with data mainly in its local memory, such an architecture will have an advantage over a classic SMP solution due to avoiding the bandwidth limits of the common system bus. The total peak memory bandwidth in this case will equal the double bandwidth of the memory modules used. Depending on the application, this may not be predictable at compile time, and task assignment may be done dynamically at run time.

Figure 3.12 illustrates two examples of AMP systems, as found in graphics processing units, which often are themselves highly parallel arrays of processing elements, and in loosely connected clusters.

In summary, from a perspective of software programmability the key questions which different memory and multi processor architectures raise are:

- How many processors are in the system and are sharing memory?
- How many memories are in the system?
- How many nodes or loosely connected clusters are in the system?
- What is encapsulated in any given cluster?
- For any given processor:
  - What is the distance to any other processor in the system?
  - What is the distance to any memory in the system?
  - How many cache levels exist and how large is each cache?
- For any given memory:
  - What is the size of the memory?
  - What processors are directly connected to the memory?

Application specificity

The fourth fundamental aspect of multicore processors and systems is the specificity of the hardware architecture to the application itself. There is a wide spectrum of application/architecture combinations; each one of them poses unique challenges from a programmability perspective.

The loosest coupling of application and architecture can be found in FPGA-like structures as shown in Figure 3.5 While the nature of the processing element itself and the communication topology determine the range of applications most suitable for the architecture, typically a wide range of applications can be implemented. The programming of the application is done independently from the hardware target and architecture-specific compilers are available to map the application programming model into the hardware architecture. One early example is the proprietary graphical programming model as had been introduced for

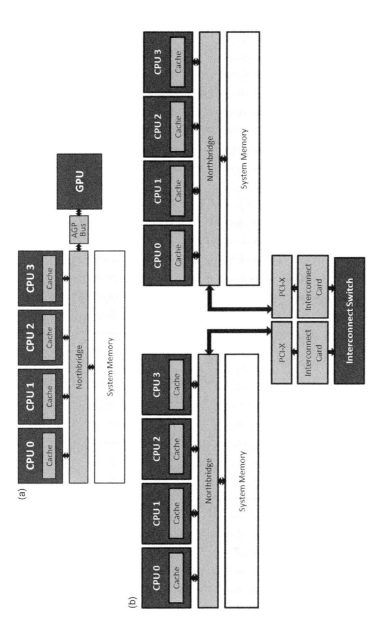

Figure 3.12

AMP: (a) graphics processing units and (b) loosely connected clusters.

the Ambric hardware; more recently, companies like Altera have started promoting OpenCL as a programming model for programming their next-generation devices.

The architecture shown in Figure 3.4 represents an array of processors but is combined with an application-specific communication structure. The "RAM Units" used in this architecture implement a very specific type of communication that is implemented as a special channel representation in the programming model used to describe the applications running on this device. This makes the programming very specific to this device, and compilers for mapping the application expressed in a specific programming model to the architecture are available.

In contrast, Figure 3.13 shows the Texas Instruments DaVinci architecture. The architecture is specific to video and audio applications, but, given the availability of two generic processing engines (ARM and TI DSP), the user still has some choice which execution engine to use for which task. There is no notion of compilation or task-dependent mapping to the processors, but multicore platforms like TI DaVinci are delivered with complex software development environments and software stacks exposing APIs for users to exploit specific application features like video and audio decoders. These software stacks do make the users largely independent of the platform, i.e. if the API remains stable, then the hardware architecture underneath can be changed without the user noticing.

The most extreme specificity of the architecture to the application can be found in designs like the application/architecture combination shown in Figure 3.14. In this case Epson used Tensilica's extensible processors to do what traditionally has been done in RTL design. The application has been partitioned into six large blocks with defined communication between them. Then, for each of the blocks (which traditionally would have been implemented in hardware), a processor with custom extensions for the particular task has been designed using the Tensilica tools.

The resulting implementation uses only very small portions of RTL. Epson claims significant time to market improvements using this design flow, part of which is achieved by doing functional verification − the traditional limiter for "signing off on RTL" − after the actual chip is

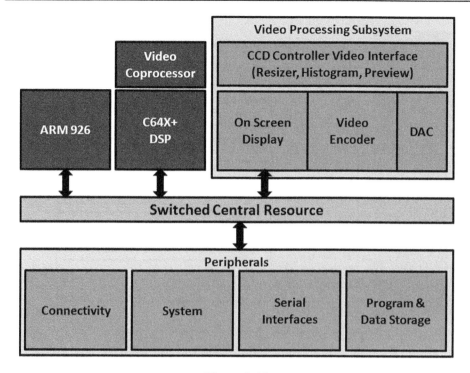

Figure 3.13
Texas Instruments Da Vinci. *Source: http://www.dspdesignline.com/showArticle.jhtml? printableArticle = true&articleId = 191600765.*

taped out. Once the architecture is defined, the software partitioning in application/architecture combinations like this one is trivial. However, to find the optimal partitioning of the application to meet performance, cost and power constraints, designers likely have to exercise an optimization loop trying out the effects of different partitioning of the application.

Application-specific platform topologies

The International Technology Roadmap for Semiconductors (ITRS) differentiates in its 2011 edition between several architecture templates, driven by the requirements of application domains.

The architecture template for networking (see Figure 3.15) satisfies the demands for scalable but largely parallelizable performance with a set of dedicated processors which can be chosen for processing at run time. The ITRS assumes the die area as constant, the number of cores

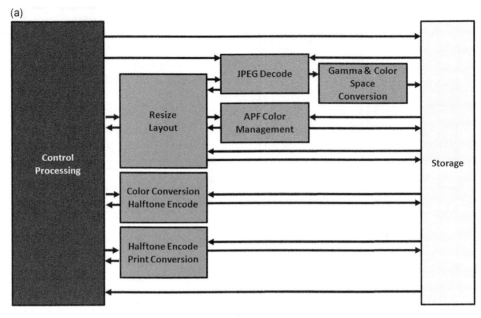

Figure 3.14

Epson REALOID application and architecture. *Source: Epson, Nikkei Electronics Processor Symposium, Tensilica, MC Expo Germany 2006.*

increasing by $1.4 \times$ /year, the core frequency increasing by $1.05 \times$ /year and on-demand accelerator engine frequency increasing by $1.05 \times$ /year. The underlying fabrics – logic, embedded memory (cache hierarchy),

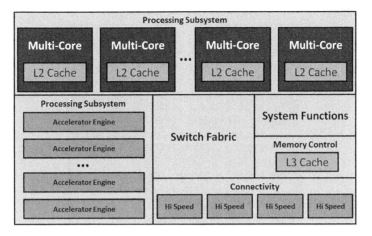

Figure 3.15
ITRS networking architecture template.

on-chip switching fabric, and system interconnect — will scale consistently with the increase in the number of cores.

The architecture template for portable consumer designs (see Figure 3.16) is most sensitive to low power due to its need to support mobility. It implements a highly parallel architecture consisting of a number of main processors, a number of PEs (processing engines), peripherals, and memories. The PEs are processors customized for specific functions. A function with a large-scale, highly complicated structure will be implemented as a set of PEs with software partitioned across the PEs. This architecture template enables both high processing performance and low power consumption by virtue of parallel processing and hardware realization of specific functions.

The Consumer Stationary SoC (see Figure 3.17) is used in a wide variety of applications in digital consumer electronic equipment, such as high-end game machines, which are assumed to be typically used in a tethered (non-mobile) environment.

It features a highly parallel architecture consisting of a number of main processors, a number of "DPEs", and I/O for memory and chip-to-chip interfaces. Here, a DPE is a processor dedicated to data processing that achieves high throughput by eliminating general-purpose features. A main processor is a general-purpose processor that allocates and

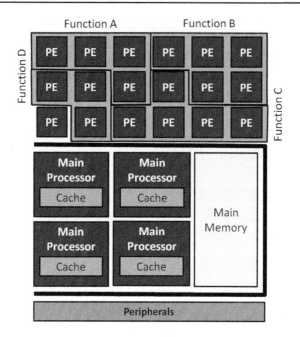

Figure 3.16
ITRS portable consumer architecture template.

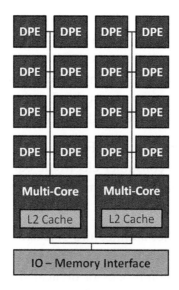

Figure 3.17
ITRS consumer stationary architecture template.

schedules jobs to DPEs. A main processor, along with a number of DPEs, constitutes the basic architecture. The number of DPEs will be determined by required performance and chip size.

Integration of multicore systems, MPSoCs and sub-systems

As discussed earlier, multicore design happens at the scope of the PCB system, the SoC, and the SoC sub-systems. Table 3.1 shows an overview of some multicore SoCs across several application domains, summarizing whether they are homogeneous or heterogeneous architecture, split between the global and potential local distribution (like a homogeneous graphics engine as a sub-system).

It also previews some of the programming challenges, i.e. whether scheduling happens at run time or compile time, and whether or not the programming is done by a general population of software developers or is done internally by the chip provider.

There are various drivers influencing the type of integration of multicore architectures:

- Application specificity: as previously shown in the section on application specificity above, the integration of processing elements in heterogeneous vs. homogeneous ways, in combination with specific communication techniques and memory organizations, varies from FPGA-like structures to specific topologies like TI's DaVinci to fully application-specific structures with point-to-point communications as shown in the Epson REALOID example. They determine the applicability for specific applications.
- Security: in the context of virtualization as described in a later chapter, specific processing elements or sub-systems can be fully dedicated to security aspects. To ensure security, they are not to be accessed from the outside or from specific parts of the design.
- Low power in networking: with low power being one of the main drivers towards multicore architectures, designers are trading off more processors at lower frequencies against a single core at higher frequency. The distribution of software across the multiple cores becomes the challenge. As a result, applications that lend themselves fairly easily to parallelization − like the increasing amount of packet

processing in networking to be scheduled at run time — use this trend extensively, resulting in the ITRS networking template as outlined in Figure 3.15.

- Low power in wireless and consumer applications: in other application domains like application processors for wireless and consumer devices, low power is equally important. As a result, heterogeneous designs have been introduced allowing the run-time switching between compatible sub-systems having different power consumption. The MPSoC in Figure 3.3 shows an ARM sub-system with a dual Cortex A7 and Cortex A15 domains, introduced as big. LITTLE sub-systems. A cache-coherent interconnect between the cores allows the switching of applications within a clock cycle between the high-performance Cortex A15 domain that consumes more power and the lower-performance Cortex A7 domain that consumes less power. This way different use cases — from high-performance gaming to lower-performance tasks like simple phone calls — can be switched between the domains at run time. To the application programmer, the switching is transparent, i.e. kept invisible, and the hardware takes care of proper cache coherency, relieving the programmer from low-level switching tasks.

- Hardware multi-threading: some processing elements provide multiple threads (sometimes called hyperthreading or hardware multi-threading). This allows for some shared elements of the execution hardware, with select bits replicated and hardware context storage for faster switching. A hyperthreading-equipped processor pretends to be several processors to the host operating system, allowing the operating system to schedule two threads or processes simultaneously. The advantages of hyperthreading are improved support for multi-threaded code, allowing multiple threads to run simultaneously, and providing an improved reaction and response time. Hardware multi-threading per core has limited scalability — bound by the saturation point of the execution units and the cost of additional threads — whereas multicore CPUs promise more potential for scalability. However, from a programmer's perspective the challenge with multicore processors is in the area of software development. Performance speed-up is directly related to how parallel the source code of an application was explicitly articulated.

Table 3.1: Example Overview of Multicore Architectures

Company	Name	# Proc	Processors	Application	Multicore Type		Software Style		
					Global	Local	Global	Local	Progr.
Infineon	Auto Platform	3	TriCore, PCP	automotive	hetero	homo	silo	CTD	internal
Freescale	SPACE	3	e200 PPC	automotive	homo		RTD		internal
Intel	Tflops	80	custom ALU	compute, graphics	homo		CTD		internal
AMD	Fusion	many	X86, PE	compute, graphics	hetero	homo	RTD	RTD	external
Azul Systems	VEGA 2	48	Custom 64 Bit	compute	homo		RTD		internal
SciCortex	SciCortex Node	6	MIPS	compute	homo		RTD		external
NVIdia	CUDA	128	custom PEs	compute, graphics	homo		RTD		external
Sun	Niagara	8	SPARC	compute, graphics	homo		RTD		external
TRIPS	TeraOps	2	custom	compute, graphics	homo		RTD		external
Toshiba, IBM Sony	Cell	9	PPC, Custom SPE	compute, graphics	hetero	homo	silo	RTD	int/ext
3DLabs	DMS-02	26	ARM, DSP	compute, graphics	hetero	homo	silo	CTD	external
Ambric	AMxx70	70	SR, SRD	multimedia	homo		CTD		external
Aspex	ASProCore	1000's	simple PE	multimedia	homo		CTD		external
Cirrus Logic	CS49300	2	custom DSP	multimedia	homo		CTD		internal
Clearspeed		64	custom DSP	multimedia	homo		CTD		external
MnD Semi	Silver Screen	8+	SPARC	multimedia	homo		CTD		internal
Stream Processors	SP16HP-G220	16	custom DSP	multimedia	homo		CTD		external
Tilera	Tile 64	64	custom	multimedia	homo		CTD		external
Xmos	Xcore	2	custom	multimedia	homo		CTD		external
ARM	ARM 11MP	4	ARM11	multimedia	homo		RTD		external
Element CXI	ECA-64	64	custom ALU	multimedia	homo		RTD		external
3 Plus 1	3P5220	5	ARM, DSP	multimedia	hetero	homo	silo	CTD	external
ARC/Synopsys	Vraptor	3	ARC750, Media Engine	multimedia	hetero	homo	silo	CTD	internal
Cradle	CT3400	12	DSP, MCU	multimedia	hetero	homo	silo	CTD	external
Freescale	MSC8144	4	Starcore, RISC	multimedia	hetero	homo	silo	CTD	external
MobileEye		14	MIPS, PE	multimedia	hetero	homo	silo	CTD	internal
Boston Circuits	gCore	16	ARC	multimedia	hetero	homo	silo	RTD	internal
Broadcom	BCM7440	2	MIPS	multimedia	hetero		silo		internal
CEVA	XS1200	1+	CEVA-X1620	multimedia	hetero		silo		internal
LSI Logic	Zevio	5	ARM, ZSP, Custom	multimedia	hetero		silo		internal
Micronas Decoder	DeCypher	2	MIPS, custom	multimedia	hetero		silo		internal
Micronas Encoder	Cypher	2	MIPS, Custom DSP	multimedia	hetero		silo		internal
Tensilica	HD Platform	6	Xtensa	multimedia	hetero		silo		internal
Texas Instruments	DaVinci DM6443	2	ARM, C64x+	multimedia	hetero		silo		int/ext
Zoran	COACH 10	5	MIPS, custom	multimedia	hetero		silo		internal
Zoran	SupraTV 160	4	MIPS, custom	multimedia	hetero		silo		internal
AMCC	Titan	2	PowerPC	networking	homo		RTD		external
Broadcom	SiByte	4	MIPS	networking	homo		RTD		internal
Cavium Networks	Octeon	16	MIPS	networking	homo		RTD		external
Intel	Xeon + E7520	4	Intel x86 IA	networking	homo		RTD		external
Raza Micro	XLR700	8	MIPS	networking	homo		RTD		external
Tilera	Tile 64	64	custom	networking	homo		RTD		external
Cisco	CRS-1	192	Tensilica	networking	homo		RTD		external
Freescale	MPC8574	34	PowerPC	networking	hetero	homo	silo	RTD	external
Epson	Realoid	7	ARM, Tensilca	printing	hetero		silo		internal
Emulex	AV150	10		storage	homo		RTD		internal
NXP	PNX952x	5	ARM, TriMedia	telematics	hetero	homo	silo	CTD	external
NEC	NaviEngine1	5	ARM11, PowerVR	telematics	hetero	homo	silo	RTD	external
Freescale	MPC5121e	2	Power, FPU	telematics	hetero		silo		external
Picochip	picoArray	300+	custom	wireless base station	homo		CTD		internal
Texas Instruments	TC16847	3	C64x+	wireless base station	homo		CTD		int/ext
Ericsson	custom ASIC	33	MCU, custom DSP	wireless base station	hetero	homo	silo	CTD	internal
STMicroelectronics	Greenside	3	ARM, ST140	wireless base station	hetero	homo	silo	CTD	internal
Sandbridge	SB3000	4	Sandblaster	wireless handset	homo		CTD		int/ext
Infineon	Music	4	custom SIMD	wireless handset	hetero	homo	silo	CTD	int/ext
Icera	Livanot	1+	DXP	wireless handset	hetero		silo		int/ext
Analog Devices		2	Sharc, ARM	wireless multimedia	hetero		silo		internal
NXP	PNX 5221	2	ARM, DSP	wireless multimedia	hetero		silo		int/ext
Qualcomm	MSM7600	4	ARM, DSP	wireless multimedia	hetero		silo		internal
STMicroelectronics	STm8815	5	ARM, ST200	wireless multimedia	hetero		silo		int/ext
Texas Instruments	OMAP3430	4	ARM, PowerVR, Custom	wireless multimedia	hetero		silo		int/ext
Zoran	Approach 5c	2+	ARM, ISP	wireless multimedia	hetero		silo		internal
Altera	NIOS	n/a	NIOS				silo, CTD, RTD		
Xilinx	MultiBlaze	3	MultiBlaze				silo, CTD, RTD		

RTD: Run Time Distributed, CTD: Compile Time Distributed

Programming challenges

Programming challenges are biggest when the application and the implementation architecture are orthogonal. Application-specific architectures offer little opportunity for software to be a contributing factor in overall system performance and optimization (once the architecture is decided on). With a fixed software layout in an application-specific architecture, software issues become one of validation.

Programming challenges can be severe when compile-time scheduling of tasks is used instead of run-time scheduling. When build-time configuration is needed for allocation of tasks to processors, automation of optimization becomes very desirable in complex systems. If the apportioning is performed when the application is decomposed or functions are coded, users need analysis tools to guide their decisions. If the scheduling is done at run time, then the operating system or scheduler assigns tasks to processors dynamically and may move tasks around to balance the load. The programming challenge is then focused on finding and expressing enough parallelism in the application.

The MPSoC user wants to develop or optimize an application for an MPSoC. However, in order to ensure the feasibility of meeting performance, constraints must be checked and the power consumption optimized, which requires the exploration of various options of partitioning the parallel software.

The MPSoC designer wants to ensure that selected applications can be run at the required performance and efficiency. This involves programming the computation-intensive parts of the applications and verification of MPSoC performance over a range of representative applications.

In each scenario, both the MPSoC user and designer need MPSoC simulation, debug and analysis. In addition, parallel programming techniques for the software need to be available in combination with efficient automation of mapping parallel software to parallel hardware. The degree of integration and the execution speed both determine the productivity achieved by users.

Application characteristics

The characteristics of the actual application running on the multicore processor influence the specific types of programming challenges. There are several issues of interest:

- Applications can be homogeneous or heterogeneous. Data parallel applications can have multiple channels of audio/video or multiple network streams. Task parallel applications can be pipelined to meet performance (video codec, software radio) or can have a mix of task types, like audio and video, radio and control, or packet routing and encryption and traffic shaping and filtering. Data parallel, homogeneous applications are likely to have a simple fixed layout, in which all CPUs do the same task or are scheduled at run time. The programming challenge is, in this case, limited to finding suitable parallelization in the application, as the actual mapping to the hardware architecture is straightforward.
- The computational demand of applications determines whether an application can be scheduled at compile time or needs to be scheduled at run time. Totally irregular applications will likely use run-time scheduling. Fixed demands normally imply a static layout and schedule, in which case parallelization of the application and the decision as to which tasks to run on which processor are done at compile time. Fixed computational demand is required for software radio and encryption algorithms; audio/video codecs are irregular but bounded (which means designers can work with worst cases). General-purpose computing with web browsers and user interfaces is completely irregular.
- The way data flows through the system determines its dependency on external memory performance and cache efficiency. When data flow is completely managed by the application using DMA engines, specific communication using FIFOs, or local memories then the influence of caches on performance diminishes. If, however, the application uses only one large shared memory, then the cache serves effectively as a run-time scheduler for data and will impact the performance of inefficiently written software. In cases where data flow is totally irregular, often the only real option for a designer is to

use a cache and let the system optimize at run time. If data flow is predetermined and fixed, like in some DSP applications, then the data flow can be statically allocated to FIFOs and local memories and scheduled. Often applications require a mixture: data streaming through the system with fixed rates (uncompressed multimedia streams, packets, wireless data) combined with fixed topologies and variable but bounded rates (compressed data) and variable irregular movement (system management functions, search algorithms). The question of how much data movement is optimized at design time and how much is decided on at run time, relying on caches and bus arbiters, has a profound impact on system performance and how to do the programming.

MPSoC analysis, debug and verification

In the majority of projects today, once the real hardware is available, the verification of software is finished by connecting single-core-focused debuggers via JTAG to development boards. Sometimes prototype boards are used in which FPGAs represent the ASIC or ASSP currently under development. More recently, designers have been able to use virtual prototypes utilizing simulation of the processor and its peripherals either in software or using dedicated hardware accelerators. All these techniques have different advantages and disadvantages.

Software verification on real hardware is only available late in the design flow and offers limited ability to "see" into the hardware. This approach does not normally take into account turnaround time in cases when defects are found that can only be fixed with a hardware change.

Prototype boards are available earlier than the real hardware, but they require the design team to maintain several code bases of the design — one for the FPGAs used as a prototype and one for the real ASIC/ASSP used later. This approach also makes it difficult to achieve proper visibility into the hardware design to enable efficient debug.

Virtual prototypes, either in software or using hardware acceleration, are available earliest in the design flow and offer the best visibility into the design, but they often represent an abstraction and, as such, are not "the real thing". This approach runs the risk that either defects are found that do not exist in the real implementation, or defects of the real implementation are not found because the more abstract representation did not allow it. Within this category there are significant differences between the time when the virtual prototypes become available and their speed. Often, abstract software processor models can be available long before RTL is verified, and they can be reasonably fast (of the order of 10 s of MIPS). However, users typically pay for this advantage by having to sacrifice some accuracy of the model. When cycle accuracy is required, models typically are available not long before the RTL, in which case hardware-assisted methods such as emulation become a feasible alternative.

Shortcomings and solutions

A fundamental shortcoming is that many solutions are single-core focused. The most pressing issues in systems running parallel software on parallel hardware require new techniques of analysis and debug. Users face issues of both functional correctness and performance. Data races, stalls, deadlocks, false sharing and memory corruption are what keep designers of multicore software awake at night (see also [2]). Multicore debug is covered in more detail in its own chapter later in this book.

MPSoC parallel programming

The choice, adoption and standardization of the right programming models will be a key trigger to the move of MPSoCs into mainstream computing. There are various advantages to using high-level MPSoC programming models that hide hardware complexity and enhance code longevity and portability.

Several programming models have been analyzed in projects under the MESCAL research program, including some dedicated to the INTEL IXP family of network processors and some as a subset of the MPI (message passing interface) standard. Other programming models focused on high-performance computing are OpenMP and HPF.

In the SoC world, ST microelectronics research is reporting on a project called MultiFlex to align more with the POSIX standard and CORBA (which has also been standardized on by the US DoD for future radios; JTRS, see [3]). Philips has been presenting an abstract task-level interface named TTL, following earlier work on YAPI. Another DSP-focused programming model is called StreamIt; even SystemC, with its concepts of channels and ports, could be viewed as a software programming model, but its adoption for software design is open to question.

Each of the different programming models offers specific advantages, often within specific application domains. In addition, the target architectures may affect the choice of programming model – for example, in non-uniform memory architectures (NUMA). This means there is a trade-off between abstraction and performance: e.g., CORBA has significant overhead and may not be suitable for high performance.

Figure 3.18 outlines graphically one possible approach in which parallel tasks – the units of work in an application – communicate with each other via channels and talk to channels via ports. Various communication modes like blocking and non-blocking can be supported, and communication can be implemented in various ways depending on the platform.

Parallel software and MPSoCs

Based on the use of models described above, one key aspect for both MPSoC designers and MPSoC users is the ability to rapidly program a

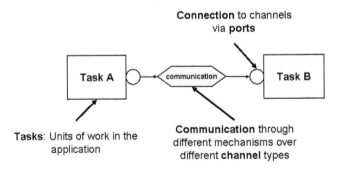

Figure 3.18
Tasks, ports and channels in a programming model.

variety of different combinations of parallel software that run on parallel hardware in an automated fashion. For automation to be possible, it is essential that the descriptions of the application functionality and the hardware topology be independent of each other and that a user have the ability to define different combinations using a mapping of parallel software to parallel hardware.

This requires a description of the software architecture in combination with the parallel programming models mentioned above. If a mechanism such as that shown in Figure 3.16 is used, in which the communication structures are separated from the tasks, then a coordination language to describe the topology is required.

In addition, a description of the hardware architecture topology is required which then allows a mapping to define which elements of the software are to be executed on which resources in the hardware and which hardware/software communication mechanisms are to be used for communication between software elements.

In the hardware world, the topology of architectures can be elegantly defined using XML-based descriptions as defined in SPIRIT. In the software world, several techniques exist to express the topology of software architectures, such as those defined in UML.

Figure 3.19 illustrates this relationship. In the upper left, the topology of a video scaler application with 13 processes is shown, which are communicating via 33 channels. The lower right shows an MPSoC topology with four processors and shared memory. If they are kept independent, then different design experiments can be set up, mapping between processes in the application and the processors executing them.

Summary

The processor power crisis can be addressed by switching to more parallel hardware with multiple processors. Several variants of multi-processor architectures exist and are determined mostly by the type of processing units they use, the methods used for communication and their memory architecture.

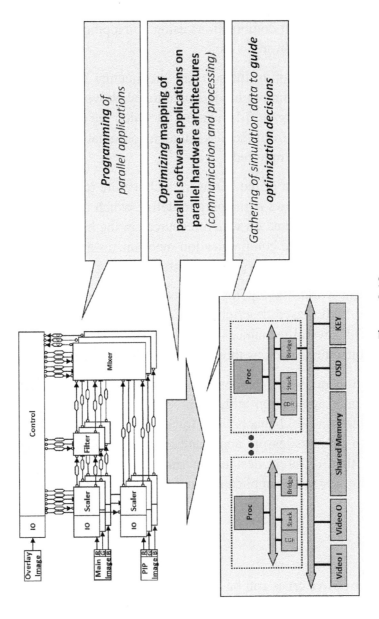

Programming *of parallel applications*

Optimizing mapping of parallel software applications on parallel hardware architectures *(communication and processing)*

Gathering of simulation data to guide optimization decisions

Figure 3.19
Application to MPSoC mapping.

The inevitable switch to multicore designs will cause a fundamental shift in design methodologies. However, the effects this switch will have on software programming and the ability of MPSoC designers and users to interact efficiently using design automation are not yet well understood and are likely to spawn a new generation of system design automation tools.

References

[1] Herb Sutter, The free lunch is over, a fundamental turn toward concurrency in software, Dr. Dobb's Journal 30(3) (2005). <http://www.gotw.ca/publications/concurrency-ddj.htm>.
[2] Max Domeika. Development and optimization techniques for multicore processors. <http://www.ddj.com/dept/64bit/192501977>.
[3] Available from: <http://jtrs.army.mil/>

CHAPTER 4

Memory Models for Embedded Multicore Architecture

Gitu Jain

Software Engineer, Synopsys

Chapter Outline

Real World Multicore Embedded Systems.
DOI: http://dx.doi.org/10.1016/B978-0-12-416018-7.00004-3

75

Introduction

The applications that will run on general-purpose computing systems are not known in advance, and the design of the memory architecture has to be general enough to support a number of heterogeneous applications with varying memory requirements. On the other hand, embedded systems are designed to run a set of pre-defined applications. The memory access patterns for these applications can be determined a priori. This information can be used to customize the memory architecture to meet the embedded system goals, whether it is to minimize cost or size, maximize performance, reduce power consumption, or support portability.

The design of the memory architecture becomes even more critical in multicore embedded systems due to the sharing of data between the multiple cores within the system. For example, how the cache hierarchy is laid out and what cache coherence protocol is used can have a great influence on the performance of the system.

In this chapter, we start our discussion with the different types of memory used in embedded systems. The memory architecture is described in section 3 with special emphasis on the cache structure. In embedded systems, the number of levels of cache, the cache size, and the cache line size can all be determined by prior experimentation and benchmarking with the target application. The embedded system designer can also choose alternate memory configurations such as DMA-controlled SRAM scratch pad memory, embedded DRAMs, and FIFOs that satisfy different needs of the application.

Multicore memory architecture options are presented in section 4. In multicore systems, memory can be configured in many different forms, ranging from memory that is completely shared among all the processors or cores, to memory that is private to each core. How information is shared and communicated between the cores varies and depends upon the memory configuration.

An overview of cache coherence protocols is given in section 5, with detailed explanation of the MESI protocol. We discuss transactional memories before concluding.

It is important to determine what type of memory and memory configuration works best for your application, and to design applications that utilize a given memory architecture efficiently. This chapter will give you a good overview of different options available to you so that you can make the right choices for your system since there is not a single memory architecture that will work well for all embedded systems.

Memory types

Memory on a low- to mid-end embedded system is usually expensive and not as abundant as on a desktop, laptop, or high-end embedded system. Given the many types of memory available for an embedded designer to choose from, you need to be aware of the differences between them so that the needs of the embedded device are met most efficiently in terms of size, storage density, cost, performance, power, ease of access and volatility.

Depending upon the size of the system, memory can be present in the form of volatile RAM, for holding variables and stack, and non-volatile memory, such as FLASH or ROM, magnetic disks, or removable memory cards, for holding permanent data. The different types of memory commonly used in embedded systems are described below:

RAM

Random **a**ccess **m**emory is used for high-speed access to temporary data used during execution of a program. Each word in memory is directly accessed by specifying the address of this memory word, and access time is independent of the word location. RAM retains its contents as long it has electrical power, i.e. it is *volatile*. The storage capacity of RAMs is limited by the bus size.

DRAM

Dynamic **RAM** is inexpensive memory that provides high storage density. Since it is capacitive storage, it needs to be refreshed periodically using a DRAM controller, which increases the power consumption. DRAMs are accessed in terms of rows, columns, and pages, which significantly reduces the number of address bus lines.

SRAM

Static **RAM** devices have extremely fast access times but are more expensive to produce than DRAMs. Many embedded systems include both types: a small block of SRAM along a critical data path and a much larger block of DRAM for everything else.

NVRAM

Non-volatile **RAM** is SRAM with battery backup so that it can retain its data when the power is turned off. NVRAM is fairly common in embedded systems but is expensive because of the battery. It is normally used for system-critical information.

DPRAM

Dual-port **RAM** is an SRAM with two I/O ports that access the same memory locations. DPRAMs are frequently used for shared memory in dual-core processors.

EPROM

Electrically programmable read-only memory is a one-time programmable, non-volatile memory. It can be field programmed once using special programming stations. **EEPROM** is an electrically erasable **EPROM**, and, unlike an EPROM, can be *reprogrammed*. EEPROMs are expensive, and writing to the EEPROMs takes longer than to a RAM. An EEPROM should be used for storing small amounts of data that must be saved when power is removed, not as your main system memory. An EEPROM is similar to *flash* memory. The main difference is that an EEPROM requires data to be written and erased one byte at a time whereas flash memory allows data to be written and erased in blocks, making flash memory much faster.

Flash

Flash memory is a low-cost, high-density, non-volatile computer storage chip that can be electrically erased and reprogrammed. Flash memory can be either *NOR-Flash* or *NAND-Flash*. NOR-Flash allows a single word to be written or read independently. It can be used for storing boot code. NAND-Flash is denser and cheaper than NOR-Flash. It is block-accessible and cannot be used for storing code. It is primarily used in memory cards, USB flash drives, and solid-state drives.

SD-MMC

A **s**ecure **d**igital **m**ulti**m**edia **c**ard provides cheap, non-volatile storage of the order of gigabytes. These cards are very compact and can be used with portable systems such as digital cameras, video game consoles and mobile phones.

Hard Disk

This is a non-volatile, random access, magnetic storage device that can be bulky and requires a disk reader to read memory locations. This memory is generally used for bulk storage without size constraints. Hard disks are only present in larger embedded systems.

Memory architecture

Embedded systems are designed to run a set of well-defined applications. You can use your knowledge of the memory access patterns of these applications to design the memory architecture of the embedded system to optimize parameters of interest, such as area, power, performance, and volatility.

Real-time embedded systems usually have heterogeneous on-chip memory architectures with a mix of data and instruction cache, DMA (direct memory access), scratch pad memory, and custom memory. In this section we will explore the characteristics of these different types of memories.

Cache

A *cache* is a small memory made of fast SRAM instead of the slower DRAM. It is attached to a processor or core to reduce access times for frequently used memory locations by storing a copy of the most recently used data or instructions.

Memory caching is effective at improving the performance of an application on the average because the access patterns in typical computer applications exhibit locality of reference — *temporal* locality if the same data is requested again very soon, and *spatial* locality if the data requested is physically close to data already in cache. Program instructions tend to exhibit temporal and spatial locality of reference as well, leading to improvement in performance when an instruction cache is used.

The cache works by keeping recently used data and instructions in a small, fast memory. A processor performing a read or write into a specific memory location first checks to see if the data from that memory location is already in the cache. If it is, it performs the read/ write on this data instead of the main memory. This is much faster than accessing the main memory, and can improve the performance of the application if used correctly.

Cache hardware decides which cache line to evict when new data is read into the cache. The cache can improve program performance tremendously, and is in fact designed to do so for the *average case*, but it is unpredictable and depends largely on the data access patterns. This can be a big problem for real-time systems where predictable performance is one of the critical requirements of the system. As a result, it has been common practice either to lock down portions of the cache (discussed later) or to simply disable the cache for sections of code where predictability is required.

The smallest, fastest cache on a computer is the *register file*. The use of the register file is controlled by software, typically by a compiler, as it allocates registers to hold values retrieved from main memory and to temporarily hold results of program execution.

Most processors have at least three additional caches: a read-only *instruction cache* to speed up instruction fetch, a read-write *data cache* to speed up data fetch and store, and a *translation lookaside buffer* (TLB) used to speed up virtual-to-physical address translation for both executable instructions and data.

The data cache is usually organized as a hierarchy of several cache levels such as level 1 (L1), level 2 (L2), and more. The memory

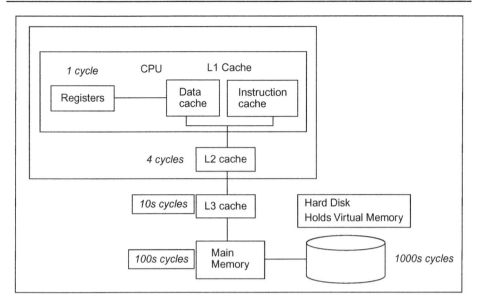

Figure 4.1
Memory hierarchy.

architecture, with the cost of data access at different levels of memory, is shown in Figure 4.1.

Translation lookaside buffer (TLB)

A **translation lookaside buffer** (**TLB**) is a type of cache used to improve the speed of virtual-to-physical address translation for systems that utilize *virtual memory* (section 3.2). The TLB is typically implemented as a *content addressable memory* (CAM) (section 3.7). The CAM search key is the virtual address and the search result is a physical address. A TLB hit occurs if the requested address is present in the TLB. The retrieved physical address is used to access memory. A TLB miss occurs when the requested address is not in the TLB. In that case the virtual address has to be translated into a physical address by looking up the page table, which is an expensive process. This virtual-to-physical address mapping is now entered into the TLB. A TLB can range in size from 8 to 4096 entries. A TLB hit takes approximately 1 clock cycle while a miss can take from 10 to 100 clock cycles. Each core in a multicore system has its own TLB.

Instruction cache

The use of instruction cache has a greater impact on performance than the use of data cache. This is because the processor is usually sitting idle while an instruction is being fetched, while it can sometimes continue doing useful work during the duration of a data fetch. Instructions also tend to have greater locality of reference, both spatial and temporal, resulting in higher hit ratios for the instruction cache. Program flow is much more predictable than data access patterns, resulting in more performance gain from pre-fetching instructions. The address of each instruction remains the same during a program's execution, leading to more predictability in instruction cache behavior. For optimal system performance, a processor needs to be busy doing computational work, not waiting for the next instruction or data to be fetched from memory.

When code executes, the code words at the locations requested by the instruction set are copied into the instruction cache for direct access by the core processor. If the same code is used frequently in a set of program instructions, storage of these instructions in the cache yields an increase in throughput because external bus accesses are eliminated. A cache read miss from an instruction cache generally causes the most delay because the processor has to wait until the instruction is fetched from main memory.

The cache hardware can be customized to exclude instructions marked for exclusion. The designer may also lock down critical instructions or sections of the code into the cache to attain predictable performance for real-time systems. Locked lines are not replaced to make room for incoming instructions.

To achieve the best performance from an instruction cache, the designer needs to make sure that the generated code is as small as possible. The designer should help the processor make good pre-fetching decisions, either through code layout or with explicit pre-fetching.

Data cache

The data cache hierarchy in typical processors is composed of several levels. The **L1**, or primary cache, is a small, high-speed cache residing in the CPU core (*on-chip* cache). It typically ranges in size from 8 KB to

64 KB. The L1 cache is normally split into two separate caches, one for instructions and the other for data.

The **L2**, or secondary cache, is bigger, slower, and less expensive than the L1 cache and, unlike the L1 cache, it usually resides outside the CPU core (*off-chip* cache). It typically ranges in size from 64 KB to 4 MB. Many processors may also include a third larger, slower and less expensive cache at the L3 level just before the main memory. The L2 and L3 caches may be shared between the processor cores.

If the data requested by the cache client (processor or core) is already contained in the cache, a so-called *cache hit*, then this request can be served by simply reading the cache, which is comparatively faster. Otherwise, upon a *cache miss*, the data has to be recomputed or fetched from its original storage location, which is slower. The percentage of accesses that result in cache hits is known as the *hit rate* or *hit ratio* of the cache. The hit rate has a direct effect upon the overall system performance.

Multi-level caches generally operate by checking the L1 cache first; if it hits, the processor proceeds at high speed. If the smaller cache misses, the next larger L2 cache is checked, and so on, before external memory is accessed. The cache controller, discussed later, handles the details of retrieving the required data from the closest level of cache (or main memory) that contains it.

There are two basic approaches to modifying data stored in a cache: *write-through*, where the data is updated in both the cache and the main memory simultaneously; and *write-back*, where only the copy of the data in the cache is updated. A modified cache block is written back to the main memory just before it is replaced by another. Write-back cache is more complex to implement, since it needs to keep track of and mark "dirty" any modified memory locations so that they can be written into main memory at some later time, when they are evicted from the cache.

A read miss on the dirty data by another cache of a multicore system may require two memory accesses to service: one for the cache holding the modified data to write it back to the main memory, and then one to retrieve the needed data into the current cache. This process of keeping the caches of different cores consistent is referred to as coherency, and will be discussed in section 5.

A cache controller is a hardware block that can dynamically move code and data from main memory to the cache memory and back. The incoming data or code replaces old code or data which is currently not being used in the cache memory. The *replacement policy* of a cache is the heuristic responsible for ejecting a cache line to make room for an incoming cache line, based upon the address of the cache line in main memory.

The replacement policy is called *fully associative* if the cache line can be placed anywhere in the cache. On the other end, if the cache line can only be placed in one place in the cache, the cache is *direct mapped.* Most replacement policies are somewhere in between, called *N-way set associative*, where an entry in main memory can go to any one of *N* places in the cache. The higher the associativity, the more chance there is of getting a cache hit as there are more cache locations in which a particular cache line can reside. This does, however, come with a cost: it takes more time and power to search for a cache line. Most caches implement a 2- or 4-way associativity as increasing the associativity beyond this is shown to have less effect on cache hit rate. See Figure 4.2.

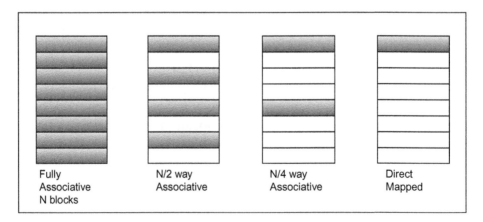

Figure 4.2
Cache associativity.

The memory of a cache is divided into cache lines. This is the size of data that is read from or written into memory at one time. Depending upon the size of the cache, there are N number of lines that can be stored in the cache at any given time, where $N = cache\ size/line\ size$.

A memory address is split into a *tag*, *index*, and a *block offset*, as shown in Figure 4.3. The *index* is used to determine in which cache line the data has been placed. The *block offset* specifies the position of the data in that particular cache line. The *tag* contains the most significant bits of the address and is stored in the cache line, as shown in Figure 4.3, and enables the cache to translate the cache address to a unique memory address. The *data* section contains the actual data. The *flag bits* are used by the cache coherence protocols and will be discussed in section 5.

For example, consider a system with 256 MB of memory and a direct-mapped L1 data cache of size 8 KB with 64-byte cache lines. There are 8 K/64 = 128 cache lines. The memory address has to be $\log_2(256\ MB) = 28$ bits long to access any location in memory. Since the cache line is 64 bytes the offset is $\log_2(64) = 6$ bits, and will point to the exact location of the memory being accessed on the cache line. The index into the cache in this case is $\log_2(128) = 7$ bits long. So the tag will have $(28 - 7 - 6) = 15$ bits. The tag is stored along with the cache line data. When a memory location is accessed, the cache controller checks to see if the cache already contains the data by using the memory index to access the appropriate cache line and comparing the tag value(s).

For 4-way associative L1 data cache, there will be 128/4 = 32 sets for the lines. This gives $\log_2(32) = 5$ different indices. Each memory access

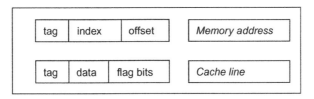

Figure 4.3
Memory address and cache line.

now needs to be checked against 4 different cache lines for a match. The memory address for a 4-way associative cache will be:

| 17 | 5 | 6 |

Cache customization

While all the cache blocks in a particular cache are all the same size and have the same associativity, typically "lower-level" caches, such as the L1 cache, are smaller in size and have smaller cache lines, while "higher-level" caches, such as the L2 cache, are larger in size and have larger cache lines.

When deciding upon a cache's total size, consider the fact that a small cache is more energy efficient and has a good hit rate for a majority of applications, but a larger cache increases the range of applications displaying a good hit rate, at the expense of wasted energy for many applications.

An embedded system typically executes just a small set of applications for the system's lifetime, in contrast to general-purpose desktops, so, ideally, we would like to tune the architecture to those applications. Two of the most important aspects for customizing a cache in an embedded system are the cache line size and the cache size.

As a rule of thumb, if the memory access pattern exhibits high spatial locality, i.e. is very regular and consecutive, a longer cache line should be used since it minimizes the number of off-chip accesses and exploits the locality by pre-fetching elements that will be needed in the immediate future. On the other hand, if the memory access pattern is irregular, a shorter cache line is desirable as this reduces off-chip memory traffic by not bringing unnecessary data into the cache. The maximum size of a cache line is the main memory page size. Estimation techniques using data re-use analysis can be used to predict the number of cache hits and misses for different cache and cache line size. Based upon the results, the best size should be selected for the cache [1].

Designers of real-time embedded systems sometimes disable the use of cache as it can lead to unpredictable access times depending upon whether the data/instruction access resulted in a cache hit or not.

An alternate solution, for hardware systems that support this, is to *lock down* lines in the cache. System software can load critical data and instructions into the cache and instruct the cache to disable their replacement. This gives programmers the ability to keep what they need in cache and to let the caching mechanism manage less-critical instructions. The chief disadvantage of this approach is that, once data and instructions have been pinned down, it is not possible to reorganize the cache contents without significant overhead. A good overview of customization techniques for the memory architecture of embedded systems is presented in [2].

A cache improves program performance for the average case, which can lead to unpredictable behavior for real-time systems with critical and consistent performance requirements. An alternate solution for real-time systems is to use *software-managed caches*, a flexible, low-overhead mechanism that allows software to steer the replacement decision in the cache [3].

For example, an instruction that accesses memory can be annotated with an attribute, called a *cache hint*, to specify whether it should be retained in the cache or not. The way this works is that the application running on the embedded system is first instrumented to determine memory access patterns. Once locality of data access patterns is determined, cache hints are annotated to the memory instructions of the original program. This can result in performance improvements without the unpredictability attached to traditional hardware-managed cache.

Virtual memory

Virtual memory allows users to store data on a hard disk, but still use it as if it were available in memory. The application makes accesses to the data in a virtual address space that is mapped to memory, whereas the actual data physically resides on the hard disk and is moved to memory for access. Virtual memory allows access to more RAM space than is physically available on the system. In a multi-tasking application, each task can have its own independent virtual address space called a discrete address space. The operating system on the computer is responsible for virtual memory management, with some support from a hardware memory management unit (MMU). Smaller embedded systems, however, do not have virtual memory or an MMU to handle the virtual memory.

Scratch pad

A *scratch pad* SRAM memory is an alternative to cache for on-chip storage. It is a small, high-speed internal memory used for temporary storage of calculations, data, and other work in progress. Scratch pads are better than traditional caches in terms of power, performance, area, and predictability, making them popular in real-time systems such as multimedia applications and graphic controllers. DSPs (digital signal processors) typically use scratch pads.

In embedded systems, the application code is known a priori, and the critical code and data, as identified by the embedded system designer, can be carefully placed in the scratch pad. This identification of which data to be placed in the scratch pad is performed while the application is being designed, and can be done either manually using compiler directives, or automatically using a compiler. This is in contrast to cache memory systems, where the mapping of program elements is done during runtime.

Efficiency of use is dictated by the degree of locality of reference, similar to the cache. Scratch pad SRAM guarantees a single-cycle access time while access to the cache is subject to cache hits and misses. The address space of data mapped onto the scratch pad memory is disjoint from the address space of the main memory. An embedded system designer has to figure out how to partition the available on-chip memory space into data cache and scratch pad memory so that the total access time and power dissipation is minimized.

Figure 4.4 shows the architectural block diagram of an embedded core processor with both scratch pad and data cache [4]. The address and data busses from the CPU connect to the data cache, scratch pad memory, and the external memory interface (EMI) blocks. On a memory access request from the CPU, first the data cache is searched, and if there is a hit the cache transfers the data to the CPU. In the case of a cache miss the scratch pad is searched and if there is a hit, the scratch pad gains

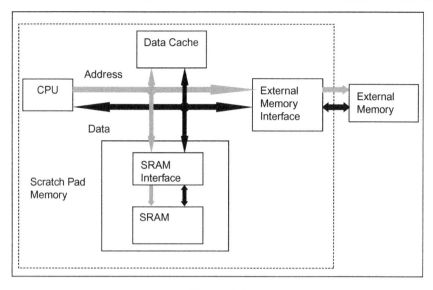

Figure 4.4
Scratch pad memory.

control of the data bus and transfers the data to the CPU. In the case of a miss, the EMI transfers a block of data equal to the cache line size from the external memory into the cache and CPU. Data transfer is managed using a *DMA* (direct memory access).

Scratch pad memory can also be used for *software overlays* as described in the next section.

Software overlays

A real-time embedded processor may choose not to have a built-in cache and cache controller. Instead, it may choose to utilize the on-chip SRAM memory as a scratch pad that can be used to store frequently used code by using *software overlays*.

Each code section mapped onto an overlay has a *run space* and *live space*. Live space is the space in the main memory where the code section resides when not running. Run space is the space in internal memory where the code section resides during execution. Software called an *overlay manager* is responsible for moving the code from live space to run space. The linker and loader tools have to provide support for generating overlay symbols for the code sections to be mapped to overlays. The overlay symbols also contain information about the run space and the live space of the overlay. This information is used by the overlay manager to move the overlays dynamically. There can be multiple overlays in a system, each having a different live space but the same run space.

The embedded system programmer is responsible for identifying mutually exclusive code sections in the application in order to be able to use this feature. The time between swaps of these sections should be high so that they don't degrade the performance by switching too frequently. The overlay section should not be larger than the run space available. Software overlays are normally used for code and rarely for data.

DMA

An efficient real-time system is one where the CPU is used for the maximum number of cycles. Normally, a CPU is fully occupied for the entire duration of a read or write operation requiring the transfer of information from or to memory.

Direct memory access (DMA) is a peripheral that allows large quantities of data to be transferred to or from memory without using the CPU. The DMA is one of the integral assets in today's embedded systems. It works in parallel with the CPU, thereby simulating a multi-processing environment and effectively increasing the CPU's bandwidth. The use of a DMA can enhance the utilization of the CPU significantly, thereby helping developers design optimized, high-performance, energy-efficient systems.

A DMA for data transfers, along with an MMU for memory mapping, can be used in multicore processors to allow computation to proceed in parallel with the data transfer. Data transfers, in which the CPU is not involved, are called *zero-copy* transfers.

The DMA can be used for the following types of data transfer:

- memory to memory transfer
- memory to peripheral transfer

- peripheral to memory transfer
- peripheral to peripheral transfer.

During a read or write operation, the CPU initiates the transfer by informing the *DMA controller*, a separate device that manages the DMA, that data needs to be transferred. Typical setup parameters, to initiate the transfer, include the base address of the source area, the base address of the destination area, the length of the block, and whether the DMA controller should generate a processor interrupt once the block transfer is complete.

After initialization, the DMA controller sends a *DMA request* signal to the CPU, asking permission to use the bus. The CPU completes its current bus activity, stops driving the bus, and returns a *DMA acknowledge* signal to the DMA controller. The DMA controller then reads or writes one or more memory bytes, driving the address, data, and control signals as if it were itself the CPU. The CPU's address, data, and control outputs are tri-stated while the DMA controller has control of the bus. When the transfer is complete, the DMA controller stops driving the bus and de-asserts the DMA request signal. The CPU removes its DMA acknowledge signal and resumes control of the bus. The DMA controller itself does no processing on this data.

DMA operations can be performed in either burst or single-cycle mode. In *burst mode*, the DMA controller keeps control of the bus until all the data buffered by the requesting device has been transferred to memory. In *single-cycle mode*, the DMA controller gives up the bus after each transfer. This minimizes the amount of time that the DMA controller keeps the processor off the memory bus, but it requires that the bus request/acknowledge sequence be performed for every transfer. This overhead can result in a drop in overall system throughput if a lot of data needs to be transferred.

The simplest way to use DMA is to select a processor with an internal DMA controller. This eliminates the need for external bus buffers and ensures that the timing is handled correctly. Also, an internal DMA controller can transfer data to on-chip memory and peripherals, which is something that an external DMA controller cannot do. Because the handshake is handled on-chip, the overhead of entering and exiting DMA mode is often much faster than when an external controller is used.

Managing data through DMA is the natural choice for most multimedia applications, because these usually involve manipulating large buffers of compressed and uncompressed video, graphics, and audio data.

DMA can be used along with an on-chip SRAM scratch pad memory to store temporary data. The on-chip memory can be divided into an L1 cache to store frequently used data and instructions, and a scratch pad for DMA transfers.

As a rule of thumb, if your application code fits into L1 internal memory, disable the instruction cache and lock the code in the internal memory. If the code does not fit, then enable the L1 instruction cache. If desired performance is not achieved, lock down the cache lines for critical code sections and use the rest of the internal memory as scratch pad for DMA transfers.

DRAM

DRAM can be used as off-chip as well as on-chip memory in embedded systems. Off-chip DRAM is used for the main memory, where code and data of a running application are stored. *Embedded DRAM* (eDRAM), as the name suggests, is integrated on the same die as the ASIC or processor.

Embedded DRAMs have performance, density, bandwidth, and power advantages that make them very attractive for use in embedded systems. Embedding memory on the chip allows for much wider busses and higher operation speeds. DRAM is four times as dense as the SRAM used for cache and scratch pad memory. eDRAM is cost effective at the system level, even though DRAM is more expensive to produce than SRAM, because the increase in density leads to area savings that offset this cost.

DRAM requires periodic refreshing of the memory cells, which adds complexity. However, the memory refresh controller can be embedded along with the eDRAM memory. Unlike commodity DRAMs, which are only available in a standard range of densities such as 4, 16 or 64 Mbits, eDRAM density can be customized to the system needs. This also leads to area and cost minimization as no memory is wasted.

Embedded DRAMs offer better memory performance through the use of specialized access modes that exploit the internal structure of data within these memories. Synthesis and compilation techniques can employ detailed knowledge of the DRAM access modes and exploit advance knowledge of an embedded system's application to better improve system performance and power.

Embedded DRAMs are used in many devices today, such as Apple iPhone, Sony Playstation, Nintendo Wii, Microsoft Xbox 360, and many more.

Special-purpose memory

In addition to general memory structures such as caches and scratch pads, there exist various other types of custom memory structures that implement specific access protocols. For example:

- A **LIFO** (last-in-first-out) memory block is present in most microprocessors and microcontrollers. It is used to store the address of an instruction in a routine when the routine is interrupted during execution by the operating system to tend to some other urgent task. Once the task is completed, the operating system returns to the LIFO to find the address of the last step in the interrupted routine so that execution can continue from where it left off. This behaves like a stack in memory.
- **FIFO** (first-in-first-out) memory chips are used in buffering applications between devices that operate at different speeds or in applications where data must be stored temporarily for further processing. Typically, this type of buffering is used to increase bandwidth and to prevent data loss during high-speed communications.
- **CAM** (content addressable memory) is a special type of memory used in certain very-high-speed searching applications. Unlike standard computer memory, where the user supplies a memory address and the memory returns the data word stored at that address, a CAM is designed such that the user supplies a data word, and the CAM searches its entire memory to see if that data word is stored anywhere in it. If the data word is found, the CAM returns the list of one or more storage addresses where the word was found. The additional high-speed comparison circuitry does, however, increase the manufacturing cost, physical size, and power dissipation of this type of memory chip.

Memory structure of multicore architecture

Memory in multicore systems can be classified into two broad categories — shared memory and distributed memory. Many modern multicore systems employ both shared and distributed memory.

Shared memory architecture

In shared memory systems, all processors or cores have direct access to common memory. All applications developed for shared memory systems can directly address and access the same logical memory locations regardless of where the physical memory actually resides. Parallel applications can share data between tasks by addressing the same memory locations. This makes data sharing fast and uniform, but the responsibility for controlling synchronization to shared data lies with the programmer. Shared memory systems have limited scalability since they share a memory bus, and increasing the number of processors or cores, or the amount of memory, increases traffic on this shared bus to access the common memory. Large shared memory systems are also expensive to produce.

Shared memory machines can be classified based upon uniformity of memory access times: *UMA* and *NUMA*.

Uniform memory access (UMA)

Shared memory multiprocessors, also known as symmetric multi-processors (SMPs), with equal and uniform access to main memory, are classified as UMA machines, as shown in Figure 4.5. If there is a cache structure, then cache coherence protocols will be needed to manage access to shared memory locations. UMA systems that have cache coherency implemented at the hardware level are labeled *cache coherent UMA* or *CC-UMA*.

Non-uniform memory access (NUMA)

NUMA machines often consist of two or more SMP blocks linked together. The memory of one SMP block is directly visible and accessible from another SMP block. Since the SMP blocks are linked via a bus the memory access times may vary depending upon the physical

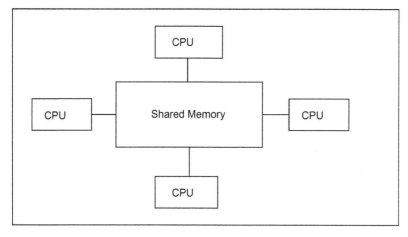

Figure 4.5
Uniform memory access multiprocessor.

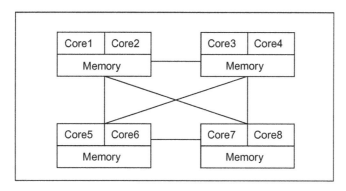

Figure 4.6
Non-uniform memory access multiprocessor.

proximity of the memory to the CPU making the request; see Figure 4.6. Like CC-UMA, if hardware cache coherency is present in the system, it is called CC-NUMA.

Distributed memory architecture

In distributed memory systems, not all cores or processors have the same access to memory; see Figure 4.7. Memory is local to a particular core or processor, with only that processor having direct access to that memory. Memory addresses in one processor do not map to another processor, so there is no concept of a global address space. Distributed memory systems

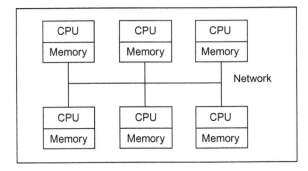

Figure 4.7
Distributed memory multiprocessor.

require a communication network to access memory on other processors. Access time to memory residing in other processors varies greatly, depending upon the proximity and the speed of the network used for data transfer. Changes made to local memory by one process have no effect on other processes, hence there is no need for memory synchronization and no overhead incurred from maintaining cache coherency. Distributed systems are scalable as it is very easy to add new processors and memory.

Cache memory in multicore chips

When considering a multicore chip, attention needs to be paid to the cache structure. Decisions such as whether the cache should be shared or private to each core need to be made. Is there a cache hierarchy, and, if so, how are the caches at the different levels managed? How is the consistency of the data in the cache and memory maintained for correct operation of the device?

Cache *shared* by multiple cores on a chip allows different cores to share data, and an update by one core can be seen by other cores with no need for cache coherence methods. Since there is now one cache per chip instead of one per core, the cache can be larger, allowing more data to be cached and leading to corresponding performance benefits. A shared cache will not add value if the threads running on the cores sharing the cache all access large amounts of data from different address spaces.

Private caches reduce contention among threads vying for space on the same cache and have the advantage that the access speed is higher since

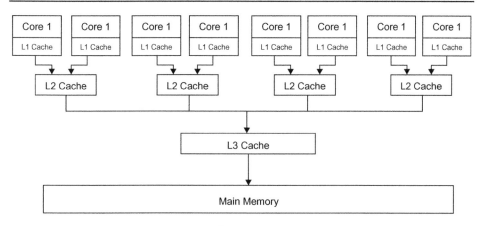

Figure 4.8
Chip multiprocessor cache structure.

the cache is closer to the core. The lowest-level L1 cache is usually private to reduce latency of frequently used data and instruction access for its core.

The highest-level cache, which is accessed just before main memory, generally should be a large shared cache to reduce the number of accesses required to main memory. For example, an eight-core chip with three levels may include an L1 cache for each core, an L3 cache shared by all cores, with the L2 cache intermediate, e.g., one for each pair of cores, as in Figure 4.8.

For private cache structures, there may be situations in which data is modified in a cache by a thread running on one core, and not yet written back to memory. A different thread running on another core may need to read that data. This kind of situation needs to be handled by the *cache coherence protocol*, discussed in the following section.

Cache coherency

Multicore systems may have several levels of memory cache as shown in Figure 4.8. To maintain consistency and validation of data, cache coherence protocols, provided by the processor, must be used. For example, consider the situation in Figure 4.9, where there are four cores

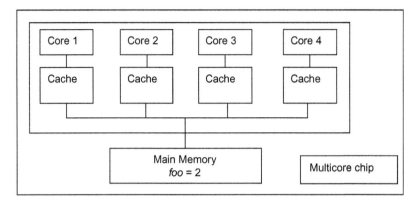

Figure 4.9
Cache coherency — 1.

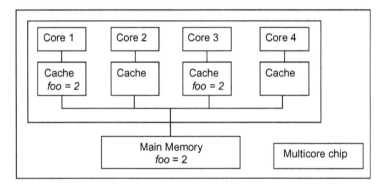

Figure 4.10
Cache coherency — 2.

with private cache memory. There is a data variable *foo* in main memory, with initial value **2**.

Both Core 1 and Core 3 read and store the value of *foo* in their cache; Figure 4.10. Core 3 now writes a new value **3** to *foo*. The value of *foo* in main memory gets updated, assuming the cache is *write-through*; Figure 4.11. Core1 now tries to read *foo*, and gets a stale, incorrect value from its cache.

Cache coherence protocols take care of situations like this.

There are two broad categories of cache coherence protocols — *directory-based* and *snooping*. Each is suited to architectures with different distributions of memory and interconnect, and with different bus bandwidth and communication characteristics.

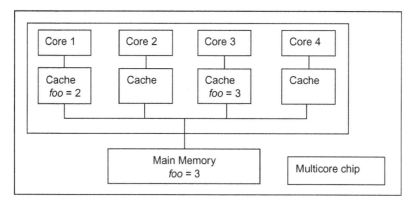

Figure 4.11
Cache coherency – 3.

Snooping protocols tend to be faster, if a common bus with enough bandwidth is available, since all data transactions are seen by all processors. Every request is broadcast to all nodes in the system, and as the system gets larger, the size of the bus and the bandwidth needed to support this protocol must grow proportionately. Thus, snoopy protocols are not scalable.

Directory schemes are slower since the directory access and extra interconnect traversal is on the critical path for cache misses. They also require dedicated memory to store the directory, which adds to the overall system cost. However, directory protocols use much less bandwidth since the messages are point-to-point and not broadcast. For this reason, they are generally preferred on larger embedded multicore systems.

Directory-based cache coherence protocol

Directory-based cache coherence protocols keep track of data being shared in an extra data structure (directory) that maintains the coherence between caches. The directory works as a look-up table for each processor to identify coherence and consistency of data that is currently being updated. The directory filters all requests from a processor to load or update memory in its cache. When an entry is changed, the directory either updates or invalidates the other caches sharing that entry. The directory can be centralized in smaller, symmetric multiprocessors and can be distributed in larger, shared-address NUMA systems or distributed memory multiprocessors.

The directory-based scheme scales better than snooping as it does not depend upon a shared bus for communication. The directory keeps track of all processors caching a memory block (cache line), and then uses point-to-point messages to maintain coherence. This allows the flexibility to use any scalable point-to-point network for communication.

Directory-based protocols fall under three broad categories:

- Full-map

    Every processor can store a copy of any block of data in its cache. Directory entries have one bit per processor, called the *Sharers* set, and a dirty bit to indicate *Exclusive* ownership. A bit set to 1 in the Sharers set indicates that a copy of the data resides in that processor's cache. If the dirty bit is set, then only one processor's bit is set. A cache has two bits of state per cache line; one bit indicates whether the cache line is valid, and the second indicates whether it has been written (dirty).

    The key issue with this protocol is that the directory memory requirements do not scale well. The number of bits needed in the Sharers set grows as the number of processors grows. Also, the larger the main memory, the larger is the directory needed. The second issue with full-map directories is that with the scaling of multiprocessor systems, the bandwidth and memory requirements of a centralized directory can become a bottleneck.

- Limited

    The directory size problem is solved by limiting the number of processors in the Sharers set, regardless of the number of processors in the system. Now the drawback is that every processor *cannot* store a copy of any block of data in its cache.

- Chained

    This protocol solves the second issue with the full-map directory protocol by maintaining a distributed version of the full-map scheme. This protocol keeps track of shared copies of data by maintaining a *chain* of directory pointers, distributed among the caches of all the processors [5].

Every time a processor loads a memory block into its cache or writes into a cache line, it sends a message to the central (or distributed) directory. This message will result in one of two actions — an update to the directory, or more messages to satisfy the request.

The directory maintains the state of each memory block; each block can be in one of three states — *Invalid*, *Shared* and *Exclusive*. The following requests are handled by the directory when a memory block is in the following state:

Invalid — No processor has cached the data.
- **Read miss**: data is read from memory into cache and the state of the block is changed to *Shared*.
- **Write miss**: data is read from memory into cache and written by the processor into the cache, but memory is not updated. The state of the block is changed to *Exclusive* and the Sharers bit vector indicates the identity of the owner.

Shared — At least one processor has cached the data; memory is up-to-date.
- **Read miss**: data is read from memory into the cache and the requesting processor is added to the Sharers bit vector.
- **Write miss**: data is read from memory into the cache and is written by the processor, but memory not updated. All processors in the Sharers set are sent invalidate messages. The requesting processor is added to the Sharers set and the state of the block is made *Exclusive*.

Exclusive — One processor (owner) has the data, the memory is out-of-date.
- **Read miss**: data request is sent to owner processor. The owner sends the data to the directory, where it is written to memory and sent back to the requesting processor. The owner's state is changed from *Exclusive* to *Shared* in the directory (it still contains a readable copy), and the requesting processor is added to the Sharers set.
- **Data write-back**: the owner is replacing the block and writes it back. The memory is updated, the Sharer set is empty and state of the block changes to *Invalid*.

- **Write miss**: the block gets a new owner. A message is sent to the old owner, which sends the value of the block to the directory, from where it is sent to the requesting processor. The Sharers bit vector is set to identify the new owner (after removing the old one) and the state remains *Exclusive*.

With all directory protocols, you have to be aware that, since communication is not instant and can vary from node to node, there is a risk that there are different views of memory at some time instances. These race conditions need to be understood and addressed by the designer.

Snoopy cache coherence protocol

Most commercial multicore processors use a cache coherence scheme called *snooping* where all caches monitor a shared inter-cache bus for incoming notifications from other caches about modifications made to shared data. There are two basic types of cache coherence protocols:

- **Write-through:** all cache memory writes are written to main memory, even if the data is retained in the cache, such as in the example in Figure 4.11. A cache line can be in two states — *valid* or *invalid*. A line is *invalidated* if another core has changed the data residing in that line. In the example above, if this technique is used, the copy of *foo* residing in cache of Core 1 is invalidated as soon as a new value is written into *foo* by Core 3; see Figure 4.12. When

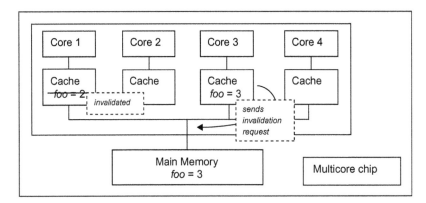

Figure 4.12
Cache coherency — invalidate protocol.

Core 1 now tries to read *foo*, it gets a cache miss and *foo* is fetched from the up-to-date copy in the main memory. This is a simple protocol but is slow and results in a lot of bus traffic. Write buffers are sometimes used to reduce traffic.

An alternative to the invalidation protocol is the *update* protocol where all updates to cached data are broadcast on the inter-cache bus; see Figure 4.13. An update event is generated for each write to data in cache, even repeated writes to the same data variable. This causes the update protocol to be slower than the invalidation protocol, which generates only one event − for the first write. For example, the invalidation protocol will need to send only one event if *foo* is written multiple times by Core 3 (since the copy in other caches is invalidated when the first event is received), but the update protocol will send an event every time *foo* is modified.

- **Write-back**: a cache memory write is not immediately written into main memory unless another cache needs that cache line. The most common write-back protocol is **MESI**, and will be discussed next.

MESI cache coherence protocol

MESI (**m**odified-**e**xclusive-**s**hared-**i**nvalid) stands for the state of each cache line at any time, maintained with two additional *flag* bits. It is based upon the data ownership model, where only one cache can have *dirty* (modified) data. When data is written, the cache modifying the

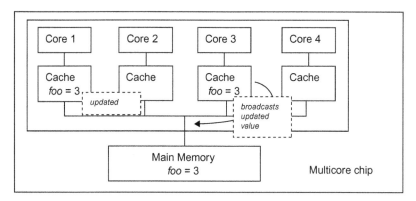

Figure 4.13
Cache coherency − update protocol.

cache line informs other caches of the fact; it does not transmit the data itself. A cache line in each cache can be in one of the following states:

- Modified: the cache line is valid and present only in the current cache. It has been modified but not written back to main memory, and is marked *dirty*. The write-back of the modified data to main memory will change the state of the line to *Exclusive*.
- Exclusive: the cache line is valid and present only in the current cache, and is *clean*, that is, it is consistent with the copy in main memory. The state of the line will change to *Shared* if another cache reads it, and it will change to *Modified* if the current cache writes to it.
- Shared: the cache line is valid and in the current cache and at least one other cache. It is consistent with main memory. If discarded the state of the line will change to *Invalid*.
- Invalid: the cache line contains no valid memory data.

Initially all cache lines are empty and in the **Invalid** state. If data is loaded for reading, the new state is changed to **Exclusive** if no other core has the data in its cache. If it does, the new state is set to **Shared**. The states of the cache line of all the other cores that contain the data are also set to **Shared** if not already so. If a write miss occurs, the state changes to **Modified** and the data is loaded into a cache line for writing. The memory is updated at the same time. **Modified** and **Exclusive** cache lines are owned by the cache in which they reside. They can be changed without informing other caches.

If a second core wants to write to the cache line, the first core sends the cache line contents and marks the cache line locally as **Invalid**. This is the *Request For Ownership* (RFO) operation. The second cache now owns the cache line, and is in the **Modified** State.

If a cache line is in the **Shared** state and the local core reads from it, no state change is necessary and the read request can be fulfilled from the cache. If the cache line in a **Shared** state is written to by its core, the state changes to **Modified** and the write operation is announced to all the cores via an RFO message so they can mark their versions of the shared cache line as **Invalid**. No bus operation is needed. If the cache line is in the **Exclusive** state, the write operation does not need to be announced on the inter-cache bus.

Table 4.1: MESI Cache Coherency Protocol

Cache line state	Modified (M)	Exclusive (E)	Shared (S)	Invalid (I)
Valid cache line	Yes	Yes	Yes	No
Memory copy status	Out of date	Valid	Valid	-
Copies exist in other caches?	No	No	Maybe	-
Action upon read	Read hit, no update from memory	Read hit, no update from memory	Read hit, no update from memory	Cache miss, update from memory, change state to E if no other cache contains line, S otherwise
Action upon write	Write hit, update cache, don't update memory	Write hit, update cache, don't update memory, change state to M	Write hit, update cache, update memory and invalidate contents of other caches, change state to E	Write miss, update memory, change state to M

The state of MESI cache lines can be summarized in Table 4.1 and Figure 4.14. The state changes are accomplished by the cores listening, or snooping, for events broadcast by other cores. The memory address of the cache line in question is visible on the inter-cache address bus so that each cache can do an associative search of its contents for a match.

From this description of the state transitions it should be clear where the costs specific to multicore operations are. Filling a cache is expensive but now we also have to look out for RFO messages. A MESI transition cannot happen until it is clear that all the cores in the system have had a chance to reply to the message. That means that the longest possible time a reply can take determines the speed of the coherence protocol. Whenever an RFO message has to be sent things are going

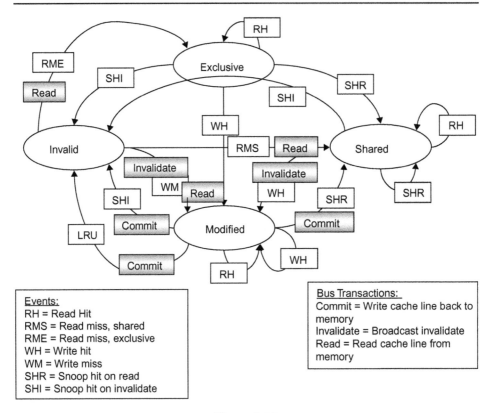

Figure 4.14
MESI cache coherence protocol state diagram.

to slow down. There are two situations when RFO messages are necessary:

- a thread migrated from one core to another and all the cache lines have to be moved over to the new core once;
- a cache line is truly needed in two different cores.

Concurrency is severely limited by the finite bandwidth available for the implementation of the necessary cache coherence protocol. Programs need to be carefully designed to minimize accesses from different processors and cores to the same memory locations. This ensures that minimal notifications are sent out over the inter-cache bus.

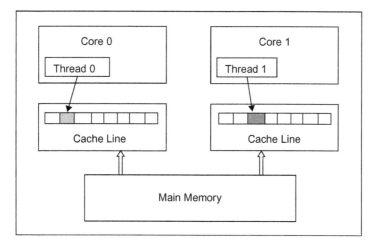

Figure 4.15
False sharing.

Some processors add a fifth state for *Shared Modified* and call it the **MOESI** protocol. A cache with the **Shared Modified** state updates the shared cache lines of other caches with current data, but does not write it back to main memory.

Cache-related performance issues

An application can have multiple threads working on a shared data structure. The most recent members of the data structure that are accessed by a thread are present in the cache of the core where the thread is running. In some cases, the data access pattern can exhibit behavior that negatively impacts the performance of the application.

False sharing and the ping-pong effect

False sharing occurs when threads from different processors modify variables that reside on the same cache line; see Figure 4.15.

For example, suppose the loop below is parallelized by assigning the loop iterations to two threads.

```
for(i=0;  i<N;  i++)

{

        C[i] = A[i]+B[i];

}
```

If round robin scheduling is used, all even iterations are assigned to Thread 0 and all odd iterations are assigned to Thread 1. Thread 0 runs on Core 0, Thread 1 on Core 1. Suppose the cache line can hold eight values of A, B or C. When starting out, A(0−7), B(0−7), C(0−7) are loaded into three cache lines for both the threads. Read operations on A and B do not cause cache invalidation. Writing to C(0) results in the cache line of Core 1 to be invalidated since it is shared. Core 1 will re-load the cache line (an expensive operation). What happens when Thread 1 writes to C(0)? The cache line for Core 0 is invalidated and re-loaded, and so on. This is called the *cache line ping-pong effect* and can lead to a significant performance hit. Notice that the threads are actually operating on exclusive data but since the data resides in the same cache line, the cache is fooled into believing that it is shared. This phenomenon is called *false sharing*. Notice also that the cache lines for A and B are not fully used by each core; only half the values are useful.

You have to take care to parallelize your application with this in mind. In the example above, a better approach would be to divide the loop iterations in half and assign the first half to Thread 0 and the second half to Thread 1. The only cache line sharing will then occur at the boundary between Thread 0's and Thread 1's data. Cache lines for A and B will now be fully utilized, leading to additional performance gain.

Processor affinity

In multicore processors, the cache structure and assignment of threads to cores can have an effect on performance. For example, assume the memory structure in Figure 4.8. If your application has two threads that share a lot of data, they should be placed on two cores, such as Cores 0 and 1, that share an L2 cache so that performance is improved by using this cache to communicate changes rather than accessing the slower memory. On the other hand, if the two threads do not share any data, and would benefit from a large cache, they should be placed in cores that do not share an L2 cache, such as Cores 0 and 2 (this works well if there are no other applications running on Cores 1 and 3.)

We can direct the operating system to assign a thread to a particular core or set of cores by using processor affinity directives supplied in most system libraries. This has a side benefit that a thread is always scheduled to run on the same core by the operating system, and there may be some data re-use due to leftover data in the cache from a previous run.

Processor affinity can also be used to grant critical processes or threads exclusive use of certain cores.

Cache locking

Applications running on embedded systems are pre-determined. Their memory access patterns can be used to assign critical code and data to cache lines and *lock them down*. This ensures that these lines are not swapped out to make room for incoming lines. The rest of the cache can be used in the normal way.

Transactional memory

A discussion of memory management in multicore processors would be incomplete without a mention of transactional memories.

The traditional approach to dealing with the data synchronization problem present in shared memory systems is to use locks to make certain critical sections of code execute without interference from other threads. This approach is difficult and error prone and has known problems such as deadlocks and race conditions that are very hard to reproduce and debug. The programmer has to account for overlapping operations on shared data in distantly separated and seemingly unrelated sections of the code.

There can be a significant performance hit when threads have to wait for a lock to get released before proceeding to execute a shared section of critical code. Locks can lead to *priority inversion*, where a high-priority thread is forced to wait for a low-priority thread to release the lock.

```
Acquire Lock {
        new_node->prev = node;
        new_node->next = node->next;
        node->next->prev = new_node;
        node->next = new_node;
}
Release Lock
```

Figure 4.16
Linked list node insertion using locks.

For example, when inserting a node in a linked list that is operated upon by multiple threads, the thread inserting the node has to acquire a lock before doing the insertion; see Figure 4.16.

Transactional memory (TM) is aimed at alleviating the difficulty of writing multi-threaded code that is scalable, efficient, and correct *without using locks*. Using this approach, the programmer specifies *what* should be done atomically, leaving the system to determine *how* this is achieved.

Parallel programs can be written by allowing a group of read/write instructions to execute in an atomic way, analogous to database transactions. TM is optimistic as it assumes that the transaction is going to succeed. The benefit of this optimistic approach is increased concurrency, as no thread needs to wait for access to a resource, and different threads can safely and simultaneously modify disjoint parts of a data structure that would normally be protected under the same lock.

Contrast transactional memory to locking mechanisms, where a thread needs to acquire a lock just in case another thread tries to perform a concurrent operation. The transaction has to be retried only when there is an actual concurrent modification to shared data by another thread. Unlike locks, transactions automatically allow multiple concurrent readers.

TM provides the following instructions for manipulating transaction state:

- **Commit** attempts to make the transaction permanent. It succeeds only if no other transaction has updated any location in the transaction's data set, and no other transaction has read any location

```
tm_atomic {
        new_node->prev = node;
        new_node->next = node->next;
        node->next->prev = new_node;
        node->next = new_node;
}
```

Figure 4.17
Linked list node insertion using transactional memory.

in the transaction's write set. If it fails, all changes to the write set are discarded.
*   **Abort** discards all updates to the write set.
*   **Validate** tests the status of the transaction. It returns "true" if the current transaction has not been aborted; otherwise it returns "false".

Using TM, developers can mark the portions of their programs that modify shared data as being *atomic*. Each atomic block is executed within a transaction: either the whole block executes, or none of it does. Within the atomic block, the program is free to read or write a shared variable as many times as it likes, without locking it.

At the end of the atomic block, during commit, the TM system checks to see if the shared data has been modified in memory since the atomic operation was started. If it hasn't, the commit writes the new value of the shared variable into memory. If it has, the transaction is aborted, and the work the thread did is rolled back. The program will usually retry the operation, without any intervention from the programmer, until it succeeds, or a pre-determined limit of failed attempts is reached.

For example, using TM, the same example of a linked list node insertion can be stated as in Figure 4.17.

Multiple threads can insert elements into the linked list without worrying about synchronization issues. If the threads are working on disparate portions of the list, all transactions will be committed, and full parallelization is achieved. Contrast this to the example with locks where only one thread can insert an element at a time, regardless of the position in the list.

TMs do have their limitations. They cannot perform any transaction that cannot be aborted and undone. This can be a problem for operations that involve, for instance, I/O. Such limitations are typically overcome in practice by creating buffers that queue up the irreversible operations and perform them at a later time outside a transaction.

TM can be implemented in both software and hardware.

Software transactional memory

Software transactional memory (STM) provides transactional memory semantics in a software runtime library or the programming language. It requires minimal hardware support (typically an atomic *compare and swap* operation, or equivalent). In its simplest form, software transactions are implemented as an *atomic* block of code which logically occurs at a single instant, as shown in Figure 4.17. When the end of the block is reached, the transaction is committed if possible, or else aborted and retried.

Each shared data variable has version information to keep track of the changes made by atomic transactions. For example, in Figure 4.18, thread T1 executes the first atomic transaction and thread T2 the second. T1 and T2 both maintain a transaction log that is initially empty. The data variables, *var1*, *var2*, *var3* and *var4*, all have some initial value and version. The initial state and transaction log are shown in Figure 4.18.

Now suppose T1 and T2 execute concurrently and read the values for *var1*, *var2*, *var3* and *var4*. Before updating the variables, T1 and T2 log

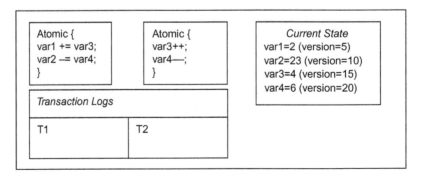

Figure 4.18
Transactional memory — initial state.

all the variables involved in the transaction and their versions. The transaction log is shown in Figure 4.19.

Now suppose T2 updates *var3* and *var4* and tries to commit the transaction. The transaction manager will check to see if the versions of *var3* and *var4* are the same as what it has in the log for T2. Assuming they are the same, the transaction will be committed to memory and the values and versions of *var3* and *var4* updated as shown in Figure 4.20.

Now assume that thread T1 has finished executing its transaction and wishes to commit the new values of *var1* and *var2* to memory. The transaction manager checks to see if the versions of *var1*, *var2*, *var3* and

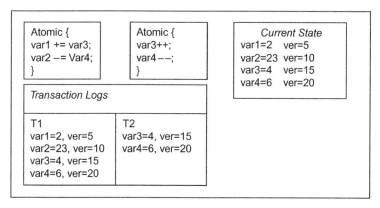

Figure 4.19
Transactional memory — state after reading.

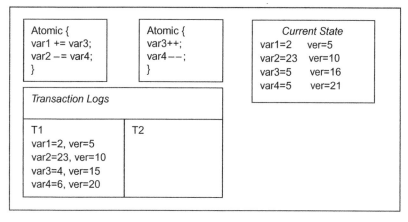

Figure 4.20
Transactional memory — state after t2 updates *var3* and *var4*.

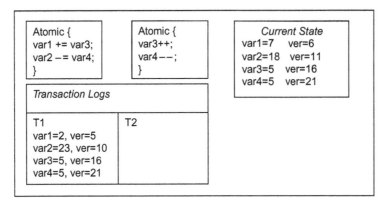

Figure 4.21
Transactional memory — state after T1 updates *var1* and *var2*.

var4 in memory are the same as those in the log for T1. It finds that the versions of *var3* and *var4* are *not the same*. This transaction will be aborted and tried again. Assume this time there is no conflict and the transaction gets committed; Figure 4.21.

TM code is atomic as seen by both TM code and non-TM code but only TM code conflicts with other TM code are detected as version information of data is only maintained for TM blocks. Changes to variables in a TM code block by non-TM code will not be detected, and may result in an inconsistent system.

There is overhead associated with TMs due to the need to maintain log operations such as load, update, store, and check version. The amount of synchronization within the application and failed attempts to commit can impact performance. Good TM systems also need to support unbounded (both in space and time) transactions where there is no limit to the size of the atomic block or the amount of time it takes to execute.

The programmer may need to provide a fallback mechanism to use traditional locks for the critical section if the transaction fails multiple retry attempts. In addition, the programming language needs to support monitoring of memory accesses for logging as well as support full TM semantics such as nesting of transactions.

The new standards for C and C++ published in 2011 add limited support for STM [6].

Hardware transactional memory

Hardware transactional memory (HTM) implementations usually make use of processor cache and cache coherence protocols to detect and manage conflicts between hardware transactions. The cache coherence protocols keep track of accesses within a hardware transaction. If the hardware transaction aborts, then the cache lines holding the tentative writes in the HTM are discarded. The hardware manages data versions and tracks conflicts transparently. Hardware transactions usually require less overhead than software transactions because hardware transactions occur entirely in hardware. Hardware transaction size may be limited due to hardware limitations (memory needed to store a transaction), whereas software transactions can handle larger and longer transactions.

HTMs are covered in more detail in the chapter on Hardware synchronization.

Hybrid transactional memory

Hybrid transactional memory provides support for bounded-size atomic transactions by coordinating between both hardware and software transactional memory. HTM is used for bounded-size transactions that do not exceed the HTM's limitations, and STM is used for the rest.

Summary

Multicore embedded systems differ from general-purpose multicore computing devices in that they are designed with a specific application in mind. The memory access patterns for this application can usually be pre-determined and can be used to customize the memory architecture of the device to optimize the embedded system for factors such as cost, power consumption, performance, and volatility of data. Many trade-offs need to be considered given the complexity of the device, such as:

- Is the memory shared or distributed among the cores? If it is shared, is the memory uniformly accessible by all the cores or does the access time differ depending upon the location of the memory? Do the cores have memory cache, and, if yes, is the cache coherent across all cores?

- Should there be a memory cache or does the system perform better without one? If there is a cache requirement, how is the cache hierarchy laid out? How many levels of cache are required? Is the cache on-chip or off-chip? What is the size of the cache and cache line? Is the cache split between data and instruction cache? Are some lines of cache locked down to meet real time performance requirements? Is the cache coherency managed by hardware or software? What is the cache coherence protocol?
- Are there other memory types that may benefit the device, such as scratch pad SRAMs and eDRAMs? Is there need for specialized memory such as LIFO, FIFO or CAM?
- Can transactional memory be supported in hardware or software, and if so, what are its size and performance limitations?

A performance analysis and detailed benchmarking of the embedded application with different memory configurations are needed to determine the optimal memory architecture. This chapter presented a synopsis of memory types and architecture commonly used in embedded processors as well as desktops, with the advantages and disadvantages of using any one approach. There is no one-size-fits-all memory architecture. You, as an embedded system designer, will have to evaluate the needs of your particular system and tailor the memory accordingly.

References

[1] C. Zhang, F. Vahid, W. Najjar, A highly configurable cache architecture for embedded systems, Proc. Int. Symp. Comput. Archit. (2003).
[2] P.R. Panda, N.D. Dutt, Memory architectures for embedded systems-on-chip, HiPC, (2002). <http://www.ics.uci.edu/~dutt/pubs/bc12-hipc02-panda.pdf>.
[3] K. Beyls and E. D'Hollander. Reuse distance-based cache hint selection. Proceedings of the 8th International Euro-Par Conference, 2002.
[4] P.R. Panda, N.D. Dutt, A. Nicolau, Efficient utilization of scratch-pad memory in embedded processor applications, ED&T (1997).
[5] D. Lenoski, J. Laudon, K. Gharachorloo, A. Gupta, J. Hennessy. The directory-based cache coherence protocol for the DASH multiprocessor. *Computer Systems Laboratory*, Stanford University, CA, 1990. <http://web.cecs.pdx.edu/~alaa/ece588/papers/lenoski_isca_1990.pdf>.
[6] Transactional memory in gcc. <http://gcc.gnu.org/wiki/TransactionalMemory>.

CHAPTER 5

Design Considerations for Multicore SoC Interconnections

Sanjay R. Deshpande

Design Manager, Freescale Semiconductor, Austin, TX, USA

Chapter Outline

Real World Multicore Embedded Systems.
DOI: http://dx.doi.org/10.1016/B978-0-12-416018-7.00005-5

Introduction

The demand for computing density has been growing unabated across the entire, ever-widening spectrum of applications, both embedded and general purpose. And, fueled by needs of more intelligent functionality, lower response time, more connectivity and higher level of security, the demand promises to continue to grow for the foreseeable future at an even higher rate. Fortunately, as was presaged by Moore's law, the reduction in silicon feature size has, over the same period, continued to offer a higher number for the gate count per square millimeter of a semiconductor chip. This was also accompanied by faster and faster device speeds. The trend encouraged employment of an increasing temporal parallelism via deeper processing pipelines for a while — and through that, achievement of higher processor frequencies and therefore of computing speeds — as a means of achieving increasing computing power. Concomitantly, this development also pointed to a way of decreasing the bill of materials (BOM) of a computing system, leading to the development of systems-on-chip which integrate, along with processors, other functions such as co-processors, memory controllers, and I/O interfaces in a single semiconductor chip.

Lately, with feature sizes approaching a few tens of nanometers, however, a semiconductor device's leakage power — dissipated as heat — has begun to increase at super-linear rate with reduction in feature size. In order to keep the leakage power down, the threshold voltage of the device has had to be kept high, which in turn has caused the device to be run at lower clock speeds in order to keep power dissipation within the same limit as before. The result has tended to discourage the use of higher clock frequencies to achieve higher performance and instead has encouraged higher levels of integration and of spatial parallelism to increase the compute density within a chip. The consequence of course has been an increasing momentum toward developing highly integrated systems-on-chip (SoCs), with large numbers of both general-purpose and special-purpose processing blocks operating in parallel, while keeping the clock frequencies at a moderate level to keep power dissipation reasonable [1,2].

SoC designs come a wide variety, depending on the applications they are intended to serve. But the common emergent theme is that of a multicore system, wherein multiple general-purpose processor cores are used as basic computing engines. This is true of SoCs meant for both embedded and general-purpose processing. In the former, these general-purpose processors provide computing flexibility, versatility, and adaptability to the application. General-purpose computing SoCs are characterized by just a few large processors, large on-chip caches, memory controllers, and a few high-speed interfaces for external peripheral connectivity. On the other hand, high-end embedded SoCs often boast a large number of small to medium processor cores, multiple caches, memory controllers, a plethora of peripheral co-processors, and an assortment of interface controllers for direct connectivity to external interconnections such as PCI Express, Serial RapidIO, USB, SATA, Ethernet, etc.

Thus any SoC is an assemblage of a number of different types of functional components which are connected to form a system. These interconnections allow the components to communicate with each other in order to perform the overall computations. The communications occur in a variety of forms and formats. These interconnections and their design are the subject of this chapter.

Importance of interconnections in an SoC

In any practical computational system, computational elements typically are separated from the main system memory holding the operands — either initial or intermediate — and from the portals which bring primary operands from outside the system or take the results of the computations out of the system. A portion of the overall computational effort and time is expended in communication, bringing those operands to computational elements and carrying results from them, these transfers being carried over the system-level interconnections. From an application performance point of view, this communication time is considered overhead. Some of this overhead is unavoidable, but the remaining portion can be swayed noticeably by the communication infrastructure of the system. The total execution time of an application is the sum of the computation time and communication overhead. The ratio of communication overhead to computation time varies from application to application and is of key significance to how well the application is judged to perform – the higher the ratio, the lower the judged performance. As a result, the performance of an application within an SoC can be influenced significantly by the system's communication infrastructure. The role and importance of interconnections in an SoC, therefore, can hardly be overstated.

Terminology

For the purpose of this chapter, unless it is necessary to differentiate, by "core" we will refer to any device that performs computations. Such a device may be a general-purpose instruction set architecture processor (herein specifically referred to as a processor) or a special-purpose fixed-function block. Examples of the latter type of core are I/O interface controllers, network interface controllers, and on-chip function-acceleration engines. In general, a core actively initiates communication activity to perform storage accesses or other types of interactions with another device.

A system also has non-core devices that might be targets of communication activity by the cores and those don't initiate communication activity on their own, but do produce response activity such as production or reception of data. Examples of these types of devices are memory controllers. Such devices will be referred to as "target" devices, in this chapter. Note that some core devices themselves may be targets of communication activity initiated by other core devices, as in the case of initiation of DMA activity by writing to registers in a DMA controller. By a multicore SoC we refer to a system on a single chip with multiple cores and non-core devices.

By interconnection, we don't just refer to the wires that carry the information, but also include the function blocks that must exist along the way to control and coordinate these transfers of information.

Organization of the chapter

In this chapter, we identify the various important aspects of these interactions and discuss how they influence the choice of system interconnections. We will ignore communication paths internal to devices within the SoC such as cores or target devices as they usually are very specific to the needs of that device and not shared with other devices. Instead, we will focus on interconnections used to carry inter-device communications.

To be sure, none of the issues encountered in defining a multicore system interconnection are new; they are faced in designing any multi-processor system. However, given that the interconnection and the devices it connects are on a single die, the solutions employed may be different. For example, the solutions can take advantage of the fact that there can potentially be more wires available to connect a device to the interconnection and can therefore carry more semantics. On the other hand, interconnection topologies that are easier to achieve in three dimensions may be harder to implement in a multicore system.

It is to be expected that a high-performance or high-throughput SoC is likely to experience much higher internal communication traffic rates than a low-end SoC. It is therefore to be expected that interconnection design for high-performance or high-throughput SoCs is correspondingly more challenging than that for a low-end SoC. So in this chapter we will focus more on solutions for the former. But whenever apropos, we will indicate solutions that are appropriate for low-end SoCs.

There are a number of excellent books already available on the subject of interconnection networks for computer systems listed in the references at the end of the chapter [3–5]. It is not the intention of this chapter to merely provide a comprehensive summary of all of the material contained therein, as such a repetition would be of little benefit. Instead, the material covered here covers the subset of concepts, ideas, and considerations that are important to designing a practical on-chip interconnection infrastructure for a state-of-the art, high-performance SoC.

In the next section, we will survey the nature of communication that occurs between different components within an SoC. We will try to understand the key characteristics of this communication with a view toward what would be needed in terms of system interconnectivity to support such communication. We will also look at recent SoC trends that put additional requirements on interconnections.

Just as not having the right roadways in the high-traffic-intensity areas of a city can cause the traffic to snarl up, so could not having an interconnection to match the SoC transaction traffic. In section 3 we will study the inherent topologies of traffic inside the SoC and what system organization choices affect it. By understanding these topologies we can fashion the appropriate interconnection to support them.

Next, in section 4, we will briefly discuss the factors that are important to the performance of applications within an SoC, and what aspects of the interconnection contribute to those factors and how. Specifically, we will highlight the important roles that latency and bandwidth play in determining application performance.

In section 5 we get introduced to how interconnections are represented graphically as well as terminology used to describe their behavioral characteristics.

A bus topology is an interconnection of choice for many an SoC. In section 6, we discuss the essential characteristics of the topology and its operation that make it such a popular choice, but which, unfortunately, also become a source of drawback when it comes to supporting a large system.

Section 7 contains a discussion on the characteristics that the interconnections for future SoCs will need. It also analyzes issues regarding transaction handling countenanced in these interconnections and approaches to solutions.

Section 8 delves deeper into the design of building blocks of large interconnection networks. It looks in some detail at some of the key related design considerations.

Section 9 introduces metrics used to quantify the inherent performance and cost profiles of different interconnection schemes and topologies.

Section 10 presents a number of interconnection topologies that are suitable for SoC applications and cites their scores on these metrics.

In real life, outside academic technology demonstrations, one hardly develops an interconnection network without a backdrop of actual SoC products. Section 11 visits some of the practical considerations that are worth paying attention to in order to create effective interconnection solutions in an industrial context.

Finally, Section 12 concludes with a perspective of what is presented in this chapter and the main takeaways from it.

Communication activity in multicore SoCs

The first step in designing interconnection infrastructure for a multicore system is to understand the communication requirements in the system. How the various devices interact determines what type of communication activity will occur.

Transaction-based communication

We will use the term "transaction" for a unit of communication. In general, a transaction is a "tuple" of fields of information sent by a device expressing the storage access or another type of service request being issued into the system. Thus a transaction is composed of sufficient information to be independently meaningful for the sender and the receiver of that transaction.

Within the communication traffic of a multicore system, one can observe many types of transactions. In particular, the following two main categories of transactions can be distinguished:

- storage-oriented transactions.
- inter-device messages.

We will discuss these in some detail to develop a deeper understanding of how they can affect SoC interconnection design.

Storage-oriented transactions

Within this category, we distinguish between addressed access transactions and non-addressed transactions.

Addressed Accesses

Cores make "Load" and "Store" accesses to storage locations in the system to bring in their input operands and to store the results of their computations, respectively. The actual information about the access is issued by the device as a "transaction". Thus a Load access can cause a "Read" transaction and a "Store" access can cause a "Read (with-intention-to-Modify)" or a "Write" transaction. Both of these transactions carry an "address" of the location being accessed. Both of these transactions also specify the amount of data being accessed. The Write, additionally, carries the data to be stored in the location. Typically, additional qualifying information (referred to as "transaction attributes") may also be sent along with the transaction to provide richer semantics (Figure 5.1).

In general, there may be many variations of Read and Write transactions. Read-Write traffic is the predominant type of transaction traffic in an SoC.

Non-addressed storage-oriented transactions

In an advanced system, there may also be many other types of address-bearing transactions, such as cache management transactions, as well as transactions that don't carry addresses to provide richer functionality, such as ordering and synchronization. These non-address transactions, too, need to traverse the same interconnection pathways as Read/Write transactions.

Messages

In a message model, a core, instead of accessing information contained in a storage location using its address, communicates directly with another device. To do so, the core must be aware of a separate name-space used to identify devices in the system. To transfer information to

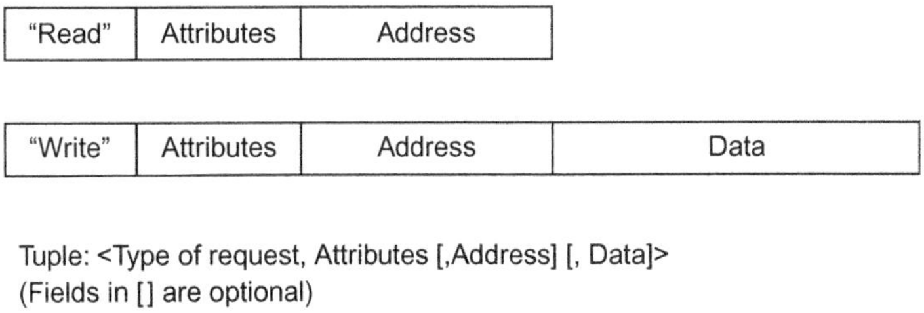

"Read"	Attributes	Address	

"Write"	Attributes	Address	Data

Tuple: <Type of request, Attributes [,Address] [, Data]>
(Fields in [] are optional)

Figure 5.1
Read and Write transactions.

another device, a core sends a "message" specifying the information or data to be transferred and the "name" or an identifier of the device that is intended to receive the information. When a message reaches its destination, hardware at the receiving device may deposit the message data into local storage locations internally addressable by the receiving device.

Typically, a symmetric operation to "receive" a message directly from another device is not easily achievable due to the practical difficulty of and inefficiency in causing the target device to produce such a message on demand. Instead, a receive message operation often checks local storage intended for depositing the message contents for arrival of a message from the intended other device. Thus, one rarely encounters an explicit "Receive" operation traversing the system, although it is not impossible to imagine one that would check whether a message is available from the target device for the device executing the receive and then return the appropriate status, along with the data, if available.

Interrupts

Special types of messages may be employed whose arrival at a device may have a side effect of causing an interrupt to the receiving device. These types of messages are also transmitted with device identifier labels. Typically, these messages do not carry data, although that is not always the case. Here, again, the system would use the device identifier within the interrupt message to route it to the intended destination.

Transactions carrying messages, including interrupt messages, are collectively referred to as "message traffic".

Concurrency of communication and segregation of traffic

In a multicore system with a large number of cores, one must expect a high rate of concurrent transaction activity, be it for storage accesses, for instruction fetches, for data accesses, or for inter-core signaling. If the applications running in the system exhibit poor cache hit rates, this activity gets amplified.

This transaction activity represents the natural parallelism available in the system. Dependency considerations and target resource constraints typically require some level of serialization of these activities, which reduces the amount of exploitable parallelism. However, to ensure efficient use of the hardware assets, the system should enable simultaneous progress of the concurrent activities that do not need to be serialized.

For example, not all concurrent activity is targeted to the same device and therefore does not inherently need serialization. However, system constraints could, nonetheless, precipitate artificial serialization and thereby reduce performance [6]. Any such serialization tends to reduce the utilization of cores. To maximize core utilization, and therefore system performance, the system will need to employ an appropriate interconnection network that helps avoid bottlenecks that could contravene the natural concurrency available.

1. Load/Store accesses themselves can often be differentiated into following types:
   - Coherent accesses by processor cores and other devices
   - Non-coherent accesses to I/O by processors
   - Non-coherent accesses by I/O to private regions in memory
   - Cache management transactions

Coherent accesses are specially marked storage accesses for which system enforces coherency semantics on the data being accessed. Accesses, especially those targeted to I/O devices or accesses by I/O devices to their private regions in memory, are typically non-coherent and do not directly interact with coherent traffic. Yet some others may consist of inter-core interrupts or messaging.

These types of transaction are orthogonal to each other. If these independent kinds of traffic can be kept from interfering with each other, it can reduce congestion and the delays they encounter, resulting in an improvement in performance. On the other hand, if they are intermingled, all devices in the system are unnecessarily subjected to all of the activity in the system, thereby wasting their valuable bandwidth.

Recent trends in SoCs

In the previous section we looked at the basic requirements for a multicore interconnection. However, it might be advantageous to keep an eye on the future when developing a multicore interconnection strategy.

Application consolidation and system partitioning

As the number of cores and other devices on a single die grows, it should be expected that a number of related applications will get loaded into a single multicore system. For example, one might consider an enterprise edge router system that sits on an intra-net LAN. Such an application may actually be composed of multiple distinct sub-applications:

— a router application with a firewall that performs packet inspection on incoming and outgoing traffic;
— an application that runs routing algorithms such as BGP, and updates routing tables for the router applications;
— an accounting application; and
— a control application that manages the entire system, handles policy decisions, and deals with exceptions and error conditions that might occur in the system.

All of these sub-applications are mostly independent of each other, but they do share data structures in a controlled manner. In fact, if the sub-applications are developed by third-party software vendors, it might be wise to ensure that they are protected from each other's erroneous behavior.

The above example illustrates a need to partition a multicore system into independent logical sub-systems and to isolate them from each other. The interconnection infrastructure of the system should enable such partitioning. Preferably, the interconnection should make it possible to take advantage of the independence of sub-applications to optimize system performance. Such inter-sub-application isolation can also enhance the security properties of the system.

Protection and security

As multiple applications are collocated in a single SoC, one immediately faces the issues of inter-partition isolation and protection. There are two types of protection that may be required: computational and performance.

Computational protection is typically achieved via the processor's MMU and the IOMMU. The MMU is used to prevent processors of one partition from accessing storage locations belonging to another partition. Similarly, the IOMMU is used to prevent IO devices or DMA operations belonging to one partition from accessing storage locations belonging to another partition.

Performance isolation aims to go beyond simple computational isolation and ensure that the partitions do not interfere with each other in terms of their execution times. Thus the goal is to prevent one partition from causing a performance loss in another partition, the extreme case being starvation.

Performance isolation in general is much more challenging to attain than computational isolation. To accomplish this, the system interconnections must provide a means of keeping the traffic belonging to different partitions isolated from each other.

Transaction service guarantees

Many embedded systems serve real-time applications. These applications deal with real-time traffic that flows in over network connections, gets processed, and perhaps results in real-time traffic that flows out via output portals – sometimes also egress network connections. All of this external traffic translates to internal traffic carrying communication activity in the system's interconnection network.

A real-time application is often characterized by strict service demands on the system. The demands are in the form of minimum bandwidth requirements for input and output traffic and maximum latencies for certain actions and processing times in the system. These requirements translate to requirements on processing rates achieved in the system, which in turn place upper-bound requirements on service times of transactions and minimum throughput requirements on transactions belonging to certain flows. The interconnection networks of such systems must be equipped with mechanisms that can achieve such goals.

Functional requirements and topologies of SoC traffic

As discussed above, a majority of transactions in an SoC are storage access related. Nominally, the SoC interconnections carry these transactions from their sources to their destinations. On the surface, this function is not unlike inter-computer networks such as a local area network (LAN) or even the Internet. But there is a difference: in the latter, the network is merely required to passively transport encapsulated information that is interpreted and acted upon only by the computer systems at the end points of such a transfer. However, inside an SoC, a transaction conveys semantic information about the desired primitive storage operation that needs to be effected in the system, potentially, with state changes occurring at multiple sites across the system. Considerable efficiency is gained if the SoC interconnections participate actively in coordinating and orchestrating the enactment of the transaction's semantics instead of passively transporting transactions from one point to another, as happens in a LAN or the Internet. To make available such efficiency, the SoC interconnections need to be made communication-content aware and an integral part of the system hardware that effects the semantics of various storage access operations and the auxiliary transaction activity supporting them. Thus it is quite important to comprehend the semantics associated with SoC transaction traffic to derive the necessary functionality the SoC interconnect should embody, before starting to design it.

Since all of the transactions generated in the system must ultimately travel over the available connectivity in the system, it is important to develop an understanding about the topology of transaction traffic in a multicore system and what factors contribute to it. In general, the more an underlying SoC interconnection topology matches the topology and capacity needs of the communication activity of an application, the better the application can be expected to perform, as the traffic has to take fewer detours and encounters lower congestion.

Understanding the native topology of SoC traffic will better enable the interconnection designer to choose the most appropriate topology and make appropriate cost-performance trade-offs. On the one hand, a complex topology of communication activity could thus prompt a complex interconnection topology, which is also more prone to being expensive to implement. On the other hand a simpler interconnection topology is likely to be easier to build and be less expensive, but also apt to yield lower performance. As will be seen later, a bus interconnection, for example, is inexpensive to implement, but also performs poorly for a large system. So a trade-off must usually be made between the complexity and performance of an SoC interconnection.

The functionality, intensity and topology of transaction traffic within an SoC is influenced to a great degree by three factors:

1 Organization of memory
2 Inter-device communication paradigms employed
3 Presence and type of hardware coherency scheme

We will look into the relevant details of each of them.

As discussed earlier, traffic in an SoC is primarily composed of addressed accesses made by core devices to storage and the auxiliary transactions associated with them. The volume of these transactions can be greatly influenced by how memory is organized and shared in the system. However, the paradigm for inter-core communication employed in the system shapes the topology, and also to quite some extent the volume, of this traffic.

Memory organization

Given that Read-Write traffic to memory is the predominant component of the overall traffic in an SoC, the organization of memory in a system dictates the topology of the traffic pattern in it. The topology of the traffic in turn hints at the appropriate topology of interconnection(s) in the SoC.

A multicore system may be organized around centralized memory or the memory could be distributed among several sites in the system.

In the centralized organization, all memory is equidistant from all cores and all the Read data accesses nominally have equal latency. Such a memory organization is called a uniform memory architecture (UMA). Figure 5.2 shows an example of a system with a centralized memory organization. Note that all accesses from cores and other devices to memory must travel through the central, global interconnection network.

As all memory locations are equidistant from all processors, the UMA model greatly simplifies the task of the software by not having to worry

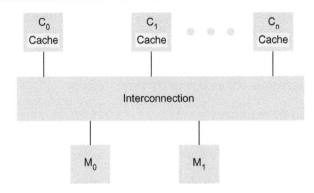

Figure 5.2
Uniform memory architecture system organization.

about data placement to achieve efficiency and performance goals. This makes the UMA model very attractive from the software perspective.

A symmetric multi-processor (SMP) is the type of structure that comes closest to implementing the centralized shared-memory model with other associated characteristics mentioned above. Over the last two decades, because of the simplicity and flexibility of the programming model and robustness of its performance with respect to placement of data, the SMP model has become popular in the industry as a basis for parallel computing platforms that are at the center of most server systems, the workhorses of the IT industry. It is also perhaps the best first model of parallelism for the nascent embedded multicore market to utilize.

In a distributed organization, not all of the memory is equidistant from every core. Some portion of the memory may be closer than other portions. Such an organization is therefore referred to as a non-uniform memory architecture (NUMA) (see Figure 5.3). In such a system, multiple clusters, each composed of cores, other processing elements and local segments of memory interconnected via a local network, are interconnected via a global network, which results in a hierarchical interconnection scheme. In a NUMA system, although all of the system memory is accessible to all cores and devices, it is expected that most of the memory accesses won't leave the cluster.

Unlike in a UMA system, in a NUMA system, the performance of the application can be very sensitive to placement of data. If access to

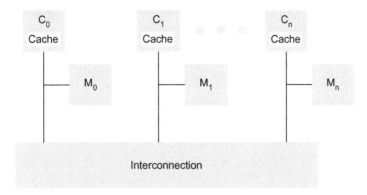

Figure 5.3
Non-uniform memory architecture system organization.

often-accessed data incurs long latencies because it is further away from its user, the performance of the application can suffer greatly. Therefore, when a significant difference exists between near and remote latencies, the operating software must be made aware of data access patterns and must take great care to place frequently accessed data proximate to the cores accessing it so that data accesses are localized as much as possible. Applications, too, need to become aware of latency differences and need to be restructured and partitioned so that their segments access nearby data more frequently than. This onus of being aware of data placement and differences in latencies increases the complexity of programming a NUMA system. Much research is being invested into ways of automating data placement and relieving this burden.

From the foregoing, it might seem like UMA is always the right choice for memory organization. However, that is not the case. As the number of devices in the system grows, the interconnection used to connect them to the centralized memory grows in physical extent. The result is that it takes longer (i.e., it takes a larger number of clock cycles) for every memory access to reach memory, and, in the case of a Read transaction, an equal number of additional cycles for the data to be returned to the requesting device. This causes the performance of every device in the system to fall. At some size, the performance loss can be large enough that it outweighs the intangible benefit of ease of programming. So for a sufficiently large system, the only option might be to employ a distributed memory organization and a NUMA programming model.

Clearly, if the trend in semiconductors continues, in a matter of a few years the system integration levels will get sufficiently high that distributed memory organization will become necessary. Already Intel and Tilera have demonstrated such highly integrated SoCs [7–9].

From the point of view of interconnection design, the choice between UMA and NUMA organization is significant. In the former case, all of the memory accesses must travel across the global centralized interconnection. In a NUMA system, since only a relatively small percentage of all accesses are expected to traverse the global network, the bandwidth demand on such a network could be expected to be relatively lower. For the same reason, latency to memory in a UMA system is far more critical than that to distant memory in a NUMA system.

Implications of inter-device communication paradigms

There are two fundamental paradigms for processes in a multicore system to communicate with each other: shared memory and message passing.

The shared-memory model

In a shared-memory model of interaction, any two communicating devices can both access a common address-based namespace region, can directly read data from and write data to locations within that region, and can thus exchange information via those locations. Inter-processor communication is primarily achieved via Read and Write operations, the same primitives that are used to access storage in a single processor environment.

While the shared-memory model represents a very straightforward and simple extension of a uniprocessor access model, accesses to shared locations by different devices need to be coordinated to maintain coherency among copies of a single data value and consistency among multiple data values in order for cooperative computations to produce correct results. This coordination adds to the basic memory access transaction traffic in the SoC. These additional transactions also need to be handled in a special manner. Figure 5.4 shows a sample set of transactions commonly found in coherent systems.

Topology of coherency activity

Coherency is an agreement achieved in a shared-memory system among various entities accessing a storage location regarding the order of values that location is observed to have taken. In the presence of caches, hardware-maintained coherency relieves the software of flushing the caches to drive updated data to the memory for other entities to see and thus provides significant performance benefit to applications. Most modern shared-memory systems provide hardware cache coherence.

Cache coherence brings with it a very unique and specialized set of transactions and traffic topology to the underlying interconnection scheme. In addition to Read and Write transactions, there are cache management transactions, some of which may have global scope, and auxiliary transactions that carry coherency responses. There may also be

Transaction	Issued by	Access type (and destination indication)	Semantics	Variations
Read	Core and other devices	Read (Address)	Read data from storage	Size, Cacheability, Coherency
Read with intent to modify	Core	Read (Address)	Read data from storage into cache in order to modify it	Coherency requirement
Clean cache block	Core	Write (Address)	Copy modified data from cache to storage	Coherency requirement
Flush cache block	Core	Write (Address)	Flush modified data from cache to storage (and invalidate the cached copies)	Coherency requirement
Write with flush	Core and other devices	Write (Address)	Flush modified data from cache to storage and then overwrite in storage	Size, Coherency requirement
Write (cache block) with kill	Other devices	Write (Address)	Write a cache block into storage and invalidate cache copies in place	Coherency requirement
Cast out (cache block)	Cache hierarchy	Write (Address)	Write the modified cache block in storage; deallocate cache block	
Synchronize	Core	Synchronization	Establish storage barrier	
Data cache block invalidate	Core and other devices	Invalidation (Address)	Invalidate a data cache block	Coherency requirement
Instruction cache block invalidate	Core	Invalidation (Address)	Invalidate an instruction cache block	Coherency requirement
TLB Invalidate	Core	Invalidation (Virtual address)	Invalidate a TLB entry	
Interrupt	Core and other devices	Interrupt (Core identifier)	Interrupt a core	

Figure 5.4

Examples of some of the common types of request transactions, their sources, and their coherency requirements.

a need to multicast or broadcast certain transactions to multiple entities in the system in order to maintain coherency among the system's caches and its memory. For example, an operation that requires invalidation of certain data from caches must be broadcast to all of the candidate caches in the system. So, while Read and Write transactions without coherency have a simple device-to-device transaction topology, coherency enforcement for the same transactions leads to a complex many-to-many communication topology, requiring gathering of individual coherency response transactions and subsequent multicasting of the combined coherency response. The coherency interconnection in the SoC must efficiently support all these activities.

As an illustration of the foregoing discussion, Figure 5.5 contrasts the salient components of communication activity for a coherent Read operation versus those for a non-coherent one. As is represented there, these functional requirements for coherency add significant complexity

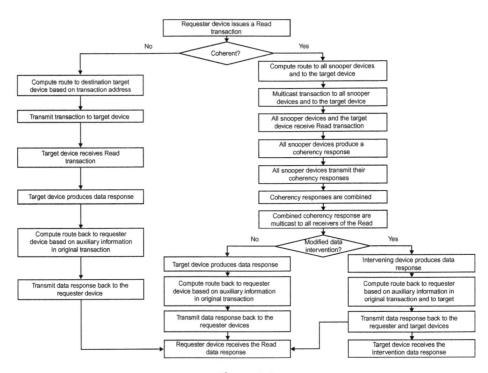

Figure 5.5
Communication activity of a coherent access.

to the topology of communication activity over what is needed in the non-coherent case.

The totality of the topology necessary for supporting coherency in the system then is the union of the topologies needed for all of the individual types of coherent accesses. Usually, the final topology is an all-to-all communication capability for address and data transfers. The ability to broadcast and multicast can provide flexibility and efficiency for address paths. For data transfers, an all-to-all topology with concurrency of individual transfers is often best suited. For response communication, there is a need to combine, which suggests a star-like topology. Traditional bus interconnections naturally support these communication topologies, and so it should come as no surprise that they have therefore been popular for multi-processor coherency in SoCs.

Snoopy or directory-based coherency protocol

Coherency protocols require transactions representing coherent operations to be presented to devices other than the two, the requesting core and the responding target, that are directly involved. The act of presenting a transaction to a device is referred to as "snooping". The two styles of coherency protocols — "snoopy" and "directory-based" — arise out the two different methods used to determine the candidate devices to snoop.

Snoopy protocols are based on the assumption of a broadcast-oriented, passive wired-bus topology of system connectivity. In such an interconnection, every transaction is visible to all of the components connected to the bus without any extra effort. Thus, all of the devices, including the intended target device, "snoop" every transaction as it is issued by a requesting device and take appropriate coherency action. This action involves every listening device updating the coherency state of the location addressed by the transaction, in concert with the type of transaction (if the location happens to be resident in the device's cache), and generation of a coherency response to ensure correct final states of the location in all caches of the system. This action may also be followed by the delivery of an "intervention" data response.

In a very large system involving high tens or hundreds, or even conceivably thousands of cores, global broadcast of every coherent transaction can create overwhelming traffic throughout a multicore system. This traffic can create massive congestion, slowing down transaction progress, and thus bringing the system performance to its knees. Directory-based protocols avoid automatic broadcasts of coherent transactions to all the devices in the system. Hardware function blocks called "directories" keep track of caches in the system that may contain a copy of any given memory location, and the coherency state of the data in them, based on previous accesses to that location [10–13]. Whenever a coherent transaction arrives into the system, the directory is looked up to identify a list of caches which may have a copy of the addressed memory location. Based on its type the transaction is then dispatched to be presented (snooped) only to the caches in that list. The generation and collection of coherency responses, combining of these responses, and subsequent distribution of the final response to the snooped devices is similar to that process in the snoopy protocol, except that it is limited to the above set of caches. This selective multi-cast of transactions results in enormous reduction in coherency traffic.

Transaction ordering support in shared-memory model

Consistency among multiple data values is fundamental to communication between cooperating processes [14]. "Producer-consumer" is a well known model of cooperation between parallel processes and is the basis of interactions between cores in an SoC. Storage operation ordering is essential in a shared-memory system for a pair of producer and consumer processes to work together correctly. For example, after a producer produces data in shared storage, it can set a flag, also in shared storage. If the consumer reads an updated flag, for correct operation it must read newly deposited data and not stale data. These are referred to as the ordering semantics of storage operations. Different processor architectures employ different means of enforcing these semantics, but they are essential for multiple cores to cooperate with each other.

Over the years, computer designers have developed various types of "consistency models" that specify ordering properties satisfied by storage operations, operating assumptions, and necessary actions among participants under which interactions between writers and readers of data work correctly and efficiently. Instruction set architectures of processors make architectural assertions about ordering of storage operations, especially to enable multiple processors to work correctly. They may also offer operations to establish, on demand, a certain ordering regimen over a set of storage operations. For example, the SPARC-V9 architecture defines three memory models that may be achieved by programs: total store order (TSO), partial store order (PSO), and relaxed memory order (RMO) [15,16]. POWER architecture, on the other hand, assumes a default weak storage ordering on most storage accesses and provides synchronization operations to establish strong ordering between storage accesses.

Processor implementations satisfy these ordering semantics internally. But as processors issue transactions into the system in response to program execution, it becomes the responsibility of the system to enforce the appropriate ordering semantics on these transactions. The system's interconnection hardware must therefore be designed to achieve the necessary ordering semantics.

Such ordering assumptions are not limited to processors' instruction-set architectures. I/O and mezzanine on-chip interconnection standards also lay down transaction ordering rules that devices connected to the interconnection can depend on for their interactions. For example, PCI defines precise rules about which transaction type may bypass which other, and which transactions types must "push" which other transactions [17]. Similar ordering rules exist for AXI interconnection [18]. Such ordering rules are also important when cores and other on-chip devices in the SoC coordinate their interactions with I/O devices off the chip. For all of these interactions to occur correctly, SoC interconnections are expected to adhere to the ordering rules and semantics associated with I/O protocols as well.

The SoC interconnections must honor and preserve these specified consistency and synchronization models and I/O protocol ordering semantics. To do so, it becomes necessary that the participating active elements of the interconnection network be designed with intimate awareness of such semantics being supported within the SoC. It is to be expected that this awareness strongly influences the management of buffering and the flow of transactions in these elements.

The message-passing model

When computing entities want to communicate with each other but don't share common accessible storage, message passing is the only means of interaction between them. In a true message-passing model of communication, information transfer is accomplished by the explicit actions of sending and receiving data as messages between communicating entities using the namespace identifying the entities. As the message-passing model avoids the latency, coherency and consistency issues associated with shared-memory-based communication, it is fundamentally a highly scalable model for inter-computational entity communication. This is what makes it an attractive model for large-scale computing environments such as supercomputers and even the Internet.

However, because of the complexity involved in defining, implementing, and managing a scalable and virtualized device namespace and associated message storage, most processor instruction set architectures don't offer native operation codes to provide generalized message passing primitives at the instruction level. (Note: an example of an ISA with native message passing is the transputer architecture. See [19].) Consequently, in most systems, message passing involves calls to software routines that package the indicated data as a message. Several standard message passing application programming interfaces (APIs) have been defined for the purpose. These include the MPI interface, the MCAPI interface, the Unix System V Pipe Interface, and the BSD Unix Sockets interface, among others [20–23]. The more common message-passing libraries are covered in more detail in the Communication and synchronization libraries chapter.

These standard middleware library packages define data structures and subroutines to send and receive messages. However, the headers and trailers in these data structures, inside which the application data is wrapped and sent, represent a per-packet overhead that reduces the efficiency of transmission. Plus, the acts of sending and receiving messages incur, either directly or indirectly, the overhead of library routines, interrupt processing, and/or execution of polling routines. All of these types of overhead conspire to make message passing an expensive inter-core communication model for all cases but those where there is no recourse. As a result, true message passing is typically not a common method of choice for inter-core communication in a multicore SoCs; instead, shared memory offers a more efficient and attractive alternative, as discussed below.

Message passing via shared memory

Shared memory can be used to advantage to implement efficient message passing.

At the software level, messages still appear to be point-to-point communications between pairs of processes which, being unaware of the underlying implementation, identify each other using a namespace that does not refer to memory addresses. But the middleware library routines map the message-passing namespace into shared data structures via which messages are exchanged.

In a shared-memory-based implementation of the message-passing model, the communicating cores interact via memory-mapped data structures. Message data is simply written to shared buffers in memory, which the receiving core reads from. The act of sending of message is achieved by merely updating a few data elements that indicate availability of a message that is ready to be received. The reception of information is also performed by reading those data elements and checking for values indicating availability of fresh message data. As a result, in a shared-memory model, the message-passing overhead turns out to be very low, of the order of only a few instructions.

Thus, when implemented via shared memory, the traffic topology of these messages at the software level becomes indistinguishable from the other memory access transaction traffic generated by the cores.

Going beyond the basic movement of data, when true decoupling among communicating processes is desired, interrupting message transactions are employed as asynchronous "doorbells" to awaken the receiver to the availability of messages.

The shared-memory model thus adds very low overhead to the actual process of communication, which decreases as the amount of data transmitted increases. Instead of moving a large amount of actual data from one core to another, only the pointer to the data can be passed to the recipient. This results in very low communication latency and potentially enormous effective bandwidth.

Communication with non-processor devices

Also, most non-processor devices, often, have no built-in ability to send and receive messages. But they do have the capability to perform Read and Write accesses. They must therefore depend on some means based on shared-memory accesses to communicate with processor cores and other accelerator blocks. The exact mechanisms can differ from device to device and a detailed discussion of them is beyond the scope of this chapter, but suffice it to say that they often use simple Read and Write transactions and utilize special values for storage locations to coordinate the communication.

Because of these many advantages, shared memory can be the method of choice for most inter-core communication within an SoC.

Performance considerations

The performance of an application is simply measured by the amount of time it takes to complete its computation. For applications in which computations are repeated for different sets of data, the same metric is expressed somewhat differently: as a rate of completion of computation, or the number of times the computation can be performed per unit time. For example, for an SoC running a network packet-processing application, the relevant performance metric would be the number of packets processed per second. The smaller the completion time, the higher is the rate of completion, and the better is the application performance.

Two factors related to SoC interconnections affect application performance: transaction latency and bandwidth.

Transaction latency

A unit computational step such as an "add" or a "subtract", or even one such as copying a value from one variable to another, can only be performed once its input operands become available. So if input operands are not available, the core must suspend its computations and idle. This idle time is detrimental to the performance of the application that is running in the core. As noted earlier, cores perform Read operations to fetch input operands from system storage if they are not locally available. The idle time of the core is then directly tied to the time it takes to complete these Read operations. Cumulatively, the performance of an application is quite sensitive to the average completion time of the application's Read operations.

A core does not commonly need to wait for its Write operations to complete. Most Write operations are "posted", meaning that as soon as the Store completes locally at the core, the core continues with its computation without waiting for the resulting Write transaction to start or complete. The hardware thus decouples Stores from corresponding Writes. Hence, the completion times of Writes do not have a first-order effect on an application's performance. Of course, if Writes get backed up severely, they can begin to interfere with the core's ability to perform Stores due to the developing scarcity of resources in the core to complete them, but we will ignore those effects here to simplify the discussion.

We refer to the time to complete a Read operation (the time duration between when a Read transaction is issued and its data response is received), when it is the only transaction in the system, as its "latency".

To appreciate the importance of Read latency to the overall performance of the application, a simple exercise can suffice.

As mentioned, the performance of an application is measured by its execution time; the shorter the time, the better the performance. The execution time of the application in a processor can be expressed as follows:

$$T_{\text{execution}} = \# \text{Instructions} * \text{CPI} * \text{Clock cycle}$$

where:

- # Instructions = the number of instructions in the application
- CPI = the effective number of cycles taken by each instruction of the application

- Clock cycle = Clock period of the clock referenced in estimating the CPI, in seconds. (It is quite common to refer to the core clock here.)

If we assume that the number of instructions that an application executes and the system clock cycle are an application-dependent constant, $T_{execution}$ of the application is proportional to the effective number of clock cycles per instruction.

There are different types of instructions that get executed in a processor, but for the sake of this discussion let us make a simplifying assumption that there are only two classes of instructions:

- Computational and Store instructions that get executed in a single cycle within the processor
- Load instructions that get serviced either within the processor's cache hierarchy or must go out in the system as Read transactions.

Again, for simplification, let us assume that 40% of the total instructions are Loads.

$$CPI_{EFF} = f_{CS}{}^*CPI_{CS} + f_{LC}{}^*CPI_{LC} + f_R{}^*CPI_R$$

where f_i is the fraction of total instructions of type i; where i is

- CS, which refers to Computational and Store instructions;
- LC, which refers to Loads that are satisfied by the cache hierarchy; and
- R, which refers to the Loads that must be sent out as Reads into the system.

Note that $f_{LC} + f_R = 0.4$ based on 40% of instructions being Loads, as assumed above.

If, for yet further simplification, we assume that $CPI_{LC} = 5$, a reasonable average number for a two-level cache hierarchy, Figure 5.6 shows the effect of CPI_R (CPI-R in the figure) on CPI_{EFF} (CPI-Effective in the figure) for various ratios of Loads that result in Reads.

Note that $CPI_{EFF} = 1$ is the best performance the application can expect. As the figure shows, there can be significant degradation in application performance as the Read latency increases: from a minimum CPI_{EFF} value of about 3 core cycles to a high of more than 40 — a more than 13-fold slowdown of the application.

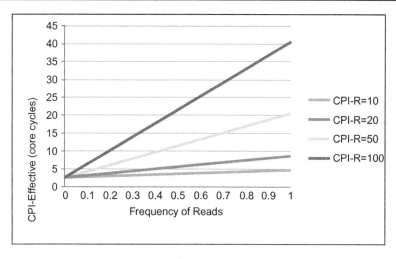

Figure 5.6
Performance effects of Read latency.

The application slowdown can indeed increase to an alarming number if the data is placed far from the requester. This is especially true of the UMA organization discussed above in which all of the memory is equidistant from the core. The latency of a Read transaction is directly proportional to the time it takes to cross the interconnection from the requester to the location's destination. The larger the interconnection, the longer will be the time needed for the Read transaction to any location and its data response to cross it, and thus the higher will be the Read latency, and the more dramatic loss in application performance as a result. Thus, even though the raw processing power of the SoC goes higher with increasing numbers of cores in it, there is a diminishing return in actual performance gain achieved on a per core basis, as the SoC interconnection also grows in size with the increasing number of cores and other devices connected to it. Beyond a certain size, the UMA model ceases to be a viable option and a NUMA model must be embraced. In the long term, a hierarchical structure that utilizes UMA nodes connected via an interconnection to form an overall NUMA system organization might become attractive.

Queuing delays

Another cause of increase in Read delay is queuing delay.

Latency refers to the minimum possible response time for the Read for a given placement of its data relative to the requester. In general, though, the Read transaction is apt to encounter additional delays along the way, perhaps because it has to wait for one or more other transactions to move out of its way or for some other events to occur. This additional delay suffered by the Read is referred to as its "queuing" delay. Thus:

Total Read delay = Read Latency + Total queuing delay for Read

The latency of a Read transaction is a function of the structure and topology of the system interconnections and of the placement of the addressed location with respect to the requester. Thus, given an already defined system structure and placement of data in the system's memories, the Read latency is fixed. However, the queuing delay experienced by a transaction is a function of the intensity of traffic along the transaction's pathways.

Queuing delay on a link

Transaction traffic in the SoC interconnection has many sources and can vary in terms of types of transactions it is composed of and its burstiness. As these transactions traverse the interconnection from their source to their destinations, they get serviced by the various communication elements along the path they take. When multiple of them arrive simultaneously at a given element, they are queued and get serviced one at a time. So it is natural to look upon an interconnection as a network of communication service elements with queues, and with transactions visiting some of these service centers as a function of the path they take [24]. The total delay a transaction experiences is the total sum of the queuing plus service delays it accumulates as it visits these service centers along its path. The total latency of a Read operation then is the sum of such latencies of the original Read transaction and its response transaction.

It is very important to perform queuing analysis of an interconnection to understand its impact on application performance, as it can potentially be large, although quite non-intuitive to gauge. A number of examples of queuing analyses of multi-processor systems and their interconnections are available in the literature [25,26]. However, frequently, those published results are specific to those interconnection constructs and the assumed traffic patterns, which are rarely representative of the actual interconnection design constraints and workloads one might encounter in the SoC being designed. Thus those results are readily applicable to an actual interconnection and application workload at hand, so much so that in practice one almost always has to resort to simulation of the specific interconnection design one has in mind. Therefore, here, in the immediate, we will restrict ourselves to a more modest yet useful goal of developing an intuitive insight into the queuing delays a transaction may face in the interconnection, and that can readily be attained by understanding the queuing delay characteristic of a single service center, such as a single link in the interconnection. By understanding the queuing behavior of such a simple system, one can develop rules of thumb to judge the adequacy of a proposed interconnection scheme in the face of traffic it must carry.

Let us assume that we can model a link as a simple queue with a single server, not unlike a single bank teller serving customers that arrive randomly and form a line to get served. In such a simple queue, the average inter-arrival delay and service times of transactions are assumed to be constant over time.

For such a system, the average time spent by a transaction in waiting to be serviced and then getting transmitted, T, is given by:

$$T = \frac{1/\mu}{1 - \lambda/\mu}$$

where μ is the average service rate for transactions, and λ equals average rate of their arrival at the service center [27].

The average utilization of the link equals λ/μ.

Intuitively, if the arrival rate is faster than the service rate, or if transactions arrive at an average interval that is smaller than the average time taken to transmit a transaction (a very slow teller during busy banking hours), there will be a growing backlog of transactions and the average time spent in the system will grow to infinity. But let us study what happens as the arrival rate approaches the service rate.

It is easier to think in terms of average service time of a transaction —
since it corresponds to its average transmission time — than the average
service rate in our discussion. We use the symbol σ to represent it in the
following.

Figure 5.7 shows Total Service Delay as a function of average inter-
arrival delay, $\tau = 1/\lambda$, for three values of average service time, σ. Note
that queuing delay encountered by a transaction is given by total service
time minus the service time for the transaction. Figure 5.7 also shows the
asymptotes of the three curves where the average utilization equal 1.

It can be observed that, for long inter-arrival delays (i.e., low arrival
rates), the total service delay tends to simply equal average service time
with very little queuing delay. But as the inter-arrival delay, τ,
approaches the average service time, σ, total service time increases non-
linearly. Notice also that the larger σ ισ, the larger the inter-arrival delay
at which the total service delay begins to grow non-linearly. This means
that on average, the longer an individual packet takes to be transmitted,
the lower is the arrival rate at which the link gets congested and begins
to exhibit long queuing delays.

A response to a Read transaction often carries a lot more data than needed
by a core's original Load operation: for example, a cache line instead of a

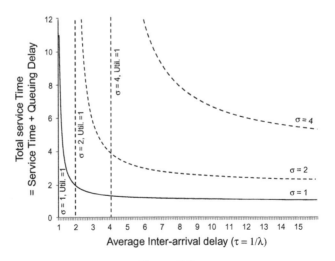

Figure 5.7
Effects of average arrival rate on total service delay.

word needed by a Load. Typically, the data needed by a Load operation is carried at the head of the transaction packet to reduce the latency for the operation. But the remaining portion of the data being transported plays a part in the traffic by contributing to the increased utilization of the links it travels on. The longer it takes to carry the remaining portion of the data in the packet, the higher the utilization and thus the higher the queuing delays on the link. This phenomenon is sometimes referred to as the "trailing edge effect" of a read data-packet's length.

From an interconnection design perspective, then, it is important to keep packet sizes from getting too long.

Notice that in all three cases, total service delay increases without limit as inter-arrival delay approaches the average service time value, or in other words, when utilization, ρ, approaches unity. So design effort should be focused on ensuring that every link in the SoC interconnection keeps its utilization well below 1.

From the foregoing, it can be seen that the queuing delay suffered at a link along the pathway is inversely proportional to the utilization of the link, which is the ratio of the total consumed bandwidth and bandwidth capacity of the interconnection link. Thus, as the traffic flowing across the link grows, the link utilization approaches unity (value of 1), and the queuing delay across it can grow non-linearly to a very high value and dominate the total delay for a Read transaction. This, in turn, causes the delay for Read transactions to grow and, consequently, causes the application performance to drop. It is therefore important to estimate the traffic across links within a proposed interconnection to ensure that the utilization of the links is below a safe level.

Bandwidth

Estimation of link utilization involves estimation of transaction traffic that will cross the link and then dividing that by the maximum traffic throughput that the interconnection link can possibly carry, or its "capacity" or "bandwidth". This traffic can be measured in terms of the number of transactions per second. In the case of a link carrying data, the number of bytes per second is often more convenient.

To estimate the traffic carried by any link in the system interconnection, we can start with goals for application performance, which is usually a primary input to the design process. From that it is possible to estimate the number of transactions issued by various devices per iteration of the application. The number of transactions per iteration that might flow on a given link of the interconnection can then be computed. This number multiplied by the performance goal of application iterations per second gives the transactions per second estimate for the link.

This latter number divided by the capacity of the link gives an estimate of the utilization of the link when the application is being run in the SoC. As we saw above, that number needs to be well below 1 to ensure that the application will not suffer undue performance degradation due to excessive congestion and consequent high queuing delays for transactions.

Interconnection networks: representation and terminology

Over the years, quite a wide variety of interconnection schemes have been proposed for organizing multi-processor computer systems. Each of the proposed schemes has attempted to improve upon certain aspects of communication that occur in a system. However, only a few of these have ever actually been implemented in real systems. Only a subset of this small number appears to be suitable for multicore SoCs. In this section, we will briefly review the taxonomy of interconnections and understand the theoretical bases for their classification.

Representation of interconnection networks

An interconnection network is often represented as a graph wherein nodes represent active communication elements, such as end-point devices and switches, and edges joining the nodes represent links between respective active communication elements. The nodes representing end-point devices are called end-point or terminal nodes. The "topology" of an interconnection network is the arrangement of its various nodes and edges. Figure 5.8 shows examples of a few interconnections. In the figure, small circles or boxes represents nodes and the lines represent edges of the graph.

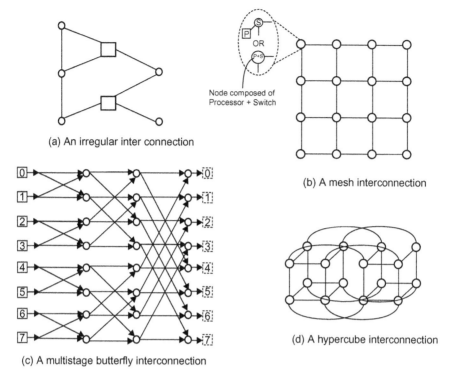

(a) An irregular inter connection

Node composed of Processor + Switch

(b) A mesh interconnection

(c) A multistage butterfly interconnection

(d) A hypercube interconnection

Figure 5.8
A few interconnection schemes.

Often, end-point devices are physically distinct from switches, but don't necessarily have to be. As illustrated by the Processor + Switch (P + S) node in the node blow-up inset in Figure 5.8(b), an end-point device might embody the functionality of transferring information between a pair of links to which it attaches and thus also act as a switch between those links.

Note that in the example of a multi-stage interconnection called the butterfly network, the edges are directed, implying that the corresponding links are unidirectional. Each of the square nodes on the left is a computational unit, which, depicted as dashed square nodes in the diagram, is also connected to the switches along the right edge of the network. Each of the switch nodes in the interconnection network contains a 2×2 crossbar switch.

Figure 5.9 shows a traditional (a) and a graph-oriented representation (b) of the address portion of a bus topology. It also shows how the

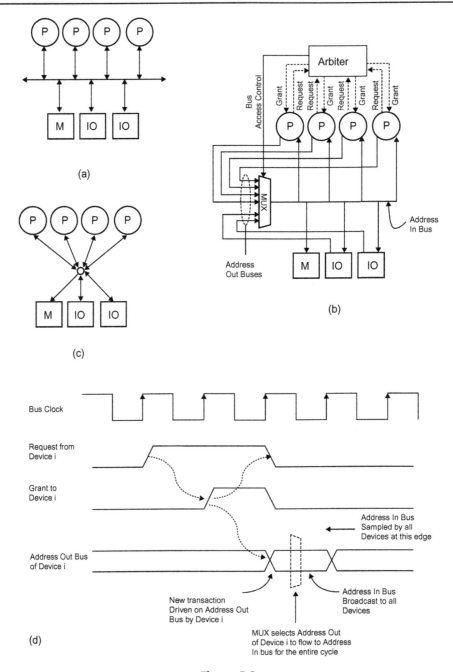

Figure 5.9
Different representations of a bus interconnection.

topology is implemented on a chip (c), without the associated control logic. The timing is shown in Figure 5.9(d). Unlike the high-z state available in board-level bus implementation allowing bidirectional wire connectivity to the processors, memory, and I/O, on-chip implementations have only two states for logic signals which require explicit multiplexing of address outputs to ensure only one source is actively driving the common broadcast address input network back to the units.

Direct versus indirect networks

Interconnection networks can be classified as either direct or indirect. In a direct network there exists a direct link between each pair of communicating end-point devices. Thus transactions exchanged between such pairs of devices do not go through an intermediate switch or another end-point device also operating as a switch. In the graphical representation, such a system would correspond to having a graph with no nodes representing purely switches. Of course, if all end-point devices need to communicate with each other, the result would be a complete graph over all of the end-points.

In practice, we find that such complete connectivity is rarely needed; it is also impractical to implement. In fact, even when complete connectivity is needed, the more practical approach is to implement a crossbar. The distinction between complete connectivity versus crossbar connectivity is illustrated in Figure 5.10. Figure 5.10(a) shows a four-node system which is redrawn in Figure 5.10(b). Figure 5.10(c) shows the connectivity needed to produce a true full connectivity that allows multiple simultaneous accesses into a node. Such simultaneous access capability is typically not available inside a node. Instead, a more practical arrangement, shown in Figure 5.10(d), that multiplexes incoming accesses into a single shared port is provided. The result actually is a crossbar-like connectivity, as depicted in Figure 5.10(e).

In the mesh configuration of Figure 5.8(b), it would appear that the end-point devices are connected directly to each other. However, in practice, each node in the graphical representation of the system is actually is computational sub-system connected to a switch that additionally connects to other switches along the edges shown in the diagram.

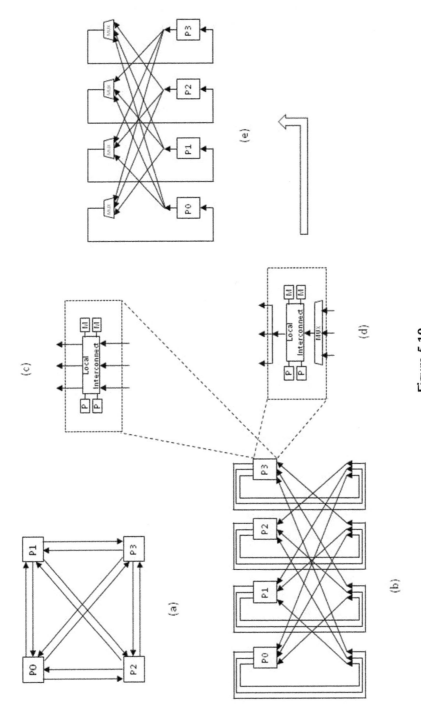

Figure 5.10

Comparison between a complete interconnection and a crossbar.

In theory, it is possible for the computational node to double as a switch node to enable communication between non-adjacent nodes. In such a node, the processor in the end-point would carry out the switching operation and relay information from one of its links to another. Such a node would be depicted as being Processor + Switch (P + S), as shown in Figure 5.8(b). With such end-point functionality, all nodes in the system could communicate with each other without seemingly needing a complete-graph connectivity. However, using a processor to do the switching function is slow and wasteful of performance, so it becomes fruitful to invest in a separate switch.

Almost all practical interconnections of today are indirect networks.

Circuit-switched versus packet-switched communication and blocking versus non-blocking networks

The switches may steer information combinationally or may operate in a sequential manner such that they receive information from a link, store it in their own internal memory elements, and forward it to a subsequent link over separate clock cycles.

In a traditional bus of Figure 5.9, the switch is simply a multiplexor, which combinationally steers the selected output and broadcasts it to the inputs of all the elements of the system. However, such a combinational broadcast operation seriously degrades the throughput of communication by requiring that the clock period be long enough for signals to propagate from the output of any of the issuers of a transaction to the inputs of all of the system elements. For this reason, most switches are designed to operate in a sequential manner.

Even with switches designed to operate sequentially, the question arises about the behavior of switches along the path of an end-to-end communication: should the switches maintain a stable end-to-end connection along the entire path for the duration of communication? In other words, should the communication be circuit-switched?

Interconnection networks were first studied in the context of telephony. Back then, switches were configured to set up circuit connections between callers and the called. A circuit was set up from an input port of the switch, to which the caller was connected, to the output port, to which the called party was connected, and was held continuously in a dedicated manner for the given connection until the call, which could last a long time, ended.

As discussed above, however, communication between end-point devices in SoCs is oriented to transactions, which are of short extent, and thus their duration of traversal through an individual switch is short. Unlike the case of telephony, setting up a connection across an entire path of a transaction is quite inefficient given the short amount of time it is occupied by the transaction. It is therefore much more efficient to transmit transactions on a switch-to-switch basis along the path from source to destination. At each stage, the transaction picks an appropriate destination output port of the current switch to hop to the next switch along the path to the destination. In this method of transmission, called packet switching, a path through the switch or between two switches is held only during the transmission of a single transaction between a sender and receiver across it, after which the path may be used for another transaction belonging to a distinct sender-receiver pair.

The circuit resources being exclusively held for call connections for a long time created opportunities for blockages to other connections involving distinct pairs of input and output ports that might have also wanted to be set up concurrently. This issue led to an exploration of non-blocking interconnection networks. A non-blocking network is a network that allows simultaneous one-to-one circuits to be set up between any two pairs of its input and output ports (and therefore, transitively, between all possible pairs of input and output ports, at any given time).

However, no such non-blocking requirement exists for packet-switched networks as a possible blockage persists only for a short duration of transaction transmission.

Base-form vs. encoded signaling

In long-range data communication, such as in gigabit or 10 gigabit Ethernet, the original data to be sent is often encoded before it is transmitted. This is done to reduce the number of wires or signal pathways needed and to increase reliability of transmission over long distances. The original data is recovered by decoding it at the destination. Such transmission always results in a significant addition to the data access and transmission latency due to the processes of encoding and decoding. Only applications that have high latency tolerance and wherein the increase in latency due to the encode-decode process adds up to only a small portion of the otherwise inherently long data transmission delay, often of the order of few to few tens of milliseconds, can utilize such a method of communication without significant loss in performance. By contrast, latency is of paramount importance in SoCs where latencies are typically of the order of a few tens of nanoseconds, and the increases in latency caused by data encoding is not tolerable. Therefore, in SoCs, the data is mostly transmitted in its native, or base, form, without any encoding.

Transaction routing

The class of storage accesses is composed of request transactions and numerous types of response transactions.

The system needs "address maps" to determine the route address-carrying transactions will take to their destinations. However, for non-address-carrying transactions, such as those for enforcing synchronization or supporting storage accesses, the system must provide an alternative method to route them. This is typically done with tags.

Given that message destinations are indicated using device identifiers as opposed to addresses, the system must have "device maps" as well as a means to translate this separate namespace into routing information for the interconnection. This namespace is distinct from the namespace for addresses of storage locations that are needed for Read-Write traffic.

Most often, messages are sent to individual devices. However, it is possible that certain message types are intended to be multi-cast to more than one device or, in the extreme, broadcast to all of the devices in the system. The multi-casting of a given message may be achieved by the sender by explicitly sending multiple individual messages, one at a time. Or, the system may ease this by providing in the device namespace collective identifiers that stand for sets of receiving devices.

Bus as an SoC interconnection

A bus interconnection is the easiest and most inexpensive way of connecting multiple communicating devices together.

In its simplest form, a bus is composed of a set of electrically passive wires to which multiple devices connect with stubs as shown in Figures 5.9(a) and 5.9(b). In some instances, the wires may be interjected with active drivers to maintain signal quality, but the overall electrical path still is logically equivalent to a wire.

In a bus interconnection, distinct sets of wires form independent busses that are used to transmit distinct types of transactions. Address or Message requests, data for Write requests or Read responses, and other types of responses, are common transaction types for shared-memory multicore systems.

Transaction information in the form of electrical signals placed on these wires is broadcast to all entities connected to the bus, but only the necessary devices that act further upon the information will pay attention to the signals and pick them up. The use of bus wires is time-multiplexed using a global clock to which all of the devices are synchronized. Information belonging to a transaction being transmitted is driven out by the sending device starting at the beginning of clock cycle and picked up by the receiving devices very near the end of that cycle, the electrical signal getting propagated to all devices within that one cycle (see Figure 5.9(d)). In the next cycle, a different device is allowed to send its transaction, or the same device is allowed to transmit another transaction, over the same set of wires. In this manner, transactions from devices are serialized and broadcast across the system, providing an all-to-all communication capability.

A device wanting to send to other device(s) a type of transaction signals to an arbiter a request to do so. The arbiter chooses one among the many requesters and signals back to it the granting of a permission to do so. The chosen device then drives its transaction on the bus during a predetermined number of cycles after receiving the grant as shown in Figure 5.9(d).

Using a single bus to connect all system components immediately fulfills the basic communication need in the SoC: any device can communicate with any other device in the system. But as we saw in earlier sections, maintaining coherency among caches and memory is also an important function of a multicore SoC interconnection.

The achievement of coherency depends on participating devices agreeing on the order in which accesses to a storage location that is subject to coherency should occur. This happens naturally in a system with a bus interconnection topology.

In a bus topology, at most one transaction can be transmitted per cycle over the bus's broadcast medium (i.e., its wires). The system "recognizes" at most one transaction at a time as "occurring in the system" by allowing it to be driven on the bus's transmission medium during a bus cycle. This establishes an unambiguous total order on all transactions occurring in the system, including the ones necessary for achieving coherency for a given location. A transaction thus occurring in the system is also made visible to all the system's devices during the same cycle it is placed on the bus through an electrical broadcast. This action accomplishes the creation of a common view of, and thus an agreement on, the cycle during which a given transaction occurred, among all devices. Together, the two actions achieve a global agreement among devices on the order of any set of transactions occurring in the system. This also helps to synchronize among the participants of the protocol the production, combination and distribution of coherency responses and the matching of them with the original transactions via simple temporal correspondence.

Note that, although some of the transactions can get "retried" and thereby get "repositioned" in the system's order of transactions, even these "retry" events are made simultaneously visible to all the devices by also broadcasting the response transactions that effect the retry, so that the global agreement about the transaction order is safeguarded in spite of the retries. Thus of the two transactions appearing on a bus, the first one to complete successfully is considered to have occurred earlier in the system and thus ordered ahead of the other.

The snoopy cache coherency protocol, which is easy to specify, implement and verify, takes advantage of the transaction serialization and broadcast properties of the bus topology. Global visibility of the common order among a system's transactions also makes synchronization among the system's devices easy.

Because of these above-mentioned properties and the simplicity of its construction, a bus interconnection has been quite popular in multicore coherent systems.

Limitations of the bus architecture

A bus-based coherency protocol serializes all storage accesses by handling one transaction at a time, and it depends fundamentally on the broadcast of a transaction to all the participating entities in a single bus cycle to enforce cache coherency.

But dependence on a single-cycle broadcast of transactions creates a scalability problem for a bus. As more entities are connected on the bus, both the capacitive load and wire length the signals have to traverse grow linearly with the number of devices connected on the bus. The propagation delay of fall time in CMOS under the assumption of constant driver current is proportional to the total capacitance of the net [6]. The cycle time of the bus must be greater than this propagation time. The result is an increase in bus cycle times as the system size grows.

As the number of cores in the system grows, the arrival rate for requests generated by these cores also increases. Both of these factors combine to cause the utilization of the bus to increase rapidly with system size. As we saw in the case of a link modeled as a simple queue (section 4.2.1), the queuing delays grow dramatically with increasing utilization. In short, the bus is easily "overwhelmed" as the system grows in size and extent.

In general, the relationship between the utilization of the interconnection resource and service delays is important. Service delays increase as utilization increases. If resources can be added, the utilization can be lowered and the service delays can be brought to linear regions. The ability of an interconnection scheme to grow its resources as its traffic requirement grows, keeping the utilization of those resources low to keep service delays in a linear region, is referred to as the "scalability" of that interconnection.

There are other disadvantages to a single-bus-based system. In a complex multicore system, there can be high level of exploitable parallelism. Relying on a single bus to convey all the transactions undermines the operational parallelism inherent in the system.

In a bus topology, all types of transactions are placed on the common bus. This reduces the amount of bandwidth available for any one specific type of activity, such as coherence or inter-processor messaging. For example, transactions implying no coherency activity can deny the necessary bandwidth for transactions requiring coherency action.

Additionally, in a simple bus protocol each transaction is presented to all entities connected to the bus regardless of whether that entity is the transaction's destination or not. This means every transaction is also snooped by every processor on the shared bus whether it needs to be or not. The effective snoop rate is the sum of access rates from all the processors. This unnecessary snooping can overburden and congest a processor's coherency management and cache pipeline and throttle its internal computations. In a shared-bus interconnection, these problems only get worse as the system grows larger.

A simple shared bus is especially unsuitable in the context of application consolidation, as it cannot take advantage of system partitioning to reduce congestion. This problem can be appreciated by considering the following example: assume we have a 16-processor system that is partitioned into four equal sub-partitions of four processors each. For the sake of simplicity, assume that each of the applications running on the system has a cache miss rate of 5% per processor, or 1 access every 20 system cycles. Each partition has a cumulative miss rate of 20%, which is not very high. However, the cumulative total across all the partitions is a whopping 80%! This level of traffic results in very large queuing delays. Also, such heavy snoop traffic into processors can seriously impede computation internal to the processors.

Fabric-oriented interconnects for larger SoCs

The performance-related shortcomings of a bus-based interconnect as the system scales in size can be mitigated by employing more fully connected, network-oriented interconnection schemes with multiple parallel transmission paths that also overcome frequency limitations. Such schemes are popularly referred to as fabric topologies [28,29]. We will briefly review their common characteristics.

Unlike a shared bus, the fabric is a point-to-point topology instead of a broadcast topology, so the fabric cycle time does not degrade as system size increases.

A fabric topology offers multiple paths for transactions to flow between devices. Thus, fabric-oriented topologies are characterized by high throughput − better matching the multicore system's requirements.

Transactions traveling to different destinations can be decoded and then can be sent along separate paths, increasing the available bandwidth and concurrency in transaction processing. This lowers congestion along any given link in the fabric, thereby keeping the average service delay along the link lower.

In a fabric topology, it becomes possible to provide separate paths for different types of activities. For instance, coherent transactions, non-coherent transactions and message transactions may all flow along different paths. This eliminates mutual interference of the three classes of transactions, thereby improving the service times for each class.

Fabric topologies can help in application consolidation without causing congestion. Fabric topologies can be logically partitioned, with each partition hosting a different sub-application. The partitioning can be achieved by employing traffic filters. Particular topologies may even allow partitions to avoid sharing many electrical paths within the fabric. Thus the sub-applications operate independently, for the most part, and their transaction traffic rarely intermingles. This inter-partition isolation exploits available parallelism, leading to better application performance. Traffic isolation also conserves processor coherency bandwidth, which is a scarce resource in a large, coherent multicore system.

Transaction isolation capabilities can also be extended to create more secure systems, preventing unauthorized transactions from reaching protected devices.

Of course, deploying fabrics involves considering all related trade-offs.

Fabrics typically are more expensive because of the additional active logic they employ to accomplish their myriad functions. This translates to more silicon area and power. However, feature size and power reductions associated with technology advances should continue to help lower these impacts.

Fabrics divide an electrical path for a transaction from source to destination into multiple pipelined sub-paths. This invariably increases the base latency faced by a transaction when there is no competition from other transactions. However, counteractive techniques such as out-of-order execution in cores, more outstanding transactions, larger processor caches, and deeper cache hierarchies, including system-level caches, can ameliorate the negative effects of longer paths through the fabric and, in fact, can improve the overall application performance.

One of biggest challenges created by a fabric-oriented topology is the implementation of a coherency protocol. The basic mechanisms that make the operation of a snoopy protocol easy in a simple bus topology are no longer available; a more sophisticated protocol is usually needed.

Transaction formats

Figure 5.1 showed the examples of Read and Write transactions, with each field shown separately and elaborated fully. We will refer to such a format as the "base format" of the transaction. When transactions are carried in the system in their base format, as is the case in most bus-based systems, all of their fields can be transmitted simultaneously and in parallel over independent wires, which leads to higher performance. However, the links carrying them, and buffering structures meant to hold them, can also become extremely wide. This can mean a high wire count and thus higher wiring congestion in the system. Wider buffering structures could also mean larger static power dissipation.

Such links and buffer structures can also become too transaction-type-specific, reducing the possibility that they could be reused somewhere else in the system for carrying and holding other types of transactions. This inflates the design effort by necessitating the creation of a wide variety of structures to accommodate different transaction types. In large systems, however, where wires can run for longer lengths and transaction may need to traverse many junction points, it becomes efficient, design effort-wise, to have common structures that can be reused in multiple places. So it helps if transactions are carried in a more uniform format.

Packetized transaction formats

Figure 5.11 displays examples of a few transaction types that have been organized into a "packet" format. The packets are formed out of units of uniform width, 128 bits in these examples, but need a different number of units depending on the information content of the transaction, although, in practice, only a few selected sizes of packets are necessary. The packet is transmitted one such unit at a time and is termed a "flit", short for <u>Fl</u>ow <u>Un</u>it, over a link of the same width.

The figure shows labels of H and C to the left of the various flits of the packets that distinguishes their contents. H stands for the "header" flit

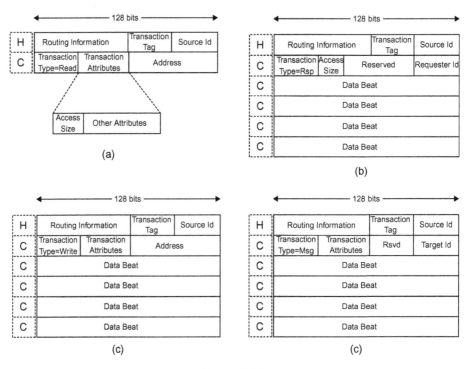

Figure 5.11
Packetized transaction formats.

and C stands for a "content" flit. All of the packets in Figure 5.11 are shown carry Routing Information in the header flit that typically is not part of the original transaction. It is added by the interconnection hardware. It is assumed in these examples that information regarding the path the packet is supposed to take through the system is computed and stored in the packet for easy access. The pre-computed path information does help save time that would be spent in computing the next step in the path and thus helps expedite the transaction to its destination. But this is not mandatory. Alternatively, the next leg of the path can be computed at each stage based on the other information available in the packet, such as Address in the Read (a) or Write (c) types of packets and Requester ID in the Response packet (b).

Although the packet formats, their widths, their particular contents, their sizes and their relative placement are shown here strictly for the purpose of illustration and can vary from one interconnect solution to another, the figure does highlight some of the common elements of a packetized transaction: first, each packet carries the entire transaction along with its identifier, the Transaction Tag, which helps correlate corresponding response transactions back to the original. A packet also carries a Transaction Type field that indicates the contents of the packet and the specific semantic action that must be taken on the transaction's behalf. The type also indicates other fields that might be carried by the packet. For example, the Write type in Figure 5.11(b) indicates that the packet is also carrying data. The Access Size field in the packet, which announces the amount of data being carried, helps in scheduling the transfers and management of buffers. Notice also that the packet for Transaction Type = Msg (Figure 5.11(d)) carries a Target ID instead of an Address.

Notice that packets carry more information than just the transaction itself. In the packets of Figure 5.11, the header flit usually contains extra-transactional information. For instance, routing information contained in the header is not germane to the transactions themselves. In practice, packets may carry even more information such as for flow control and error-detection codes. This additional information represents communication overhead, which lowers the effective transaction bandwidth compared to transactions transmitted in their base format.

Note that the width of the packets is significantly smaller than the number of bits needed for the whole transaction. This leads to narrower data paths in the interconnection and lower static power dissipation. The narrower data paths, typically, can be run at higher frequency to regain the bandwidth lost by packetization of transactions. As the silicon feature sizes decrease and the static power portion of the overall power dissipation increases non-linearly, narrower data paths at higher frequency are often a good trade-off.

Transporting transactions in packetized form also has the advantage of being able to create interconnections that are uniform throughout the system. This means that the interconnections can be assembled out of common building blocks. This ability to reuse building blocks is of great practical benefit not only for the design process, but also in verifying the operation of the overall interconnection, which could be quite extensive for large systems.

The formats shown in Figure 5.11 are 128 bits wide, suitable for traversing 128-bit wide links. It is quite possible to define alternative formats that are narrower or wider. However, there is a three-fold danger in defining a format that is too narrow: firstly, by definition, the width of the link will be narrower, which, while better from the point of view of congestion, does mean lower bandwidth for the same data transmission frequency. Secondly, the packet takes a larger number of flits to get delivered. This in turn increases the latency for transaction delivery. Thirdly, the same packet will now occupy the link for a larger number of cycles. That is, its service time over the link goes up. The result usually is more queuing before getting transmitted and longer service delays, possibly leading to lower application performance.

Transaction routing

Since each transaction is an independent unit of communication, it needs to be routed as a unit and is conveyed from its source to destination independent of other transactions. Given a transaction, the SoC must determine its path through from source to destination.

For a part of a transaction's journey, the path may be a default one. For example, memory access transactions issued by I/O devices may travel along inbound paths in a tree-like interconnection structure toward its root, which is unique for the tree. But when the transaction arrives in the system where there is an option regarding which path to choose, the system must make a calculated routing decision.

This calculation typically uses look-up tables for "name-spaces", which simply are sets of identifiers carried by transactions to identify their destinations or targets. These tables must be set up at least minimally ahead of the arrival of the first transaction, and they may be more fully populated later by software. SoCs carry different look-up tables for different namespaces that might be referenced by transactions.

The namespace being accessed is often implied by the type of transaction. For example, storage access transactions carry an "address" of the storage location being referenced, which simply is a positive binary integer, starting at zero. SoCs have address maps against which the transaction's address is compared to identify its destination and path through the system interconnect. Message transactions, on the other hand, refer to a different namespace that associates a unique identifier to each device or to a group of devices capable of receiving such transactions.

A few types of transactions may not refer to a specific address or destination device. Such transactions often imply broadcast to all relevant devices. Examples are synchronization transactions that are broadcast to all devices within a coherent shared-memory subsystem.

Building blocks of scalable interconnections

In this section we look deeper into the building blocks of SoC interconnections. We will delve to develop a deeper understanding of the ubiquitous crossbar connectivity and the trade-off between cost and performance. We will discuss issues of routing, queuing, arbitration, and transmission of data across the network. We will touch on issues of transaction priorities and virtual channels.

In the most generic description, an interconnection network can be looked upon as a connected sub-system of "switches" and "links". Computational or storage devices in the system, referred to as "end-point" or "terminal" devices, attach at the periphery of this sub-system. The interconnection sub-system is intended for transmission of information between end-point devices that produce and/or receive it.

Links typically are bundles of passive wires that provide pathways for electrical transmission, while switches are the active elements in the interconnection system.

Links

A link connects a pair of switches, a pair of end-point devices, or a switch and an end-point device. A link is a wire-set that propagates "symbols" of information via correlated electrical signals. A switch is an intermediate digital function block that carries and steers information from one link to another.

The links themselves could be unidirectional, half-duplex bidirectional, or full duplex bidirectional.

A link could be divided into unidirectional or bidirectional sub-links made up of subsets of wires. A sub-link is divided further into single or multi-bit signals that carry symbols of information. For example, a link at the memory subsystem could be composed of a unidirectional address sub-link and a bidirectional data sub-link plus other response sub-links.

In most common cases of an on-chip interconnection network, the temporal correlation among different values propagating over wires of a given sub-link is achieved via a clock. Values on all of the wires of a sub-link change or are captured in synchrony with such a clock.

All of the symbols defined by the stable values of the signals of a sub-link during a given clock cycle form a tuple of information, which, in the present discussion, is a complete transaction or a flit of a packet. A complete packetized transaction may need one or more such flits to be transferred.

Transaction flows from one terminal device to another over an interconnection network by passing through a sequence of one or more links and zero or more switches until it gets to its destination device.

Clocking considerations

The clock edges to which a sub-link's signals are synchronized may be those of a globally shared clock. For a small system, these common edges can be delivered with tight skew control at both sender and receiver ends of a sub-link. In this case, the sub-link can be run synchronously.

As the size of the system grows, distribution of a global clock presents small difficulty. But maintaining tight skew and low drift between the common edges of the clock at far away points becomes an increasingly harder undertaking. The phase difference between corresponding edges of the clock at the two ends of a link can become too great and too uncontrollable for the link to be reliably run synchronously at a high rate. In this case, the source clock is transmitted with data and used to clock the data at the receiving end. Such data must then be retimed and synchronized to the receiver's clock. This requires extra logic and may lead to some loss of bandwidth. Yet, if the same clock can be supplied without independent intermediate PLLs to the two link partners, the data can be recovered quite efficiently at the receiver by clever tuning of the received signals [30].

If, on the other hand, the same master clock is used to feed independent PLLs to supply clocks to the two link partners, the two clocks will have the same nominal frequency but could differ in phase from each other slightly at any given instant. The sub-links must then be run either asynchronously or "plesiochronously". In the latter mode, the fact that the clocks at the sender and receiver, although out of phase, have the same nominal frequency lends itself to a more efficient implementation and is thus preferred.

Switches

In practice, switch designs can vary widely based on the particular needs of the system. However, it is useful to visit the basic concepts and canonical switching structures and their purpose. We will look at some of the more important ones.

Switch data paths

Figure 5.12 shows a schematic representation of a simple $m \times n$ crossbar switch, where m is the number of input ports, 0 through $m - 1$, and n is number of output ports, 0 through $n - 1$, of the switch. The figure shows the data paths through the switch. The control paths that select the settings of the various port multiplexors (muxes) at various times and the details of the queuing structures are not shown, but will be discussed later. In the data path, every input buffer has a path to each of the n output multiplexors dedicated per output port of the crossbar. Each multiplexor selects and services at most one of the m inputs requesting transfer of a transaction at a time.

Figure 5.13 also shows an $m \times n$ switching structure. The DES block at each input port merely deserializes the incoming packet before it is stored in the central shared storage. This is done to match the transfer

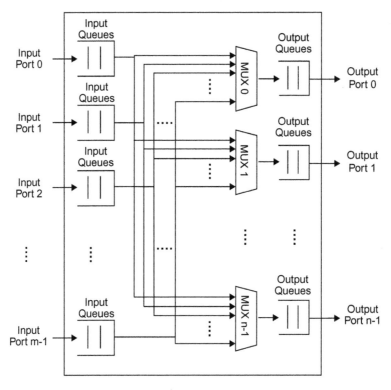

Figure 5.12
Internal structure of a basic $m \times n$ crossbar switch.

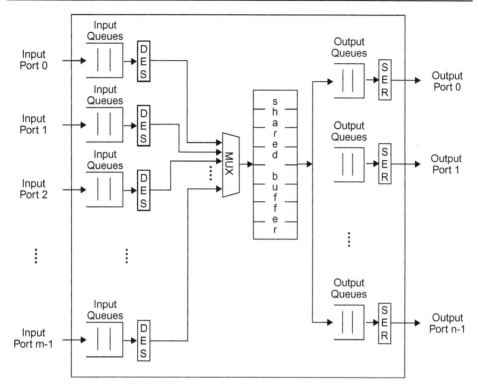

Figure 5.13

Structure of an $m \times n$ crossbar switch with internal deserialization and shared storage.

rate of the switch to the sum of the arrival rates at the input ports. The SER block at each output port re-serializes the packets on the output link.

In this crossbar structure, there is a single multiplexor that chooses one input port at a time out of m input ports and transfers a transaction from the queue of that port to the queues of one of n output destination ports. Note that the shared buffer in the middle is not a requirement and could instead be replaced by a direct wire connection between its input and output.

The two structures are functionally equivalent, but represent contrasting performance, area and power trade-offs. The crossbar structure in Figure 5.12 offers potential parallelism in information transfer across all of its multiplexors at the cost of higher area and power. On the other hand, in the structure of Figure 5.13, the transactions can only be transferred serially, between one pair of ports at a time. Notice too that both structures could perform a broadcast or multicast operation given appropriate additional control.

The commonality of a crossbar between Figures 5.12 and 5.13 is worth paying attention to. The latter can be viewed as being composed of two parts: an $m \times 1$ crossbar switch followed by a delivery structure that enables delivery of a transaction to one or more of the output queues at a time.

The basic crossbar structure is ubiquitous and is found not only in complex interconnects, but variations of it are found even in bus structures. Notice the topological similarity of the bus organization in Figure 5.9(b) and the central switching structure in Figure 5.13. If, as shown in the latter figure, one accounts for some storage at the input ports, the bus has an $m \times 1$ crossbar structure followed by a transaction distribution structure. The difference is that, in a bus, the information is always expected to be broadcast from an input port to all of the output ports, which is often not the case in a switch.

Switching control

Typically, a switch is a highly parallel logic circuit, and to use it efficiently requires that it be operated in a sustained manner so as to exploit that parallelism. Four categories of activity occur in the switch and in the operation of an interconnection network comprised of switches and links. They are:

- **Routing** of transactions along the desired path through the interconnection;
- **Arbitration** when shared resources are being contended for by multiple transactions;
- **Flow control** to coordinate the data transfers so that transaction data is moved from one point to the next through the interconnection only when there is storage space to hold it in the next spot so that the data is not dropped or overwritten; and
- Actual **transfer of transaction data** from one place to the next along the path from source to destination.

The control of a switch involves the cycle-by-cycle coordination of data transfer across it. It is comprised of the following basic actions:

- At each input, make a choice regarding which transaction from the input queue to forward to one of the output ports. (Arbitration)
- Determine, for each such transaction, which output port it is destined to and generate a request for that output port. (Routing)
- Make a choice, per output port multiplexor, depending on the fullness of the queue at the output port, as to if and which of the requesting input ports to service in the current cycle. (Arbitration)
- Indicate to the individual input ports if their request will be honored next. (Flow control)
- For each input queue that requested a transfer and was selected by the output port multiplexor, transfer out the selected transaction. (Data transfer)
- For each output port multiplexor that has accepted a request, transfer the transaction from the input port across the multiplexor to the output link queue. (Data transfer)
- For each output port that received a request for a transfer and issued a selection to an input port, transfer in the selected transaction. (Data transfer)
- For each input queue that is transferring a flit of a transaction out during this cycle, compute and communicate updated flow-control information to the upstream device or switch connected to it. (Flow control)
- For each input queue that has space for more transactions, transfer in a flit of a transaction if the upstream device or switch has transferred one over the input link. (Data transfer)
- For each output queue that is transferring in a flit of a transaction in during this cycle, update its current queue full/empty status to be used by its port multiplexor. (Flow control)
- For each output queue that has transactions to transmit to the next downstream switch or device, make a choice regarding which one to transmit. If the output link buffer is a combined one, the transaction is simply the one at the head of the single common queue. (Assume that, once a transaction is selected for such a transmittal, all of the

flits of that transaction are transmitted before another transaction is selected for transmission.) (Arbitration)

- For each output queue that has a transaction flit to transmit to the next downstream device or switch, if that device's or switch's flow control indicates availability of buffering space in its input queue, transmit the flit. (Data transfer)

The labels in the parentheses at the end of the items above indicate their categories.

In the case of transactions that have multiple destinations, as is the case for a coherent memory-access transaction being snooped across a multicore system, for instance, the list needs to be augmented to include actions needed to achieve multi-casting of the transaction so that copies of the transaction get created and sent along the multiple links necessary. As should be expected, such capabilities add significantly to the internal complexity of the switch design.

Performance of the interconnection is intricately tied with the above activities. Let us look at some of the important aspects and issues associated with these activities in more detail.

Routing

Routing involves making a decision regarding which one of the potentially multiple possible immediate next destinations to transfer the transaction to in order to eventually carry the transaction to its intended destination. As indicated earlier in this chapter, the type and the address − or some other label in one of the namespaces defined in the system − being carried by the transaction uniquely identify the destination(s) of a given transaction.

A "map" − such as an address map − is commonly set up in programmable registers to specify address ranges and the paths the transaction must take to reach them. When a transaction arrives, its address is compared against the entries in the map to extract the path information.

It is indeed possible, in theory, that the routing decision is based on looking up a map at each switch to determine the next link to traverse in a scheme called "destination routing". However, for a large system with 10 s or 100 s of cores and devices, such maps can become quite elaborate and complex to look up, making the maps too expensive in terms of area and power to replicate at each switch input or even on a per-switch basis.

In some such extended systems, it can become more economical to implement a map at the site of each source device. In a method called "source routing", each entry in such a map then contains routing information indicating which link to follow at each switch along the entire path that the transaction would take. For example, in a mesh network (see Figure 5.8(b)) such information can be a string of 3-bit fields, each of which identifies the next link to take by indicating whether the transaction should take a link to one of the four directions, east, west, north, or south, or to the local node, in which case the transaction has reached its destination. Such information is extracted and carried along by the transaction, with each switch along the way looking up its portion of the information.

Note that in the above routing scheme, a switch is not aware of the overall topology of the interconnection and the path taken across the interconnection is fixed at the source. Such a scheme is therefore called an "oblivious" routing scheme. But in the presence of faulty or congested links, oblivious routing can be a problem. If the interconnection has multiple possible paths to the destination, an "adaptive" routing scheme that chooses the links based on the state of the various links can become very useful [31].

Preventing head-of-line blocking

Arbitration at the input port queue of a switch chooses a transaction to transport through the switch. If the queue is strictly first-in first-out (FIFO queue), then the choice is simple: the oldest transaction in the queue is the chosen candidate. But such a choice can be fraught with problems, some of which we shall now discuss.

One problem is that of "head-of-line blocking". If the output port that the oldest transaction is headed to is busy with other transactions, then the oldest transaction could be stuck in the input queue for quite some time. It is possible that one or more transactions behind the oldest transaction are headed to some other output ports which are not busy, and these would have been able to make progress through the interconnection were it not for the stuck transaction ahead of them that is blocking their progress.

One way around this problem is to move the oldest transaction out of the input queue even if the output port it is headed to is temporarily busy. This is done by providing an internal buffer that a transaction can always be moved to from the input queue even if its desired output port indicates that it is busy.

Another way is to allow the transactions in the input queue to leave the queue out of order as long as there is no specific ordering requirement between those transactions. (Note that it is not altogether unusual in an SoC to find such inter-transaction ordering dependencies. For example, posted memory transactions arriving from a PCI hierarchy are required to be performed in the system in a strict order, independent of which portion of system memory the individual transactions access.)

To enable out-of-order egression of transactions, the input queue analyzes more than one transaction that may reside in it to determine their destination output ports. All the output ports so identified are sent requests. All such requests that are awarded grants by their respective output ports result in the corresponding transactions being extracted from the queue and forwarded to their intended output ports.

Clearly, the ability to analyze multiple input queue entries and the ability to extract them out of order comes only with added logic complexity, which creates timing challenges and adds to area and power, since it is replicated per switch input. The compromise is to analyze only the first few entries of the queue, but that diminishes the effectiveness of the analysis.

Another method to achieve the same effect is to organize the input queue into physically separate sub-queues, one per output port, in a scheme called "virtual output queuing" (see Figure 5.14). As the transactions arrive at the input port, their destinations are analyzed and placed into one of the virtual queues. The virtual queues now independently make requests to their respective destination output ports.

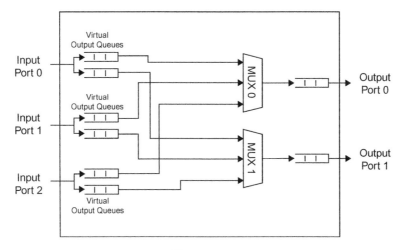

Figure 5.14
Structure of a 3 × 2 crossbar switch with virtual output queues. Alternatively, in a scheme called "virtual input queuing", each output port maintains a separate queue for each input port. Both these schemes improve the performance of the switch, albeit with a significant increase in both control and data path complexities.

Quality of service considerations

The primary elements of quality of service in an SoC are end-to-end transaction latencies and bandwidth guarantees on a per-device basis.

Latency guarantees

Queuing delays not only can significantly increase the effective latency of an access, as we have seen earlier, but they can make those latencies unpredictable as well. Unpredictable latency makes it difficult to determine the size of buffers a network interface must carry.

Predictable latency can be critical also in the above buffer overflow situations, for example. A network interface may fetch a descriptor in order to determine where the incoming frame is to be stored. If the descriptor cannot be read in time, the buffer itself may be overrun by subsequent packets arriving at the network interface. A similar hazard exists on the transmission side too. If the data being read from system memory does not arrive in time, a packet data under-run occurs, and the malformed packet gets transmitted on the network.

One obvious avenue for correcting the high latency for such Read transactions is to treat them and their responses with higher priority so that the Read responses will reach the requester sooner than otherwise. A higher-priority transaction is selected to be served in the system and to move forward through the interconnection ahead of other lower-priority transactions that might also be contending for service at various points along the transaction's path. This helps reduce the queuing delay the transactions would otherwise face. However, just raising the priority of these Reads and their responses is not sufficient.

Head-of-line blocking in the interconnection can undermine this attempt to reduce the latency of such transactions. If the input queues at a switch are handled in a strict FIFO order, high-priority transactions can be stuck behind low-priority transactions. Virtual queuing at the input ports of switches based on priority of the transactions provides an effective solution to this problem. Figure 5.15 shows a 3 × 2 crossbar switch with two virtual queues at input ports. Each virtual queue corresponds to a

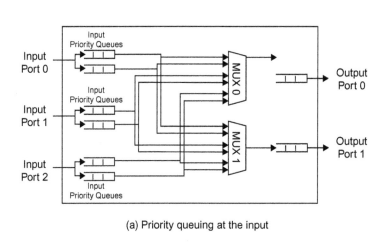

(a) Priority queuing at the input

Priority 0, Output 0
Priority 0, Output 1
Priority 1, Output 0
Priority 1, Output 1

(b) Combined priority and virtual output queuing

Figure 5.15
Example of a crossbar switch with virtual queues to avoid head-of-line blocking.

unique priority level and holds all of the transactions with that priority level traversing that port.

For the virtual queue method to be effective in reducing latencies, though, it is important that transactions of different priorities issued by the same source are free of mutual ordering requirements and are able to traverse the interconnection out of order if necessary. If such a mutual ordering is occasionally required, some separate means would need to be employed to achieve the required ordering between transactions, such as the requester ensuring that the previous transaction is complete before issuing the next one into the system.

Bandwidth guarantees

Bandwidth guarantees are needed to ensure that the applications achieve their performance goals. (Conversely, as we saw earlier, the latter implies bandwidth expectations per device involved in the computation.)

Bandwidth guarantees are necessary to avoid loss of data due to buffer overrun conditions at the network receive interfaces. These interfaces run at a fixed frequency and potentially bring in packets at the network's line rate. The contents of these packets are temporarily held in the network interface block (such as an Ethernet controller) buffer before being transferred to the SoC system memory for temporary storage or directly to another block for further processing. Over-the-network flow control is usually too slow to respond to a "buffer full" condition by stopping the flow of packets and thus preventing an input buffer from being overrun. Instead, the SoC must ensure that the arriving packets are transported out of the buffer at a speed as high as the line rate of the interface.

To provide bandwidth guarantees, the arbiters in switches have to be programmed to perform fixed time division multiplexing among requesters' transactions as well as reserve buffer resources in queues on a per-requester basis to ensure that the transactions are not denied progress, and therefore throughput, through the interconnection.

Deadlock prevention

Most common architectural sources of deadlocks in SoC interconnections are transaction completion requirements or inter-transaction ordering demands interplaying with resources in the system's actual implementation held by transactions to create circular event dependency chains. We illustrate the latter issue via an example involving PCI.

Figure 5.16 depicts a simple system involving a processor, memory and a PCI device, all connected through a central switch containing a common shared buffer for transactions, as shown. The buffer itself has two parts: one pool of buffers dedicated to request transactions and the other for data response transactions. (We will ignore other types of transactions in this simplified illustration.)

In the figure, the processor has issued a large number of Read requests (R) to the PCI device so as to have backed up into and completely occupied the shared request buffer of the switch and to back up into its own internal request queue. Concurrently, the PCI device is depicted to have issued Posted Memory Writes (W) to memory followed by data responses (RD) to the processor's Read requests.

If, when the Writes from the PCI device arrive at the input of the switch, the shared request buffer is full of Read requests headed to the device

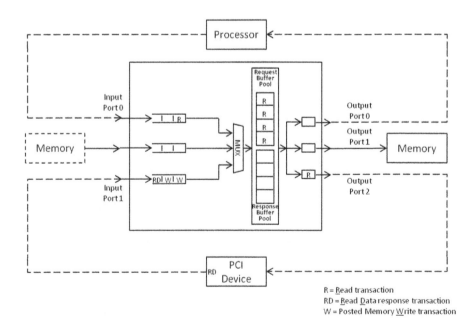

Figure 5.16
Example of a PCI-related deadlock.

itself, a deadlock occurs as shown in the diagram. The deadlock occurs because the Posted Memory Write requests are not able to make forward progress. Per PCI ordering rules, Read responses coming from PCI cannot bypass the Posted Writes arriving thence and are therefore also not able to make progress either, even though there are response buffers available in the switch.

A deadlock such as this can be resolved using the switch structure shown in Figure 5.17. There the switch has been modified to create "virtual channels", or effectively multiple parallel pathways, across it. The port that connects to the PCI device carries two virtual queues, one (Virtual Channel 0) for carrying the PCI device's own Read requests to system memory and the second (Virtual Channel 1) to carry its own Posted Memory Writes and Read data responses for processor's requests. The lone queue from the memory carries Read responses. Additionally, locations in the shared request buffer are reserved for Posted Memory Write requests that ride in virtual channel 1. Because of this buffer differentiation, the Writes from PCI are able to get into the request buffer and make progress toward the memory. That then makes way for

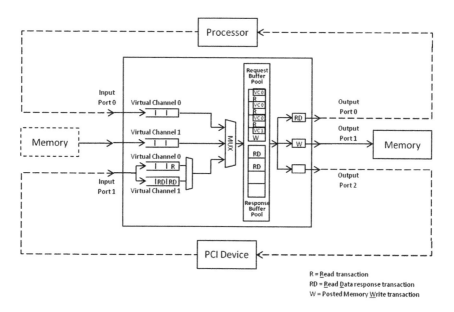

Figure 5.17
Resolution of the PCI-related deadlock using virtual queues.

the Read data responses, too, from PCI to progress toward the requesting processor.

Virtual queues at the two ends of a link similarly enable creation of virtual channels across it. This way end-to-end virtual channels can be established across the entire interconnection. Transactions can then be assigned to the appropriate virtual channels along their path to prevent circular resource-based dependencies.

In general, virtual channels offer a general mechanism to avoid such circular dependencies in the interconnection and prevent deadlocks [32].

Queue storage

As we have just seen, switches need storage at many sites throughout their design. Storage is needed for both the data path and control algorithms. Depending on the transaction sizes, the throughput rate and delays in receiving flow control credits, the storage needed can be small or large. However, the amount of storage needed for the former usually far surpasses that for the latter.

Storage can be implemented in logic flip-flops or memory arrays, the former simply inferred by the compiler from the Register Transfer Level specification of the switch logic. Flip-flops offer highly parallel access to the information contained and are better suited for the control portion of switch's logic.

Memory arrays are regular and are constructed out of optimized storage circuit cells which are significantly more area-efficient. But memory arrays have significant per-instance overhead associated with addressing logic, error correction and built-in self-test logic. Nonetheless, beyond a certain number of bits, memory arrays still turn out to be a better trade-off for storage associated with the switch's data path.

However, unlike flip-flops, memories limit the parallelism of access. Typically, a compiled memory has single input and output ports, allowing at most one read and one write command at a time. This limitation constrains the way a switch can operate. For example, to implement a queue using a single memory array to minimize overhead, but still achieve the properties of virtual queuing, the control logic must maintain separate virtual queues, implemented with flip-flops, containing pointers to the locations of transactions in the memory array and other information such as priority. The queue control logic must now contain an arbitration function based on priority and other criteria to select at most one transaction at a time to exit the queue by reading out of the memory array, as shown in Figure 5.18.

Arbitration

Arbitration occurs in two types of sites inside a switch: first, at an output port of the switch, the multiplexor must choose among the requesting input ports the next port that would be allowed to transfer its transaction or packet. Second, at a queue, the next transaction to leave the queue must be chosen.

As we have discussed, a queue could be simple, or it could be organized as multiple virtual queues for various reasons. If the queue is simple, then the arbitration decision is trivial: by definition, the queue works in a first in, first out manner, and so the transaction in the front of the queue is always the one to leave next. In the case of virtual output queuing, each of the virtual queues is again a simple queue and the arbitration is on a per-virtual-queue basis. However, if the queue is composed of multiple virtual queues separated for priority or by channel for deadlock prevention, but is implemented within a single physical structure such as

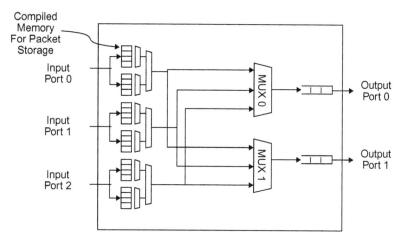

Figure 5.18
Virtual queuing using compiled memories for packet storage.

a memory array, as we have just seen, then the arbitration decision is more complex — one of the many potential contending transactions must be chosen to move forward. This decision is similar to the one that must be made at the output port multiplexors. We will review considerations for arbitration next in some more detail.

In the absence of any other overriding requirement, the property desired of a multiplexor is "fairness". By fairness we may simply demand that the multiplexor not starve a transaction by permanently refusing its request to move to the next switch level. This type of fairness can easily be achieved via a "round-robin" scheme of selecting the next transaction. In this scheme, every contending queue with a candidate transaction registers its request for forward progress by writing to a bit within a ring of bits. The multiplexor control logic continuously goes around the ring for asserted requests and services them one at a time. The logic guarantees that none of the requests, if present, is skipped, thus preventing starvation of transactions that have registered their request.

But the above scheme at times lets a recently arrived transaction out ahead of a transaction that has been waiting for longer time. That could be considered "unfair". To rectify this situation, the arbitration could be based on the "age" of the transaction — age being defined as the amount of time the transaction has been waiting at the multiplexor. In the improved scheme, then, the arbiter lets the oldest transaction go first.

If some transactions carry higher priority than others, the priority-based arbiter chooses the highest-priority transaction first. But this can lead to starvation of lower-priority transactions, which are continuously superseded by higher-priority transactions. Special steps must be taken in the arbiter to avoid such starvation. Starvation prevention is also an important attribute to achieve when deadlock prevention is the goal for arbitration.

Flow control

Although some communication protocols such as Internet Protocol (IP) do drop data in the presence of packet congestion, in SoCs loss of data that has arrived on-chip is not acceptable unless the executing software application drops some data as part of its policy. Thus the SoC interconnection is required to carry out internal communications reliably and without any loss of data. For this reason, progress of transactions from one point to the next throughout the interconnection must be coordinated between the sending and receiving entities so that no loss is incurred during the data transfer. Primarily and specifically, the design must ensure that a transaction transmitted by the sending entity can be received and stored by the receiving entity. (We shall ignore hardware-induced errors such as bit flips, etc., in this discussion.) This is the purpose of flow control signaling, which usually is composed of a signal by the sender to request permission to send a transaction or packet on to the receiver, and a corresponding signal back from the sender granting a permission to send.

When the transactions or packets are of uniform size, the request-grant signals can be simple, single-bit signals. But in practice, this is rarely the case. Typically, for the sake of storage efficiency, there are a few different-sized packets. The flow of different-sized packets gets reflected in the flow control signals, which must now be extended to distinguish between requests and grants to send specific-sized packets of information.

Similar embellishments to flow control signaling are required to indicate different priorities of packets and to account for flows along distinct virtual channels.

Such flow control signaling needs to be implemented both inside a switch between input ports and output ports as well as between switches that are partners on a given link. If the link is split into multiple independently operating sub-links, separate flow control signaling must be defined for each of them.

Evaluating and comparing interconnection topologies for future SoCs

Under the assumption that the scaling of semiconductor devices per, or nearly per, Moore's law will continue in the future, the systems on a chip will continue to scale in size. Under that scenario, the SoC interconnection networks will be called upon to connect tens, hundreds, or perhaps even thousands of processing elements. One would then expect the interconnections to be a significant portion of the total silicon cost of the SoC. Since it will be more cost-effective and efficient to assemble the interconnections out of replicated modular building blocks than building them in an ad hoc manner [33], it is important to understand how the various topologies behave under scaling.

Metrics for comparing topologies

Over the years, interconnection network topologies have been compared and contrasted using many different characteristics [34], but in the context of their incorporation within future SoCs, the three most important metrics are: diameter, bisection bandwidth, and VLSI complexity. The first two metrics are related to the performance criteria of latency and bandwidth and have been adopted from [3], and the latter is related to the growth rate of the area necessary for implementing the network logic and wiring. (Note: comparison using similar considerations also appears in [35].)

For all of these metrics, we will be interested in how they scale as the number of devices connected to the interconnection, N, scales. We are interested in asymptotic rates of scaling as a function of N. We will not necessarily insist on an incremental rate, i.e., the rate as one connection is added at a time.

We will look at what each of these metrics measures.

Diameter

In an interconnection, a path, or route, from a source node to a destination node may pass through multiple other nodes and edges. There could be multiple paths between a given source-destination pair. Since each node and edge crossed along the path adds to the delay a transaction going from source to destination experiences, the length of this path is thus a good nominal measure of the delay performance of the path.

The length of a path is measured in terms of number of edges crossed, called "hops". A minimal path between two nodes i and j in a given network is the one encountering the least number of hops. Let that be represented by $H_{\min}(i, j)$. The diameter of the network is the maximum of the $H_{\min}(i, j)$s for all source-destination pairs $\{i, j\}$ in the network.

In general, given the importance of latency to performance, we should like to see that the diameter grows as little as possible as the number of end-point nodes grows.

Bisection bandwidth

A network N is defined to be "connected" if there exists a path from every end-point source node to every end-point destination node.

Let the number of nodes in a connected network N be represented by $|N|$. A "cut-set", $C(N_1, N_2)$, of the interconnection network, N, is a set of edges in N that, when deleted, partitions the network into two connected sub-networks, N_1 and N_2, with numbers of nodes in the respective sub-networks being $|N_1|$ and $|N_2|$.

The bandwidth of an edge equals the bandwidth of the link in the interconnection it represents. The bandwidth of the cut-set $C(N_1, N_2)$, BW_C, is the sum of the bandwidths of each of the edges in the set.

A "bisection cut-set" of a network N is a cut-set, $C_B(N_1, N_2)$, such that $|N_1|$ and $|N_2|$ are very nearly equal and N1 and N2 are such that they also contain nearly equal numbers of end-point nodes.

There can be multiple bisection cut-sets $C_B(N_1, N_2)$ possible for a given network. The one that is important as a metric is the cut-set that has the least BW_C.

The 'bisection bandwidth", BW_B, = Minimum $\{BW_C\}$ for all possible $C_B(N_1, N_2)$.

In general, we should like to see that BW_B grows as the number of end-point nodes grows. Otherwise, the traffic could get more and more congested as the interconnection grows, increasing the utilization of edges. As seen earlier, this leads to a non-linear increase in the queuing delays experienced by transactions and resultant loss of performance.

VLSI design complexity

The primary purpose of an SoC is computation. In that light, an interconnection is looked upon as a necessary overhead. Its cost in relation to the silicon area dedicated for computation should be minimized.

VLSI design complexity deals with the growth rates of the area of silicon that would be needed for implementing the interconnection as its design scales to connect N devices [36]. In determining design complexity, it is assumed that the topology scheme of the interconnection remains uniform.

Insofar as an interconnection is considered an overhead cost in the SoC, it is desirable that its growth rate remain as low as possible while still providing adequate communication performance.

A survey of interconnection networks suitable for future SoCs

Over the years, a plethora of multi-processor interconnection topologies have been proposed and analyzed in the literature. Many of them require full use of the three spatial dimensions for their implementation and are therefore not suitable to be implemented in SoCs, which are still manufactured from what is primarily a two-dimensional piece of silicon. Apart from a crossbar that we present as a reference point, we discuss below a few topologies that qualify to be considered for SoCs by their virtue of being implementable in two dimensions.

A pipelined bus

The primary problem with a standard bus is its cycle time, which doesn't increase as the number of devices attached to it increases. This problem can be alleviated by pipelining the bus. Thus an electrical broadcast of signals within a single cycle is no longer required; a signal can take multiple cycles to propagate from source to destination.

A pipelined bus improves the bandwidth of the interconnection, but is not able to partition the traffic. The costs of fan-in and fan-out stages for a pipelined bus grow proportional to $N*\log N$, where N is the number of devices attached to the bus [37].

As the size of the system grows, the delay through the pipelined bus grows at least proportional to $\log N$, which should be considered as its diameter.

The bisection bandwidth of a single pipelined bus remains constant under system scaling. Thus although a pipelined bus scales better than a traditional single-cycle broadcast bus, it is still limited by its maximum bandwidth, which does not scale with the number of devices attached.

Multiple buses

Instead of a single bus, multiple buses might be used and transactions distributed among them. Multiple buses allow the traffic to be partitioned. If the number of buses is B, the cost of de-multiplexers at each device will be proportional to $\log B$. If these multiple buses are pipelined, the cost of the interconnection will grow proportional to $B^*N^*\log N$, where N is the number of devices attached to the bus.

Assuming these multiple busses are also pipelined, the delay through them, and therefore their diameter, is also proportional to $\log N$.

The bisection bandwidth of a multiple bus structure is a constant B times the bandwidth of a single pipelined bus. So it too, does not scale with N, but it has a better constant value when compared with a single bus.

A ring

In this topology, the devices form nodes and are connected in the form of a ring with point-to-point wire connections between neighboring devices as shown in Figure 5.19. For example, a ring topology is used in IBM's Cell processor and in NetLogic's XLP332 processors [38] (see Figure 5.20). In a ring topology, transactions flow around the ring from device to device. Because the propagation of signals is limited to be between neighboring devices, the ring can be operated at a high frequency.

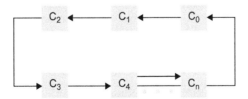

Figure 5.19
An example of a ring topology.

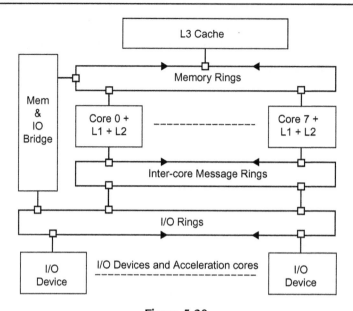

Figure 5.20
NetLogic XLP332E block diagram demonstrating the use of ring topology.

The latency is not uniform across devices in a ring; some are farther than others, with the average being $N/2$, where N is the number of devices connected in the ring. So the diameter of the ring interconnection is proportional to N.

The bisection bandwidth of a unidirectional ring is constant at 1 and does not scale at all with the number of devices connected. This can be a serious problem for a ring network.

The cost of a ring interconnection grows proportional to N. If more than one memory sub-system is independently connected to the ring, a NUMA model is created, since, for a given core, the multiple memory sub-systems will not be equidistant.

A crossbar

Note that a crossbar (Figure 5.12) is also a three-dimensional topology and does not scale very well. Yet we study it because it is an important structural component of other interconnection structures.

In a crossbar, during any cycle, there can be N simultaneous pair-wise connections active from any of the N devices to any other device. The bisection bandwidth of the network is proportional to $N/2$. This enables high throughput across the interconnection. Depending on the construction, the interconnection can also support broadcast or multicast functionality.

A transaction can get from one node to another in a single hop. That would imply the diameter of a crossbar interconnection is 1. However, that is misleading, as the delay through the crossbar is proportional to N [39]. So it is more appropriate to consider the diameter to be proportional to N, since, to maintain a fixed frequency, the path would have to split into a number of pipeline stages proportional to N.

The biggest drawback of a crossbar is its cost, which grows proportional to N^2, where N is the number of devices. However, for small values of N, cost could be acceptable. Depending on implementation, a crossbar can be amenable to partitioning. A crossbar supports the UMA model.

Mesh topology

In a mesh topology, a device sits at every grid point of the mesh. (See Figure 5.8(b) and Figure 5.21.) A mesh is a two-dimensional topology, and is well suited for implementation on a die [29,33,40]. The cost of a mesh clearly is proportional to N, where N is the number of devices in the system.

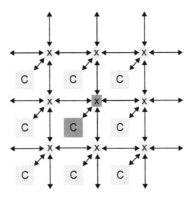

Figure 5.21
An example of the mesh topology.

Like the ring, the latency is not uniform, its average being proportional to $\sqrt{N}$, which then would also be its diameter. Because of high latency, a mesh is more suitable for a NUMA model with the core at each grid point carrying its own local memory unit. A mesh also exhibits high throughput capability. Because of its properties, a mesh is an attractive choice for a scalable interconnection to support a very large number of cores [6].

The bisection bandwidth of a mesh network is proportional to $N/2$.

Some practical considerations in designing interconnections

In addition to the theoretical aspects of interconnection networks reviewed in the last section, it is necessary to pay attention to some practical considerations that are important to vendors of SoC products. We will examine these in this section.

Hierarchies of interconnections

As has been pointed out in the preceding sections, while most traffic in an SoC is composed of Read-Write traffic, different traffic has different other functional requirements. For example, traffic from I/O devices may have very different ordering requirements than traffic generated by processors. Different I/O devices may adhere to different transaction ordering frameworks. Similarly, not all I/O traffic may need enforcement of coherency.

It therefore stands to reason that the overall traffic be separated into sub-classes such that each sub-class has similar handling requirements. Each of these sub-classes can then be put on a separate sub-interconnection of the system. Of course, bridges may be required across these sub-interconnections for transactions that need to receive additional handling.

Scalability in implementations

It is the number of devices in the system and the characteristics of their interactions that determine the choice of interconnect scheme that should be employed in the system. The maximum number of cores that can be physically accommodated in an SoC can vary greatly based on how sophisticated and large the cores are. The simpler and smaller the cores, the more of them can be accommodated on a die. However, simplicity of the core could translate to lower performance per core.

The number of cores in a system is also a function of the parallelism inherent in the intended application. Thus the optimal number of cores is one that maximizes the cumulative performance of the application, which can vary from application to application. While some exploratory scalable multicore systems with a few tens of cores have been demonstrated recently, the current state of the art for practical systems of interest varies from under ten to low tens of cores.

The development of an SoC is an expensive project. To reduce cost, SoCs embrace the concept of reuse, wherein many of the subsystems are off-the-shelf components, including cores. Each SoC may require a different number of these off-the-shelf components. The task of designing the SoC then becomes one of system integration. Such a methodology requires the concept of a "system platform" to which all of the components attach. The central and substantial ingredient of a platform is the interconnection network. In order to minimize the effort of designing many different SoCs, it is advantageous if the platform, and therefore the system's interconnection network, is scalable such that different-sized platforms can be easily produced.

Performance and cost are two metrics of importance when it comes to scaling the interconnection. As we saw in Section 10, different interconnection topologies vary significantly on these two metrics.

Summary

We started by observing that, in order to design an SoC interconnection, it is essential to take into account the communication traffic anticipated within the system. Looking at traffic types exposes opportunities that could be exploited to improve application performance. Understanding them is the first step in coming up with a solution. Communication traffic is composed of diverse types of transactions, but it is commonly dominated by the memory access traffic from cores and other devices.

An effective SoC interconnection solution attempts to match the topology of the traffic. The native topology of the transaction traffic is molded by the computation model employed in the multicore system: shared memory or message passing, or a combination of the two, and also the organization of the system memory itself. The need to maintain coherency brings very special transactions and communication topology requirements. The interconnection supporting coherency activity needs to be designed with intimate knowledge of the coherency protocol and ordering semantics expected by the instruction set architecture of the cores involved, thereby making it specialized in topology and core-ISA-aware functionality.

Beyond the basic computational needs are additional forward-looking considerations such as security, application consolidation, and quality of service that may also be needed to be kept in sight.

Multicore systems are characterized by high-level transaction concurrency. A well-performing system must not counteract the inherent exploitable parallelism. Partitioning traffic among multiple concurrent interconnections organized hierarchically might prove to be a good option.

An interconnection is central to the operation of a multicore SoC, and its significance to application performance is critical. Specifically, low storage Read access latency is of key importance to good application performance, to which the interconnection directly contributes via delays along its links. Queuing delays along the interconnection pathways can add significantly to the latency of memory operations. These delays grow non-linearly as link utilizations approach unity. So the interconnection link bandwidths must be provisioned to keep their expected utilization low.

We presented common terminology used to classify interconnection networks and introduced representations used to study them. We also touched on some topics related to their operation.

We then understood why a bus topology has been so popular in early multicore SoCs. However, while simple to design, it is not very scalable, an attribute of increasing consequence in SoC interconnection design.

The larger the system scales, the more entities are connected to the bus. The wires are longer and the fan-out is higher, resulting in longer propagation times and lower bus frequency. In addition, more entities request service from the bus, lengthening queuing delays, which increases latencies and lowers throughput.

After the discussion of the bus topology, we examined how a fabric-oriented topology can remove the performance barriers of a bus-oriented design and offer other benefits. A fabric-based multicore interconnect gives designers the opportunity to create systems that more efficiently deliver the inherent performance advantages of multicore solutions. And, just as importantly, fabric technologies enable systems that can be more easily scaled with additional cores and more functionality to keep up with the demands of the communications equipment marketplace.

Next we studied in some depth the building blocks of fabric-oriented interconnection networks and examined design issues and choices related to them.

We then introduced the diameter, bisection bandwidth and VLSI design complexity metrics for evaluating and comparing interconnections. We looked at a number of fabric topologies that are a suitable choice for a multicore SoC interconnection. We found that they each display different performance characteristics and cost metrics.

Last, we discussed some practical issues that must be paid attention to when developing an interconnection solution in an industrial context: in order to minimize the cost of development, one must choose an interconnection topology that will be suitable for a number of SoCs, and also develop it with an eye for reusability of building blocks. Appropriate parameterization should be incorporated in the design to make this possible.

As the system grows, choosing one interconnection scheme over another can have a major impact on the overall performance, capabilities and scalability of a multicore system. Each should be evaluated based on the specific requirements of the embedded application(s) for which the multicore system is being designed.

References

[1] J.D. Owens, Research challenges for on-chip interconnection networks, IEEE Micro September-October (2007).

[2] W. Huang, et al. Scaling with design constraints: predicting the future of big chips, IEEE Micro July-August (2011).

[3] W.J. Dally, B. Towles, Principles and Practices of Interconnection Networks, Morgan Kaufmann, 2004.

[4] J. Duato, et al. Interconnection Networks, An Engineering Approach, Morgan Kaufmann, 2002.

[5] S. Pasricha, N. Dutt., On-Chip Communication Architectures, Morgan Kaufmann, 2008.

[6] J. Neil Gunther. A new interpretation of Amdahl's law and geometric scaling, Performance Dynamics Company, 2002.

[7] Tera-scale Computing. Q3, 2007, download.intel.com/technology/itj/2007/v11i3/vol11-iss03.pdf.

[8] M. Baron., Tilera's cores communicate better, Microprocessor Report (2007).

[9] Many Integrated Core (MIC) Architecture. Intel Corporation. <http://www.intel.com/content/www/us/en/architecture-and-technology/many-integrated-core/intel-many-integrated-core-architecture.html>.

[10] D. Lenoski, et al. Stanford DASH multiprocessor, IEEE Comput. March (1992).

[11] D. Chaiken, et al. Directory-based cache coherence in large-scale multiprocessors, IEEE Comput. June (1990).

[12] A. Agarwal. An evaluation of directory schemes for cache coherence, ISCA '88 Proceedings of the Fifteenth Annual International Symposium on Computer Architecture.

[13] D. Chaiken, et al. Directory-base cache coherence in large-scale multiprocessors, IEEE Comput. June (1990).

[14] S.V. Adve, K. Gharachorloo, Shared memory consistency models: a tutorial, WRL Res. Rep. September (1995).

[15] R. Marejka. Atomic SPARC: Using the SPARC Atomic Instructions, ORACLE: Sun Developer Network.

[16] Available from: http://developers.sun.com/solaris/articles/atomic_sparc.

[17] PCI Local Bus Specification, Revision 3.0, February 2004.

[18] AMBA AXI Protocol, Version 1.0. ARM Limited, 2003−2004.

[19] P.H. Stakem, The Hardware and Software Architecture of the Transputer, PRB Publishing, 2011.

[20] M. Snir, S. Otto, S. Huss-Lederman, D. Walker, J. Dongarra, MPI: The Complete Reference, MIT Press, Cambridge, MA, 1995.

[21] W.Richard Stevens, UNIX Network Programming, Prentice Hall, 1990.

[22] S.J. Leffler, M.K. McKusick, M.J. Karels, J.S. Quarterman., The Design and Implementation of the 4.3 BSD Unix Operating System, Addison-Wesley, 1989.

[23] Multicore Communications API Specification Version 1.063 (MCAPI). The Multicore Association^{TM}.

[24] E. D. Lazowska et al. Quantitative System Performance: Computer System Analysis Using Queuing Network Models, Prentice Hall.

[25] Quing Yang, et al. Analysis and comparison of cache coherence protocol for a packet switched multiprocessor, IEEE Trans. Comput. August (1989).

[26] C. Kruskal, et al., The Performance of multistage interconnection networks for multiprocessors, IEEE Trans. Comput. December (1983).

[27] L. Kleinrock, Queuing Systems; Volume 1: Theory, John Wiley and Sons, 1975.

[28] T.R. Halfhill, TI boosts base-station processors, Microprocessor Report April (2012).

[29] M.B. Taylor, et al. The RAW microprocessor: A computational fabric for software circuits and general-purpose programs, IEEE Micro February (2002).

[30] C. B. Stunkel et al. The SP1 high-performance switch, Proceedings of the Scalable High-Performance Computing Conference. 1994.

[31] B. Abali, et al. Adaptive routing on the new switch chip for IBM SP systems, J. Parallel Distr. Comput. 61 (2001) 1148−1179.

[32] W.J. Dally, C.L. Seitz, Deadlock-free message routing in multiprocessor interconnection networks, IEEE Trans. Comput. May (1987).

[33] W.J. Dally, B. Towles, Route packets, not wires: on-chip interconnection networks, Design Automation Conference, ACM, 2001.

[34] T.Y. Feng., A survey of interconnection networks, IEEE Comput. December (1981).

[35] D.N. Jayasimha, et al. On-Chip Interconnection Networks: Why They are Different and How to Compare Them, Intel Corporation, 2006.

[36] D. Langen, et al. High Level Estimation of the Area and Power Consumption of On-Chip Interconnects, 13th Annual IEEE International ASIC/SOC Conference (2000).

[37] V. Lahtinen, et al. Comparison of synthesized bus and crossbar interconnection architectures, Proceedings of the 2003 International Symposium on Circuits and Systems (ISCAS) (2003).

[38] T. R. Halfhill. NetLogic broadens XLP family, Microprocessor Report. July 2010.

[39] M.A. Franklin, VLSI performance comparison of Banyan and crossbar communication networks, IEEE Trans.Comput. April (1981).

[40] T. Krishna, et al. SWIFT: a swing-reduced interconnect for a token-based network-on-chip in 90 nm CMOS, *IEEE International Conference on Computer Design* (ICCD) (2010).

Further reading

A. Windschiegletal. A wire load model considering metal layer properties. Ninth International Conference on Electronics, Circuits and Systems, 2002, vol.2, no., pp. 765−768.

D. Krolak. Unleashing the cell broadband engine processor. http://www.ibm.com/developerworks/power/library/pa-fpfeib

CHAPTER 6

Operating Systems in Multicore Platforms

Bryon Moyer
Technology Writer and Editor, EE Journal

Introduction

Embedded systems have a wide range of ways of providing low-level system resource and service support for software applications through the operating system (OS). They can have no support at all, or they can

Real World Multicore Embedded Systems.
DOI: http://dx.doi.org/10.1016/B978-0-12-416018-7.00006-7

provide rich system services of the order of what desktop computers provide, and even services not found on the desktop. One of the challenges of embedded systems is specifically that range: there's no standard way of configuring an OS for an embedded system. "Over-configuring" can introduce unnecessary complexity, overhead, and cost, while "under-configuring" can leave too much work for the application.

There are different high-level ways of approaching the problem, however, and they tie into some of the general architectural definitions mentioned in the Architecture chapter. In particular, they tie into the memory organization, since one of the defining aspects of the classic multicore architectures is how the operating system views and manages memory. The alternative configurations all represent different balances between the need for high performance and predictability on the one hand and the desire for some entity that can isolate an application program from low-level system resources and, to some extent, from other application programs on the other hand. As a general rule, more services and isolation tend to mean lower performance. Exactly what that trade-off means varies by system, and, as new architectures are devised, coupled with increasing processor and memory speed, the relative costs have diminished over time. But there are costs nonetheless.

We will discuss three fundamental ways of looking at a multicore system through the eyes of an OS: symmetric multiprocessing (SMP), asymmetric multiprocessing (AMP), and lightweight or bare-metal systems (the latter of which has its own chapter). For this discussion, we will assume "simple" or "plain" instances of OSes that directly access the hardware that hosts them. It's possible to add further abstraction through virtualization; that will be the subject of the following chapter.

It can be helpful to clarify a few concepts that become important when discussing OSes. We introduced the generic concept of "tasks" in the concurrency chapter. In this chapter, we need to be more specific. A "process" defines the scope of what we conveniently think of as "a program". From a practical standpoint, it circumscribes the namespace that a typical program can access without any special effort. It is possible to get two processes to talk to each other, but it takes extra work, and the compiler of one won't understand the names defined in the other. (Of course, it may be possible for one process to access the

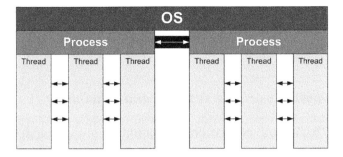

Figure 6.1

Process and threads. Threads can easily intercommunicate; inter-process communication must be explicitly built.

memory space of another through the magic of pointer arithmetic in the C language, but that's generally considered a bug — and a weakness of the C language.) Figure 6.1 illustrates this relationship.

A "thread", on the other hand, is part of a process. It's a path of execution that can conceptually proceed independently of another path. That possible parallelism remains conceptual in a single-core system, which offers no true concurrency. But in a multicore system, these threads can proceed at the same time — meaning that, at times, thread-safety issues that remained under the radar in a single-core machine may suddenly rise up to cause problems in a multicore system.

Even though threads have a conceptual level of autonomy, they are sub-units of a single process, so they understand the same namespace and share the same overall scope. Whether running on a single core, or on a multicore system, some operating systems support only one process, while others support multiple processes. Those that support one process simply manage a set of threads within a unified address space. Those that manage multiple processes manage multiple address spaces, and multiple threads within each such space.

When discussing high-level concurrency issues, we tend to refer to both processes and threads generically as "tasks". But when discussing OSes and, in particular, scheduling, they are often also referred to as "threads," or "contexts". When an OS decides to move one task or thread out of execution and swap another one in, a "context switch" has occurred. That context might reflect a different thread or task from the same process as the one that got swapped out, or it might reflect a different process altogether. At this level, it doesn't matter.

Symmetric multiprocessing systems and scheduling

SMP (Figure 6.2) is the easiest configuration from the standpoint of a programmer. In an SMP architecture, all processors are identical (homogeneous), and all can access a common memory. They may also have access to "private" memory as well, but what makes a system SMP is homogeneous processors with access to a common memory. Since all processors are identical, they can all execute the same code from memory, even at the same time. This enables code to be run on any processor at any time, providing an attractive parallelism and transparency to the programmer. To the programmer, an SMP system effectively makes many cores look like a single core — although it doesn't hide concurrency issues. The SMP OS, on the other hand, must manage the use of each core at all times, and this scheduling operation can be completely hidden from the application, which might have no explicit knowledge or control over which thread is running on which core at any time.

An OS that can handle multi-threading on a single-core system needs to have a scheduling facility in place, so the move to SMP does not introduce the need for a scheduler. What changes is the level of complexity: instead of scheduling threads and processes on a single core, it must do so across

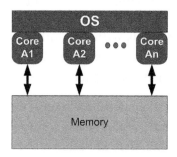

Figure 6.2
SMP configuration. All cores are identical and see the same shared memory.

several cores, and how it does that can have an enormous impact on system performance. Likewise, an application composed of multiple processes or threads that has been developed to run on a single core may be able to run on a multicore system without need for re-structuring. At the same time, re-structuring may be desirable to avoid concurrency issues, and/or to achieve desired levels of performance.

Real-time systems generally try to achieve responsiveness and throughput, in some application-dependent balance. Some applications are primarily concerned with responsiveness, and throughput is not the issue. Others may be designed to achieve high throughput, and responsiveness is consciously compromised to achieve higher throughput. The most fundamental challenge is balancing the desire to achieve higher throughput by keeping all cores busy against the trap of introducing unproductive operations that occupy cores – keeping them busy – without achieving productive results. This impacts many properties of a scheduler, but paramount among them is controlling the "affinity" between a given thread and a given core. The OS can allow full migration of threads across all available cores (no affinity), or force certain threads to run on certain cores (full affinity), both methodologies having benefits under certain application-dependent circumstances.

Generally, in multi-threaded systems, a programmer assigns each thread a priority that determines which of several threads that might be "READY" to run (not blocked or waiting for some external event) is actually given the opportunity to run. In a multicore system, this situation extends to each core. In a priority-based, preemptive scheduling system, the highest-priority thread that is READY to run is the one that will be allowed to run on a given core. But adherence to a strict priority scheduling algorithm might result in excessive context switches, and introduce significantly undesirable overhead.

The frequency of shifting may depend on overt policy or may be more subtly determined by system parameters. Specific apportionment is possible through policies like "round robin" or "weighted round robin" scheduling algorithms, where each thread or process gets a designated slice in turn. Such policies tend to be used more in real-time operating systems (RTOSes) due to their focus on deterministic program behavior.

Absent such policies, scheduling decisions boil down to the fundamental question of how long a stall merits a context switch, and how high a priority thread is waiting, READY to run. Switching contexts (other than the simple need to give each context a chance to proceed if there are more contexts than cores) is both about enabling important threads to have a chance to do their job and avoiding dead time in the processor if other work awaits. So if one thread initiates an action that will take some time to complete, without that thread's direct involvement (such as an I/O to a device), then, instead of waiting around for the device to complete its operation, another thread's context can be swapped in and given the use of the processor in the intervening time.

But it takes time to swap contexts — and how much time it takes depends on the nature of the system. Each context has a state that must be preserved. On a simple core, that state is located in memory, so swapping contexts means moving data into and out of memory — which can take a nominal amount of time (and will presumably make use of fast local memory). On a core with hardware hyper- or multi-threading, however, provisions have already been laid for multiple contexts, so little data movement is necessary, and a switch between two contexts which both have hardware resources can happen very quickly. Again, if more contexts than dedicated hardware resources exist, then context switch overhead becomes an issue.

Moving threads from one core to another, however, is an entirely different matter. In an SMP system, it doesn't really matter where any single thread executes, which means it could, in theory, stall on one core and then, if it's having a hard time getting going again, be moved to another core that's not so busy. That makes for more comprehensive, but not necessarily more efficient, core use. This is due to the fact that, as illustrated in the memory chapter, the context has to move, and the cache on the new core has to be refilled with the useful things that were already in place on the old core.

All of this takes time, so, while the program should perform correctly at a functional level regardless of where the threads execute, performance suffers if threads are moved around too much. The caching structure may also make it desirable to favor one core (say, one that shares an L2 cache) over another if a context has to be moved. This illustrates an example of when directing the OS to restrict a thread to a subset of cores might pay dividends.

This becomes part of the heuristic nature of scheduling algorithms, and it is why, in situations where a designer knows very specifically how threads and processes will behave in a dedicated program, he or she might benefit from a scheduler that allows the user to "tune" its performance in light of these considerations, or even to replace a stock scheduler with a custom one that incorporates that knowledge. Customizing the scheduler, however, while possible, is not common. Rather than taking that drastic a step, it's usually possible to guide the OS in scheduling operations; this will be discussed below.

The SMP configuration is widely supported by operating system providers, both in large-scale and real-time OSes. The larger OSes, of which Linux is most prevalent, tend to provide a richer array of services and options. The better-known options and providers as of writing are shown in Table 6.1.

Applications requiring a higher degree of determinism and/or a smaller footprint or lower service overhead than is provided by large-scale OSes will turn to RTOSes instead, and many of them have provisions for multicore. They're illustrated in Table 6.2.

There is some debate as to whether Linux has a true real-time configuration. One company provides a Linux package that they claim to be suitable for real-time use. Opinions vary; in general, it may indeed work well for a number of different applications, but the challenge comes for systems demanding hard-real-time that can be proven through

Table 6.1: Large-Scale OS Support for SMP

OS	Provider
Linux	Wind River (Intel)
	MontaVista (Cavium)
	Mentor Graphics
	Enea
Windows Embedded	Microsoft
Android	Wind River (Intel)
	Mentor Graphics

Table 6.2: RTOS Support for SMP

OS	Provider
DDC-I	DEOS
INTEGRITY	Green Hills
Lynux Works	LynxOS
Neutrino	QMP
Nucleus	Mentor Graphics
OSE	Enea
QNX	Momentics
ThreadX	Express Logic
VxWorks	Wind River (Intel)

calculations. Linux is a complex OS, and it's doubtful that all of the possible paths of execution could be traced to guarantee a worst-case latency. This remains an open question.

While there are many benefits to the SMP model from the standpoint of the programmer who doesn't wish to delve too deeply into the complexities of concurrency, it has its limitations. Most significantly, it doesn't scale. As discussed in the memory chapter, if too many cores try to access the same memory, at some point the bus chokes on the traffic, and no further performance improvement is possible regardless of the number of cores added — in fact, performance can decrease due to bus thrashing.

Having a single shared memory space can also leave one thread or process open to corruption by another thread or process. Separation between spaces is virtual, not physical, so errant pointer arithmetic, for example, can easily play havoc with memory that ostensibly belongs to a different context. While virtualization and "trusted zone" kinds of architectures provide the strongest solution for this, it is possible to improve things without them.

And finally, heterogeneous and NUMA architectures cannot take advantage of SMP since they violate the assumptions of "all cores are alike" and the single shared memory block.

These OSes typically come with a set of tools to help develop the system and diagnose issues. Those tools are described in more detail in the Tools chapter later in this book.

Assymetric multiprocessor systems

An AMP system is somewhat harder to define than an SMP system is. In essence, if it's multicore and it's not SMP, then it's AMP. AMP systems may have a number of different looks. We'll look at some common examples, but these by no means define what AMP is. In fact, there are different views on what constitutes AMP vs. SMP. One view is that these are hardware distinctions, with SMP being synonymous with a homogeneous set of cores with an UMA memory configuration; anything else is AMP. The challenge is that a homogeneous/UMA system can be configured in a way that's not SMP, by running a separate OS instance on different cores (per the following section). The other view includes the OS in the picture, combining that with the hardware to define SMP and AMP. According to that view, what makes a system AMP is the fact that there is more than one OS instance in the system.

OS-per-core

The simplest AMP configuration is one where each core has its own independent instance of an OS (Figure 6.3). If the set of cores is homogeneous with a shared physical memory, then it looks similar to SMP, except that all cores do not necessarily have access to the same shared memory. If you prefer the hardware-only definitions of SMP/AMP, then this specific case is the "exception": it's SMP configured as AMP. Note that separate virtual memory spaces may be separate regions within one physical memory, or they may be different physical memories.

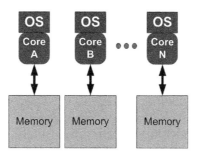

Figure 6.3
AMP configuration. Each core has a separate OS and memory.

From a software standpoint, the big difference between AMP and SMP is the fact that each core in an AMP system has separate processes; no process can be distributed over multiple cores in the way that SMP allows, since the cores may be of different architectures and execute a different set of instructions. If a multi-threaded process is moving from an SMP to an AMP architecture, it means that the original monolithic program has to be split into multiple independent programs, each one compiled and linked into an executable for the specific architecture of the core on which it will run. The cut lines may or may not coincide with the thread boundaries in the original program.

This exposes the fact that a process has a common namespace. A global variable is no longer accessible from all cores; it is only understood by the OS hosting the code where it's defined. Of course, you would never get to the point of an actual running program failing to find that global variable — the independent programs would never compile correctly in the first place. The main point is that distributing a program over multiple cores in an AMP setup requires a fair bit of work just to get things to run again. Whether they run effectively or not depends on the partitioning chosen, which is a topic covered in its own chapter.

Applications that are naturally segregated make ideal AMP candidates, where each type of processing in the program can be configured to run on a processor that excels at that type of processing. Further, it can run under an OS that is also ideal for that type of processing. This can make for very efficient systems where hardware and software are tailored to fit the individual needs of separate aspects of a program. A common configuration of this type is one incorporating a RISC CPU and a DSP on an SoC, with the RISC processor handling general application tasks, perhaps running Linux, while the DSP, perhaps running an RTOS, handles the computationally demanding signal processing or math functions somewhat independently. The two subsystems require only minimal interaction and synchronization, enabling each to run efficiently doing what it does best.

Other partitioning considerations aside, it's usually impossible to pull apart a program that was not designed to run on an AMP system in a way that completely isolates variables in one process. If the original program has data that is widely accessed by different portions of the program, then, after splitting the program apart, those separate entities are still going to need to access shared data. This requires implementation of inter-process communication (IPC), a topic covered in the synchronization libraries chapter. It's likely that you would want to minimize the amount of such communication, however, as it entails software (and potentially hardware) overhead. The partitioning process effectively cuts lines of communication, and, collectively, they constitute cutsets along the boundaries. As a general rule, you want to minimize the size of those cutsets.

Ideally, a system would be designed from the start for operation on a specific type of AMP system, and IPC considerations could then be factored into the program structure without risk of creating a web of access that would be difficult to unravel into a segregated system after the fact.

Figure 6.4 provides an illustration of this point. The program shown is split up into four partitions that intercommunicate.

If these partitions are to be split amongst two cores, then there are numerous ways to group them. As will be made clearer in the Partitioning chapter, part of the consideration is the amount of computing in each core; that should ideally be balanced, from the perspective of achieving the minimal maximum execution time for any core. But reducing communication is important as well; from that

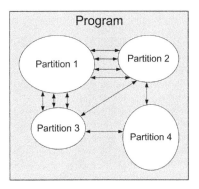

Figure 6.4
A program split into four natural partitions that intercommunicate.

standpoint, Figure 6.6 is preferable to Figure 6.5, since the latter cuts six lines of communication, while the former cuts only two.

Note that these figures are merely intended to illustrate the overall concept. In a real partitioning scenario, it's not just how many lines of communication there are, but how much data needs to be transferred, and how often, that also impact the decision. Such details are covered in the Partitioning chapter. The point is to give careful consideration to the communication, as well as computational needs of the program, and if possible tailor the hardware resources (processors and memory) to suit the program's needs. Whether six lightly used communication channels are

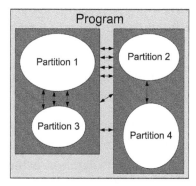

Figure 6.5
The partitions grouped in a way that requires six lines of communication.

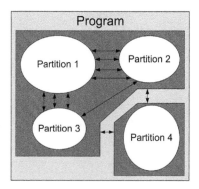

Figure 6.6
The same program split in a way that requires only two lines of communication.

preferable to two heavily used channels depends entirely on the nature of the application, and neither one is inherently superior to the other.

For these reasons, neither moving from SMP to AMP nor developing for AMP in the first place is to be done lightly, as it has fundamental implications for the structure and implementation of the program. Programs originally conceived for AMP may include things like cellphones, with a heterogeneous architecture (e.g., RISC plus DSP) requiring a separate OS (if any) per core, and pipeline architectures – although the latter are often implemented in something closer to a bare-metal configuration, to be covered below (and in its own chapter).

Multiple SMP

Another AMP implementation keeps SMP in the picture, except that, instead of all the cores being under the dominion of a single SMP instance, the cores are partitioned into one or more groups, each of which gets an SMP instance (Figure 6.7). This may also be referred to as *hybrid SMP-AMP*.

This scheme provides one way of more easily controlling access to resources between programs while still taking advantage of the benefits of an SMP environment. In an extreme instance of such an architecture, if an SMP system were reduced to one core, then the result would be an AMP system. Conversely, any AMP system can be generalized as being a collection of *n*-core SMP subsystems.

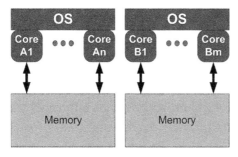

Figure 6.7
Hybrid SMP-AMP configuration.

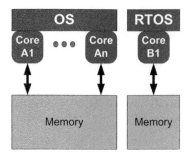

Figure 6.8
An SMP configuration with an RTOS on one core.

SMP + RTOS

Another very common practice, briefly mentioned earlier, couples the simplicity of a heavyweight OS with the deterministic nature of an RTOS (Figure 6.8). The idea is to quarantine the time-critical elements to operate under the RTOS while giving the non-time-critical portions the services of a rich OS. If it's just a two-core system, then each core gets a single-core instance of its OS. But either the full OS side or the RTOS side or both could have more than one core, in which case SMP instances could be used. As noted, such configurations might use Linux for the OS and an RTOS alongside for the demanding real-time activities.

SMP + bare-metal

Some functions are so performance-critical that they can't tolerate any system overhead. In those cases, designers try to operate with no OS at all. That's a lot of work, and often counter-productive. For example, operating with no OS requires the application to manage all the activities of the program within its own code. That code is not likely to operate as efficiently as a fine-tuned RTOS scheduler, resulting in additional overhead.

Some providers like Cavium and MontaVista (now part of Cavium) offer thin "executives" that provide a bare minimum of services with a focus on efficiency. As mentioned, this is a topic of a separate chapter, but such setups are often combined with at least one other core (often referred to as the "host") that's running a full-up OS like Linux. This is

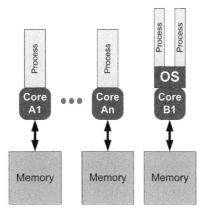

Figure 6.9
A set of bare-metal cores plus a "host" core running a full-service OS.

yet another AMP configuration (Figure 6.9, with multiple bare-metal cores). In this case, the "thin executive" is simply a type of RTOS, perhaps proprietary, and perhaps tuned to a specific architecture.

This is most common in packet-processing applications. The bare-metal cores are optimized for speed, constituting the "fast path" of the "data plane" for the most common packets that need to be handled as quickly as possible. The core handles both "control plane" activities and the "slow path" of the data plane. The cores may be identical, or, as in the example of Intel's IXP network processors, use micro-engines for the fast path and a larger ARM XScale core for the host. The IBM/Sony "BE" incorporates one central core and eight peripheral cores that generally run a small RTOS and small network stack, while the central core runs Linux.

Many OSes claim "support for AMP". Yet when used in a non-SMP setup, they're simple single-core OS instances. So what does "AMP support" mean, since the OS in such a mode knows nothing of multiple cores?

The answer boils down to initialization: the system needs an organized way to boot up, and this typically means that one core is considered the master; it coordinates bring-up of the entire system. So AMP support more or less reflects the availability of a board-support package (BSP) that can initialize an AMP system. We'll discuss more about boot-up later in this chapter.

Virtualization

AMP setups use multiple OS instances, but the cores define the boundaries of those instances. One might, however, want to share the resources of a given core between two OSes. Or you might want to have some entity outside the individual OSes that manages shared resources like I/O amongst all of the OSes, none of whom know anything about the existence of the others. Or you might want to strengthen the isolation between processes and cores and resources to make the system more robust and secure.

This is the realm of virtualization, which explores new architectures and ways of layering services and programs. Virtualization is the topic of the next chapter.

Controlling OS behavior

Various operating systems will give different levels of control of their scheduling and other operations to the user. Different applications will require different levels of control. Processes that aren't performance-critical running under an SMP OS are likely to let the OS do what it thinks is best. By contrast, in a safety-critical application with hard real-time requirements, you need to know and control how the thing will run in as much detail as possible.

OS behavior can be controlled in a number of different ways, and the following sections describe some of the more common ones.

Controlling the assignment of threads in an SMP system

As mentioned, SMP scheduling algorithms can be complex, using priority, fairness, time-slicing, and "affinity" − in one sense − to determine what to schedule where. Affinity in this context refers to the general desire to keep a suspended thread on its original core, since moving it means you have to reload a new cache, which hurts performance.

One way of taking control is to assign threads to cores manually. An SMP OS may offer the application the ability to decide for itself when to restrict certain threads to certain cores, and conversely, when to prevent certain threads from running on certain cores. You might assign every thread, or you might merely assign critical threads to, say, one core, leaving the OS to assign the other threads to the remaining cores as it sees fit. This process of assigning threads is variously referred to as "pinning," "binding," or "setting thread affinity".

Threads can be assigned to cores statically by means of the start-up configuration file, or they can be assigned when they're created, or they can be assigned in real time using an OS call. Every OS has some facility for managing this, although the details may vary between OSes.

For a given thread, you may be able to assign it to a specific core, or you may be able to give it a bit mask that gives the OS flexibility in assigning it to one of a few specific cores that you have specified. Conversely, the OS (or RTOS) may enable the developer to restrict the thread from running on certain cores, enabling those cores to be dedicated to running certain other threads and only those threads. Note that, for some OSes, the assignment is sometimes treated more as a "suggestion" − the OS tries to honor your assignment, but may not. With RTOSes, the assignments are guaranteed.

For example, the OS may define a bitmap of cores within a 32-bit word, or multiple words for larger systems. Each bit represents a particular core in the system. Then, for each thread, a bitmap specifies which cores that thread is allowed to run on and which it cannot run on. By using such a mechanism, threads can be "locked" to a single core, while other threads are allowed to share remaining cores. In systems where there are more threads than cores, then "multi-threading" strategies also come into play. In those cases, several threads must share a core, and context switching is a necessity if all threads are to run. Context switching must be highly efficient, lest more overhead gets introduced than is saved by parallel execution!

Controlling where interrupt handlers run

Interrupt handlers can also be pinned, although it is much more typical that this would be done statically, at boot-up. It's much less common for these to be moved in real time.

The flexibility you may have for pinning and interrupt handling depends very much on the system. Some multicore systems dedicate specific interrupt lines to specific cores, in which case you can't move them (unless you use indirection to "bounce" the handling from one core to another, which causes additional latency). For example, on some ARM cores — particularly those using their TrustZone architecture (covered more in the next chapter) — fixed/fast interrupts are hard-wired to core 0.

How you actually make the interrupt assignment varies by system. It's typically an API call, BSP that defines the board configuration, and it might even require some assembly-language code changes to low-level initialization runtimes to implement.

For systems where multiple devices share a single interrupt line, note that reassigning the core that will receive the interrupts will move all of the devices together; you can't move only some of the devices without reassigning the physical connections to different interrupt lines.

Partitions, containers, and zones

Some OSes have a notion of combining certain processes together for isolation purposes. Green Hills and QNX implement what they call *partitions*; Linux has a *containers* feature, which is similar to what Solaris and BSD/OS refer to as *zones*.

Partitions allow the grouping of processes into a partition. Only one partition can be active at a time, meaning that processes in different partitions can never interact. The benefit is better predictability for safety-critical applications. It comes at the cost of reduced concurrency, since, by definition, only some of the processes can be executing concurrently.

On some newer OS versions, this concurrency limitation is relaxed to some extent by allowing one partition to be assigned to one set of cores and another partition to a different set of cores. In this manner, more things can be running at the same time, while maintaining the isolation between partitions. Communication among partitions is carefully managed through secure services of the OS, rather than an open region of memory that might become corrupted for one partition due to the actions of another — a fatal flaw in the separation of partition requirement.

Containers or zones act like OSes within OSes, allowing processes in a container to have their own scheduler and dedicated access to resources. This is sometimes referred to as "OS virtualization", although it's not true virtualization.

A broader discussion of OS separation and virtualization are provided in the following chapter.

Priority

Priority is a critical parameter associated with threads and interrupts; it has a critical impact on how the scheduler decides what tasks or threads to run when. In multicore systems, it works pretty much the same way as it does in single-core systems. In fact, you could say that multicore alleviates, to some extent, conflicts between tasks having the same priority, since there's some chance that they can both run concurrently on different cores.

However, many single-core systems that operate with priority-based preemptive scheduling, which includes most RTOSes, enable developers to become (consciously or subconsciously) dependent on the principle that, in a multi threaded system on one core, a task or thread can rely on the fact that no lower or equal priority thread is running at the same time. Running such an application on a multicore system, under an OS that allows threads to run when "READY", can lead to catastrophic results that are hard to debug.

Some RTOSes offer the user the means to run the SMP system in a "single-threaded" mode, where priority order is maintained, or to consciously relax those restrictions and enable threads of different priority to run on different cores at the same time. Presumably, using this mode requires the user to be confident that such a mode of operation will not introduce any race conditions or priority conflicts.

Priority can be set statically or on the fly. Some systems allow a weighting in addition to the priority so that the scheduler can be told to give more time to specific tasks in a "time-sliced" mode. Some systems also have the concept of a "maximum" priority: in addition to the "current" priority, a maximum can be statically set. Then, if the priority of that task is raised, it can never be raised above the maximum, and the task can't spawn threads with priority higher than the maximum.

Another priority variation found in Express Logic's ThreadX RTOS is "preemption-threshold", which optionally sets a separate, higher, priority threshold that any preempting thread must satisfy. These scheduling policies and services enable the developer to achieve desired results under application-specific operational situations.

Kernel modifications, drivers, and thread safety

One of the thorniest issues bedeviling systems ever since the concept of multi-threading came along is that of thread safety, especially when monkeying with low-level OS code and drivers. For operating systems that have a kernel mode and user mode, most programming is done in user mode. That keeps the OS intact, and if the program crashes, it only affects the individual program, not the entire OS.

But, historically, when customizing or building critical low-level capabilities for embedded systems, it's been common to add or patch modules within the kernel both for better performance and, essentially, to make it look like the OS has new or modified services. These so-called "kernel mods" can be extremely tricky to debug, and failures can bring down the entire system.

There are different views at present as to what the best approach is here. One is simply that there is no longer a real performance issue to working in user mode, and that all programming should be done there. Another approach moves things from a monolithic OS to more of a "microkernel" arrangement; this will be discussed in the next chapter.

Thread safety has always been a critical determiner of the quality of a driver, and that's no more or less true with multicore than it is with single-core multi-threaded. Of course, all code must be thread-safe, but drivers can wreak more havoc than application code if they fail. The issues that lead to unsafe threading are the same for multicore as for single-core, except that the risks are higher due the fact that code can truly be concurrent.

In addition, some of the "fixes" used for single-core systems won't work for multicore systems. One of the more common threats to thread safety is the unpredictable firing of an interrupt while some critical region of code, which is intended to run uninterrupted (Figure 6.10a), is executing. The interrupt handler will start running, bumping whatever task was in progress regardless of where it was (Figure 6.10b). If the critical region and the ISR that interrupts it modify and read the same memory locations, the system could end up in an inconsistent state.

The easy solution for single-core systems is to disable interrupts while executing critical sections of a program. That way you know you'll get through the delicate parts with no interruptions; you re-enable interrupts after. But this won't work on a multicore system. When you disable interrupts, it only affects the one core. The interrupt may fire on a different core. Rather than interrupting the critical code that's running on one core, the interrupt handler will actually run concurrently on another core (Figure 6.10c). As a result, the mischief it can create may be different from what would happen in a single-core system, but it will be mischief nonetheless.

Some RTOSes address this issue through implementation of a policy that disables the scheduler, as well as all interrupt processing, when working in a critical section. Protection of critical sections from interrupt processing is expanded to protect against multi processor access. This protection is accomplished via a combination of a global test-set flag and

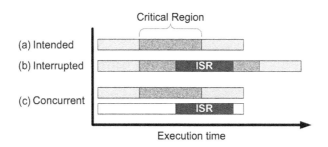

Figure 6.10

(a) Uninterrupted critical region; (b) critical region interrupted on single core; (c) ISR running concurrently with critical region.

interrupt control for the core holding the test-set flag. This solves the problem, but introduces potentially significant overhead by "freezing" all other cores while a critical section is being modified by one core.

With multicore systems, the OS must take care to protect the resources whose integrity is critical. Programmers can take control of some of this protection by using locks (or lock-free programming techniques) within the application itself. Different kinds of locks, as well as lock-free approaches, are discussed in the synchronization chapters that follow.

System start-up

One of the things that distinguish embedded systems from their general-computing relatives is the wide variety of configurations. In particular, peripherals and I/O can be dramatically different between systems. The OS has to know what's in the system, where to find it, and what code to run in order to exercise the different peripherals.

This holds true for single-core and multicore systems. But with multicore systems, you have one more element to deal with: getting the cores themselves in synch. As built, all the cores are usually the same; no one core is any more important than any other one. But someone has to be in charge of bringing the system up, so one core is allocated the role of "master", at least for the purposes of boot-up.

During system initialization, there are a variety of tasks that must be performed on a system-wide basis; each core may also have some initialization work to do that only affects the individual core. This means that each core has initialization code to run, but the master core will have more to do, since it's in charge of configuring the global portions of the system.

This is made possible through the use of "barriers". A barrier is a software construct that establishes a kind of meeting point or rendezvous. When any core reaches that part of the code, it holds there. When all of the cores have arrived at the barrier, then the barrier is lifted, and they can all move forward.

How this plays out during boot-up is that all the cores can start executing their initialization code, but the master will take longer since it has more

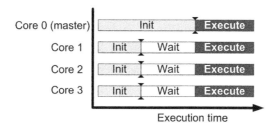

Figure 6.11
Four cores booting up; when all four have reached the barrier, then execution
can begin.

to do. So there is a barrier at the end of each initialization routine, and, once everyone has finished — and, in particular, the master has put the system into a stable starting state, then the barrier is released and the cores can start doing their work. This is illustrated in Figure 6.11.

You'll see a specific example of initialization code in the bare-metal chapter towards the end of the book.

All of the information the system needs to boot up is typically put into a configuration file that is accessed during start-up. This is a critical part of the BSP for a given system. For the most part, these files and their formats are different for each system or OS being used.

An example of such a file can be found in one standard that was set by the Power.org group; it's called ePAPR [1]. The contents include a device tree, which defines the topology of all of the chips and other "devices" on the board. It also specifies the image formats for "client" programs as well as their requirements: identifying the master CPU, specifying the entry point for the CPUs, and various other requirements and considerations for both SMP and AMP configurations. Details on device bindings and virtualization are also included. All in all, files like these provide the OS with all of the information it needs to manage the system.

Debugging a multicore system

Fundamental to the processing advantages of a multicore system is the unfortunate complexity of the software that must run on it. Both the OS and the application are subject to challenges introduced by the fact that code is being executed on multiple processors at the same time. From a software perspective, this introduces several complications to the debugging process, not the least of which is the difficulty of determining just what the system is doing at any point in time.

In a single-core system, a breakpoint can be used to pause the system at any instruction (high-level code or assembly), and at that point, all system resources can be examined: registers, memory, application data structures, thread status, and the like. This can paint a clear picture of the system's status, and the developer can use this information to determine whether the system is performing correctly or not, and also to enable it to perform better.

But in a multicore system, the picture on one core must be correlated with what is going on with all other cores, and this can be challenging. Breakpoints alone do not do a very good job of painting such a clearer picture. Fortunately, there are tools, as shown in Table 6.3, that help a great deal in unraveling the otherwise confusing array of system activities, and painting a clear picture of what the system is doing at any point in time. Operating systems such as Linux and RTOSes such as VxWorks, INTEGRITY, and ThreadX provide such tools, and they all operate in a similar manner (Table 6.3). Of course, each is unique, and features vary from one to another, but for the purposes of this brief discussion, they are sufficiently similar to discuss their common technology generically.

Table 6.3: Example Trace Tools for Debugging

Operating system	Company	Trace Tool
Linux	FSF	LTT
VxWorks	Wind River	WindView
INTEGRITY	Green Hills	EventAnalyzer
ThreadX	Express Logic	TraceX

The information gathered

In a hardware trace system, such as a Logic Analyzer, Emulator, or a debugger that captures information from ARM's Embedded Trace Macrocell, a stream of instructions is generated and can be captured, saved, and analyzed. A software tool, analogous to such a program trace, is what is generically called an "event trace". An event trace can capture system and application information that is helpful in understanding what the software is doing in real time. A software "event" is generally any system service, state transition, interrupt, context switch, or any other activity the application would like to capture upon its occurrence. It might be when a particular value exceeds a certain threshold, for example.

While the software events occur in real time and can be represented at points in real time, they typically are gathered for post-mortem analysis. Some implementations do offer real-time display of these events, but they generally happen too rapidly for human interpretation as they occur. A post-mortem analysis enables the developer to delve into the details that occur over the course of a few milliseconds in order to discover exactly what has occurred.

The operating system is responsible for capturing and saving relevant information about each event. This information, for example, might include the type of event (e.g., message sent to a queue), which thread sent it (e.g., thread_1), when was it sent (e.g., timestamp from the system real-time clock), and other related information. The OS will save this information in a circular buffer on the target, so it will always contain information from the last "n" events, depending on how large an area of memory has been allocated for the buffer. Then, either as the system is running, at a breakpoint, or system termination (crash), the buffer contents can be uploaded to the host and displayed.

Uploading the information

In order to upload the captured event information from the target to the host, a variety of mechanisms can be employed. If the data is to be uploaded as the system is running, an Ethernet or USB connection might be used. If uploaded at a breakpoint, the JTAG debugger can be used to read the buffer's memory. Most debuggers have the capability to upload a memory region to a host file. By using this ability, the trace buffer can be loaded onto the host as a file, readable by the trace tool itself. Once on the host, then the events can be read, interpreted, and displayed graphically to show the developer what has been going on.

Painting the picture

The event trace tool runs on the host, reading the uploaded event buffer and generating a graphical or textual display of the information it contained. The buffer will contain a data "packet" for each event that has been recorded by the target OS. By digesting all of the packets, the event viewer can determine the exact time span that has been captured, and every event that occurred during that span. It also can associate each event with the thread that initiated the event; and can produce histograms of common events, an execution profile for each thread, each core, and system activities such as interrupt servicing. A timeline of events can be constructed, very similar to what is shown on a logic analyzer for signals over a period of time. Figure 6.12 shows an example of how this might look for a period of time in a two-core SMP system.

In this example, application threads are listed on the left, and this list is repeated for each core in the system – in this case, two cores. Across the top is a time scale of ticks, where each tick represents a period of real time that is dependent on the resolution of the clock source used. In this example, notice that the "Sequential View" tab has been selected, and the tick marks are simply a sequential counter, not

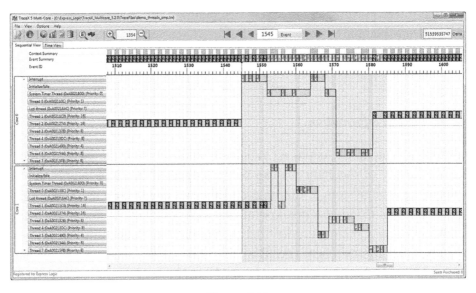

Figure 6.12
An event trace example.

correlated to real time. The correlation with real time could be seen in the "Time View" tab.

To the right of each thread, ordered from left to right, are the events that that thread has initiated, each depicted with an icon and a symbol to identify it. Vertical black lines indicate context switches between threads, and horizontal green lines indicate that the thread was running, but not initiating any events during that period. Note, for example, that thread_1 sends messages to a queue ("Queue Send" or "QS" event), and thread_2 receives those messages from the same queue ("Queue Receive" or "QR" event). Also shown are timer interrupts and activity from threads that are activated by the timer expiration (e.g., thread_0 on Core-1, and thread_6 on Core-0).

By examining the events for each thread and the events on each core, a much clearer picture of system activity can be seen than without such a tool. The developer can then determine whether system behavior is as designed, or if something has gone wrong. Further, by understanding the exact sequence of events, errors in logic or race conditions can be detected and corrected more easily.

Debugging a multicore system is indeed challenging, but the reward is achievement of greater performance. By utilizing all available tools for development and debugging, it is possible to simplify the challenge and shorten development time, while achieving desired levels of system performance.

Summary

The operating system can play a critical role in the operation of a multicore system. There are numerous ways of configuring one or more – or no – operating systems, and the best arrangement will depend on the application.

An SMP configuration has the most "hands-off" potential, where the OS scheduler can take care of making sure all tasks are executed. There are mechanisms available, such as the ability to bind processes or threads to specific cores, that allow some control over operation and the allocation of resources.

An AMP configuration allows much more flexibility and provides much more control, but it also means that everything must be managed explicitly. The interactions between cores, processes, and threads may not proceed transparently. But for applications requiring critical timing, the ultimate in performance, or guaranteed safety, the various AMP arrangements generally provide a better solution than SMP. AMP systems can also scale further than SMP systems can.

SMP is best suited to applications where the load is very dynamic — so much so that the application cannot easily keep all cores utilized in an AMP mode. AMP gives the best potential for performance if the load is more statically defined.

Virtualization adds yet more possibilities while addressing some of the vulnerabilities of systems running traditional operating systems. This is the topic of the next chapter.

Reference

[1] "Power.org™ Standard for Embedded Power Architecture™ Platform Requirements (ePAPR)", Version 1.1. <https://www.power.org/resources/downloads/Power_ePAPR_APPROVED_v1.1.pdf>.

CHAPTER 7

System Virtualization in Multicore Systems

David Kleidermacher

Chief Technology Officer, Green Hills Software, Inc., Santa Barbara, CA, USA

Chapter Outline

Real World Multicore Embedded Systems.
DOI: http://dx.doi.org/10.1016/B978-0-12-416018-7.00007-9

What is virtualization?

With respect to computers, virtualization refers to the abstraction of computing resources. Operating systems offer virtualization services by abstracting the details of hard disks, Ethernet controllers, CPU cores, and graphics processing units with convenient developer APIs and interface, for example file systems, network sockets, SMP thread scheduling, and OpenGL. An operating system simplifies the life of electronic product developers, freeing them to focus on differentiation.

Of course, this environment has grown dramatically up the stack. Instead of just providing a TCP/IP stack, embedded operating systems are increasingly called on to provide a full suite of web protocols, remote management functions, and other middleware algorithms and protocols.

Furthermore, as multicore SoCs become more widespread, application and real-time workloads are being consolidated — so the operating system must support general-purpose applications while responding instantly to real-time events and protecting sensitive communications interfaces from corruption.

In essence, the typical operating system abstractions — files, devices, network communication, graphics, and threads — are proving insufficient for emerging multicore embedded applications. In particular, developers need the flexibility of abstracting the operating system itself. The networking example of control- and data-plane workload consolidation, with a Linux-based control plane and a real-time data plane, is but one use case. Mixed criticality applications are becoming the norm; a general-purpose operating system with its sophisticated middleware and open source software availability is being combined with safety, availability, security, and/or real-time critical applications that benefit from isolation from the general-purpose environment.

System virtualization refers specifically to the abstraction of an entire computer (machine), including its memory architecture, peripherals, multiple CPU cores, etc. The machine abstraction enables the execution of complete general-purpose *guest* operating systems. A *hypervisor* is the low-level software program that presents and manages the virtual machines, scheduling them across the available cores. The hypervisor appropriates the role of SoC hardware management traditionally relegated to the operating system and pushes the operating system up into the virtual machine. The guest operating systems are permitted to access resources at the behest of the hosting hypervisor.

This chapter is dedicated to the discussion of system virtualization in embedded systems, with an emphasis on modern multicore platforms. To capture theoretical multicore processor performance gains, the first obvious approach is to use fine-grained instruction-level or thread-level parallelism. The approach has been the subject of research and investment for years — with parallelizing compilers, parallel programming languages (and language extensions, such as OpenMP), and common operating system multithreading. These fine-grained parallelization approaches have achieved limited success in unlocking the full potential of multicore horsepower. However, it has been realized that embedded system architectures in the telecommunication, networking, and industrial control and automation segments are already naturally partitioned into data, control, and management planes. This separation has driven strong interest amongst original equipment manufacturers (OEMs) to map the entire system to a multicore system-on-chip (SoC) in a manner that simply integrates the previously stand-alone functions onto a single device. System virtualization is the obvious mechanism to realize this mapping.

A brief retrospective

Computer system virtualization was first introduced in mainframes during the 1960 s and '70 s. Although virtualization remained a largely untapped facility during the '80 s and '90 s, computer scientists have long understood many of the applications of system virtualization, including the ability to run distinct and legacy operating systems on a single hardware platform.

At the start of the millennium, VMware proved the practicality of full system virtualization, hosting unmodified, general-purpose "guest" operating systems, such as Windows, on common Intel Architecture (IA)-based hardware platforms.

In 2005, Intel launched its Virtualization Technology (Intel® VT), which both simplified and accelerated virtualization. Consequently, a number of virtualization software products have emerged, alternatively called virtual machine monitors or hypervisors, with varying characteristics and goals. Similar hardware assists for system virtualization have emerged in other popular embedded CPU architectures, including ARM and Power.

While virtualization may be best known for its application in data center server consolidation and provisioning, the technology has proliferated across desktop and laptop-class systems, and has most recently found its way into mobile and embedded environments. This evolution is no different from, and is related to, the migration of multicore processors into the embedded and mobile space. System virtualization enables developers to make better use of multicore processors by enabling the hosting of a wider range of concurrent workloads.

Similarly, multicore processors enable virtualization to be more practical to implement from a performance efficiency perspective. Multicore and virtualization are certainly symbiotic technologies. The availability of system virtualization technology across a wide range of computing platforms provides developers and technologists with the ultimate open platform: the ability to run any flavor of operating system in any combination, creating an unprecedented flexibility for deployment and usage.

Yet embedded multicore systems often have drastically different performance and reliability constraints as compared to multicore server computing. We will also focus on the impact of hypervisor architecture upon these constraints.

Applications of system virtualization

Mainframe virtualization was driven by some of the same applications found in today's enterprise systems. Initially, virtualization was used for time-sharing multiple single-user operating systems, similar to the improved hardware utilization driving modern data center server consolidation. Another important usage involved testing and exploring new operating system architectures. Virtualization was also used to maintain backwards compatibility of legacy versions of operating systems.

Environment sandboxing

Implicit in the concept of consolidation is the premise that independent virtual machines are kept reliably separated from each other (also referred to as virtual machine isolation). The ability to guarantee separation is highly dependent upon the robustness of the underlying hypervisor software. As we will soon expand upon, researchers have found flaws in commercial hypervisors that violate this separation assumption. Nevertheless, an important theoretical application of virtual machine compartmentalization is to isolate untrusted software. For example, a web browser connected to the Internet can be sandboxed in a virtual machine so that Internet-borne malware or browser vulnerabilities are unable to infiltrate or otherwise adversely impact the user's primary operating system environment.

Virtual appliances

Another example, the virtual appliance, does the opposite: sandbox, or separate, trusted software away from the embedded system's primary operating system environment. Consider anti-virus software that runs on a mobile device. A few years ago, the "Metal Gear" Trojan was able to propagate itself across Symbian operating-system-based mobile phones by disabling their anti-malware software. Virtualization can solve this problem by placing the anti-malware software into a separate virtual machine (Figure 7.1).

The virtual appliance can analyze data going into and out of the primary application environment or hook into the platform operating system for demand-driven processing.

Workload consolidation

As microprocessors get more powerful, developers of sophisticated embedded systems look to consolidate workloads running on discrete processors into a single, more powerful platform. System virtualization is a natural mechanism to support this migration since complete operating systems and their applications can be moved wholesale into a virtual machine with minimal or no changes to code.

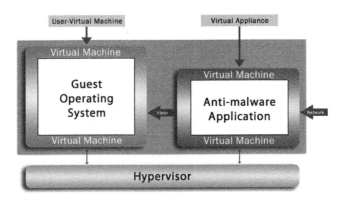

Figure 7.1
Improved robustness using isolated virtual appliances.

Operating system portability

Operating system portability is one of the original drivers of system virtualization. IBM used virtualization to enable older operating systems to run on top of newer hardware (which did not support native execution of the older operating system). This remains an important feature today. For example, system designers can use virtualization to deploy a known "gold" operating system image, even if the SoC supplier does not officially support it. In some high criticality markets, such as medical and industrial control, developers may need to be more conservative about the rate of operating system upgrades than a consumer-oriented market.

Mixed-criticality systems

A by-product of workload consolidation is the increased incidence of mixed criticality computing. Critical functions include hard real-time, safety-critical, and/or security-critical applications that may not be suitable for a general-purpose operating system such as Linux. System virtualization enables mixed criticality by providing sandboxes between general-purpose environments and highly critical environments (such as that running on a real-time operating system). Later in this chapter we provide a case study of mixed criticality automotive electronics.

Maximizing multicore processing resources

Many of these consolidation and mixed-criticality system requirements map well to multicore platforms. SMP-capable hypervisors enable multiple SMP-capable guest operating systems to share all available cores, making it far more likely that those cores will be put to good use. Furthermore, when the overall system workload is light, the multicore-aware hypervisor can dynamically change the available number of cores available to guest operating systems, enabling unused cores to be turned off to save power.

Improved user experience

System virtualization can improve user experience relative to a traditional stovepipe distributing computing approach by breaking down physical component and wiring barriers. For example, consider modern automotive head-units (the center dash computer that typically includes radio, navigation, and other infotainment functions) and clusters (the driver information system that includes speedometer, odometer, and other gauges). In a consolidated design that incorporates both head-unit and cluster functionality, designers may find it easier to mix different types of information in a heads-up display (part of the cluster) because the head-unit and cluster systems are running on the same platform. Communication between the cluster and infotainment systems is extremely fast (shared memory), and the consolidated architecture will encourage automotive OEMs to take a more central role in a synergistic design, instead of relying on separate hardware implementations from "Tier 1" companies — that is, those that supply the OEMs directly — that are not co-developed.

Hypervisor architectures

Hypervisors are found in a variety of flavors. Some are open source; others are proprietary. Some use thin hypervisors augmented with specialized guest operating systems. Others use a monolithic hypervisor that is fully self-contained. In this section, we'll compare and contrast the available technologies.

Hypervisors perform two major functions: resource management and resource abstraction. The management side deals with allocating and controlling access to hardware resources, including memory, CPU time, and peripherals. The abstraction side corresponds to the virtual machine concept: manifesting an illusion of an actual hardware platform. Typically, a general-purpose operating system executes within each virtual machine instance.

Type 2

The early PC hypervisors — VMware Workstation, Parallels, and User-Mode Linux — took advantage of the fact that PC operating systems made virtualization relatively easy: the underlying host operating system provided both resource management (CPU time scheduling, memory allocation) as well as a means to simplify resource abstraction by reusing the host operating system's device drivers and communications software

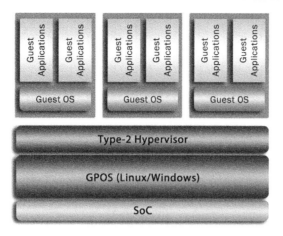

Figure 7.2
Type-2 hypervisor architecture.

and APIs. These early hypervisors are now known as Type 2 hypervisors: running on top of a "conventional operating system environment". Type 2 architecture is shown in Figure 7.2.

Type 1

The original IBM mainframe hypervisors implemented both resource management and abstraction without depending on an underlying host operating system. In the PC server world, both VMware ESX and Xen apply this same architecture, so-called Type 1 because they "run directly on the host's hardware". This architecture is shown in Figure 7.3. The primary motivation for Type 1 hypervisors is to avoid the complexity and inefficiency imposed by a conventional host operating system. For example, a Type 2 hypervisor running on Linux could be subverted by one of the multitudinous vulnerabilities in the millions of lines of code in the Linux kernel.

However, the traditional Type 1 architecture has its own drawbacks. Type 1 has no ability to provide a non-virtualized process environment. Whereas a Type 2 environment can mix and match native applications running on the host operating system with guests, a Type 1 hypervisor is purpose-built to execute guests only. Another issue with legacy Type 1 hypervisors is that they must reinvent the device drivers and hardware

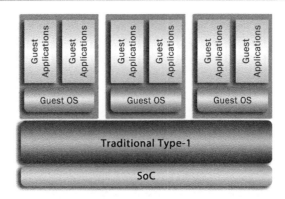

Figure 7.3
Type-1 hypervisor architecture.

management capabilities that conventional operating systems already provide, tuned for performance and robustness across applications and years of field deployment. Some Type 1 hypervisors employ a specialized guest to provide device I/O services on behalf of other guests. In this approach, the size of the software base that must be trusted for overall system robustness is no better than a Type 2.

Paravirtualization

System virtualization can be implemented with full virtualization or paravirtualization, a term first coined in the 2001 Denali project [1]. With full virtualization, unmodified guest operating systems are supported. With paravirtualization, the guest operating system is modified in order to improve the ability of the underlying hypervisor to achieve its intended function.

Paravirtualization is sometimes able to provide improved performance. For example, device drivers in the guest operating system can be modified to make direct use of the I/O hardware instead of requiring I/O accesses to be trapped and emulated by the hypervisor. Paravirtualization often includes the addition of specialized system calls into guest operating system kernels that enable them to request services directly from the hypervisor. These system calls are referred to as *hypercalls*. For example, if the guest kernel running in user mode (because it is executing in a virtual machine) wishes to disable interrupts, the standard supervisor mode instruction used to disable interrupts will no longer work. This instruction will fault, and the hypervisor can then attempt to discern and emulate what the kernel was trying to do. Alternatively, the kernel can make a hypercall that explicitly asks the hypervisor. Paravirtualization may be required on CPU architectures that lack hardware virtualization acceleration features.

The key advantage to full virtualization over paravirtualization is the ability to use unmodified versions of guest operating systems that have a proven fielded pedigree and do not require the maintenance associated with custom modifications. This maintenance saving is especially important in enterprises that use a variety of operating systems and/or regularly upgrade to new operating system versions and patch releases.

Note that this section compares and contrasts so-called Type-1 hypervisors that run on bare-metal. Type-2 hypervisors run atop a general-purpose operating system, such as Windows or Linux, which provide I/O and other services on behalf of the hypervisor. Because they can be no more robust than their underlying general-purpose host operating systems, Type-2 hypervisors are not suitable for most embedded systems deployments and have historically been avoided in such environments. Thus, Type-2 technology is omitted from the following discussion.

Monolithic hypervisor

Hypervisor architectures most often employ a monolithic architecture. Similar to monolithic operating systems, the monolithic hypervisor requires a large body of operating software, including device drivers and middleware, to support the execution of one or more guest environments. In addition, the monolithic architecture often uses a single instance of the virtualization component to support multiple guest environments. Thus, a single flaw in the hypervisor may result in a compromise of the fundamental guest environment separation intended by virtualization in the first place.

Console guest hypervisor

An alternative approach uses a trimmed-down hypervisor that runs in the microprocessor's most privileged mode but employs a special guest

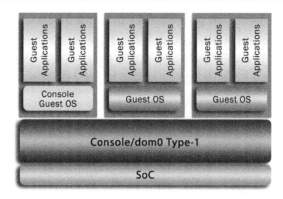

Figure 7.4
Console guest or Dom0 hypervisor architecture.

operating system partition called the "console guest" to handle the I/O control and services for the other guest operating systems (Figure 7.4). Examples of this architecture include Xen, Microsoft Hyper-V, and Red Bend VLX. Xen pioneered the console guest approach in the enterprise; within Xen, the console guest is called Domain 0, or Dom0 for short. Thus, the console guest architecture is sometimes referred to as the Dom0 architecture. With the console guest approach, a general-purpose operating system must still be relied upon for system reliability. A typical console guest such as Linux may add far more code to the virtualization layer than found in a monolithic hypervisor.

Microkernel-based hypervisor

The microkernel-based hypervisor, a form of Type-1 architecture, is designed specifically to provide robust separation between guest environments. Figure 7.5 shows the microkernel-based hypervisor architecture. Because the microkernel is a thin, bare-metal layer, the microkernel-based hypervisor is considered a Type-1 architecture.

This architecture adds computer virtualization as a service on top of the trusted microkernel. In some cases, a separate instance of the virtualization component is used for each guest environment. Thus, the virtualization layer need only meet the equivalent (and, typically, relatively low) robustness level of the guest itself. In the microkernel

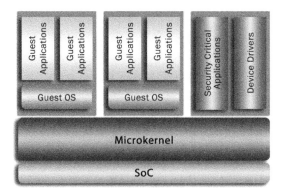

Figure 7.5
Microkernel-based Type-1 hypervisor architecture.

architecture, only the trusted microkernel runs in the highest privilege mode. Examples of embedded hypervisors using the microkernel approach include the INTEGRITY Multivisor from Green Hills Software and some variants of the open standard L4 microkernel.

Core partitioning architectures

Another important aspect of hypervisor architecture is the mechanism for managing multiple cores. The simplest approach is *static partitioning* (Figure 7.6), where only one virtual machine is bound permanently to each core.

Dynamic partitioning (Figure 7.7) is the next most flexible implementation, allowing virtual machines to migrate between cores.

Migration of virtual machines is similar to process or thread migration in a symmetric multiprocessing (SMP) operating system. The ability to migrate workloads can be critical to performance when the hypervisor must manage more virtual machines than there are cores. This situation is similar to server-based virtualization (for instance, cloud computing) where there are more virtual servers than cores and the hypervisor must optimally schedule the servers based on dynamically changing workloads. In embedded systems, the ability to schedule virtual machines provides obvious advantages such as the ability to prioritize real-time workloads.

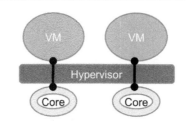

Figure 7.6
Static partitioning.

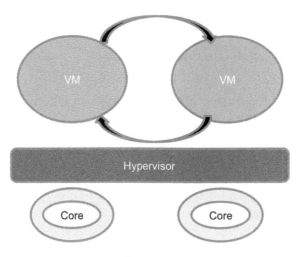

Figure 7.7
Dynamic partitioning.

Dynamic partitioning also adds the ability to time-share multiple virtual machines on a single core. This approach is well suited to the previously mentioned microkernel hypervisor model because of its ability to schedule and time-share processes as well as virtual machines. Thus, specialized applications can be prioritized over virtual machines and vice versa.

Dynamic partitioning can be critical for power efficiency. Consider the simple example of a dual-core processor with two virtual machine workloads, each of which requires 50% utilization of one core. Without sharing, each core must run a virtual machine and use half of the available processing resources. Even with dynamic frequency and voltage scaling, the static energy of each core must be expended. With shareable partitioning, the hypervisor can determine that a single core

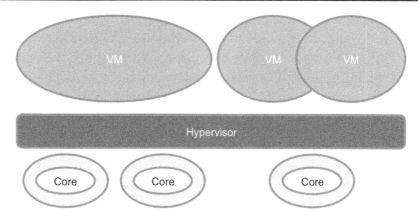

Figure 7.8
Dynamic multicore partitioning.

can handle the load of both virtual machines. The hypervisor will then time slice the two virtual machines on the first core, and turn the second core off, reducing power consumption substantially.

Dynamic multicore partitioning adds support for multicore virtual machines. For example, the hypervisor can allocate a dual-core virtual machine to two cores, host an SMP guest operating system, and take advantage of its concurrent workloads. Single-core virtual machines can be hosted on other cores (Figure 7.8).

A hypervisor may support dynamically switching the number of cores allocated to a virtual machine. This is another method for improving overall performance efficiency. With Linux guests, the Linux hot plug functionality can be employed by a hypervisor to force the guest into relinquishing or adding cores, based on some application-specific policy.

Leveraging hardware assists for virtualization

The addition of CPU hardware assists for system virtualization has been key to the practical application of hypervisors in embedded systems.

Mode hierarchy

Many processor architectures define a user and supervisor state. The operating system kernel often runs in supervisor mode, giving it access to privileged instructions and hardware such as the memory management unit (MMU) while applications run in user mode. The hardware efficiently implements transitions between applications and the operating system kernel in support of a variety of services such as peripheral I/O, virtual memory management, and application thread control.

To support hypervisors, hardware virtualization acceleration adds a third mode to run the hypervisor. In a manner analogous to the user-kernel transitions, the hypervisor mode supports transitions between user, supervisor, and hypervisor state. Another way of thinking about this tri-level hierarchy is that the hypervisor mode is analogous to the old supervisor mode that provided universal access to physical hardware state, and the guest operating systems now run in the *guest supervisor* mode and the guest application in *guest user* mode. While an additional mode of increased privilege is the lowest common denominator for hardware virtualization assistance, numerous other hardware capabilities can dramatically improve the efficiency of system virtualization and may vary dramatically across CPU architectures and implementations within an architecture.

Intel VT

Intel VT, first released in 2005, has been a key factor in the growing adoption of full virtualization throughout the enterprise-computing world. Virtualization Technology for x86 (VT-x) provides a number of hypervisor assistance capabilities, including a true hardware hypervisor mode which enables unmodified guest operating systems to execute with reduced privilege. For example, VT-x will prevent a guest operating system from referencing physical memory beyond what has been allocated to the guest's virtual machine. In addition, VT-x enables selective exception injection, so that hypervisor-defined classes of exceptions can be handled directly by the guest operating system without incurring the overhead of hypervisor software interposing. While VT technology became popular in the server-class Intel chipsets, the same VT-x technology is now also available in Intel Atom™ embedded and mobile processors.

A second generation of Intel VT adds page table virtualization. Within Intel chipsets, this feature is called Extended Page Tables (EPT). Page table virtualization enables the guest operating system to create and manage its own virtual memory mappings. A guest operating system virtual address is mapped to a guest physical address. The guest sees only this mapping. However, guest physical addresses are mapped again to real physical addresses, configured by the hypervisor but automatically resolved by hardware.

Power architecture ISA 2.06 embedded hypervisor extensions

In 2009, the Power Architecture governance body, Power.org, added virtualization to the embedded specification within the Power Architecture version 2.06 instruction set architecture (ISA). At the time of writing, Freescale Semiconductor is the only embedded microprocessor vendor to have released products, including the QorIQ P4080 and P5020 multicore network processors, supporting this embedded virtualization specification. A detailed case study of Power Architecture virtualization and the P4080 can be found later in this chapter.

In 2010, ARM Ltd. announced the addition of hardware virtualization extensions to the ARM architecture as well as the first ARM core, the Cortex A15, to implement them. Publicly announced licensees planning to create embedded SoCs based on Cortex A15 include Texas Instruments, Nvidia, Samsung, and ST Ericsson.

Prior to the advent of these hardware virtualization extensions, full virtualization was only possible using dynamic binary translation and instruction rewriting techniques that were exceedingly complex and unable to perform close enough to native speed to be practical in embedded systems. For example, a 2005-era x86-based desktop running Green Hills Software's pre-VT virtualization technology was able to support no more than two simultaneous full-motion audio/video clips (each in a separate virtual machine) without dropping frames. With VT-x on similar-class desktops, only the total RAM available to host multiple virtual machines generally limits the number of simultaneous clips. General x86 virtualization benchmarks showed an approximate doubling of performance using VT-x relative to the pre-VT platforms. In addition, the virtualization software layer was simplified due to the VT-x capabilities.

ARM TrustZone

An often overlooked and undervalued virtualization capability in modern ARM microprocessors is ARM TrustZone. TrustZone enables a specialized, hardware-based form of system virtualization. TrustZone provides two zones: a "normal" zone and a "trusted" or "secure" zone. With TrustZone, the multimedia operating system (for example, what the user typically sees on a smartphone) runs in the normal zone, while security-critical software runs in the secure zone. While secure zone supervisor mode software is able to access the normal zone's memory, the reverse is not possible (Figure 7.9). Thus, the normal zone acts as a virtual machine under control of a hypervisor running in the trust zone. However, unlike other hardware virtualization technologies such as Intel VT, the normal-zone guest operating system incurs no execution overhead relative to running without TrustZone. Thus, TrustZone removes the performance (and arguably the largest) barrier to the adoption of system virtualization in resource-constrained embedded devices.

TrustZone is a capability inherent in modern ARM applications processor cores, including the ARM1176, Cortex A5, Cortex A8, Cortex A9, and Cortex A15. However, it is important to note that not all SoCs using these cores fully enable TrustZone. The chip provider must permit secure zone partitioning of memory and I/O peripheral interrupts throughout the SoC complex. Furthermore, the chip provider must open the secure zone for third-party trusted operating systems and applications. Examples of TrustZone-enabled mobile SoCs are the Freescale i.MX53 (Cortex A8) and the Texas Instruments OMAP 4430 (Cortex A9).

Trusted software might include cryptographic algorithms, network security protocols (such as SSL/TLS) and keying material, digital rights

Figure 7.9
ARM TrustZone.

management (DRM) software, virtual keypad for trusted path credential input, mobile payment subsystems, electronic identity data, and anything else that a service provider, mobile device manufacturer, and/or mobile SoC supplier deems worthy of protecting from the user environment.

In addition to improving security, TrustZone can reduce the cost and time to market for mobile devices that require certification for use in banking and other critical industries. With TrustZone, the bank (or certification authority) can limit certification scope to the secure zone and avoid the complexity (if not infeasibility) of certifying the multimedia operating system environment.

A secure zone operating system can further reduce the cost and certification time for two main reasons. First, because the certified operating system is already trusted, with its design and testing artifacts available to the certification authority, the cost and time of certifying the secure zone operating environment is avoided.

Secondly, because the secure zone is a complete logical ARM core, the secure operating system is able to use its MMU partitioning capabilities to further divide the secure zone into meta-zones (Figure 7.10). For example, a bank may require certification of the cryptographic meta-zone used to authenticate and encrypt banking transaction messages, but the bank will not care about certifying a multimedia DRM meta-zone that, while critical for the overall device, is not used in banking transactions and is guaranteed by the secure operating system not to interfere.

TrustZone-enabled SoCs are able to partition peripherals and interrupts between the secure and normal states. A normal-zone general-purpose

Figure 7.10
TrustZone virtualization implementation with metazones within the trust zone.

operating system such as Android can not access peripherals allocated to the secure zone and will never see the hardware interrupts associated with those peripherals. In addition, any peripherals allocated to the normal zone are unable to access memory in the normal zone.

ARM Virtualization Extensions

With the Cortex A15, ARM Ltd. has added a general hardware hypervisor acceleration solution to ARM cores. This technology is known as the ARM Virtualization Extensions, or ARM VE. Similar to the virtualization extensions for other CPU architectures, ARM VE introduces a hypervisor mode, enabling guest operating systems to run in a de-privileged supervisor mode. ARM VE is orthogonal to TrustZone: the hypervisor mode only applies to the normal state of the processor, leaving the secure state to its traditional two-level supervisor/user mode hierarchy. Thus, the hypervisor executes de-privileged relative to the trust zone security kernel. ARM VE also supports the concept of a single entity that can control both the trust zone and the normal-zone hypervisor mode. One example of a commercial multicore ARM SoC that supports ARM VE is the Texas Instruments OMAP5 family (based on the Cortex A15).

ARM VE provides page table virtualization, similar to Intel's EPT.

Hypervisor robustness

Some tout virtualization as a technique in a "layered defense" for system security. The theory postulates that, since only the guest operating system is exposed to external threats, an attacker who penetrates the guest will be unable to subvert the rest of the system. In essence, the virtualization software is providing an isolation function similar to the process model provided by most modern operating systems.

As multicore platforms promote consolidation, the robustness of the virtualization layer is critical in achieving overall system goals for performance, security, and availability.

However, common enterprise virtualization products have not met high robustness requirements and were never designed or intended to meet these levels. Thus, it should come as no surprise that the theory of security via virtualization has no existence proof. Rather, a number of studies of virtualization security and successful subversions of hypervisors have been published. While they look at robustness from a security perspective, the following studies are equally applicable to any multicore embedded computing system with critical requirements (real-time, safety, security, availability).

SubVirt

In 2006, Samuel King's SubVirt project demonstrated hypervisor rootkits that subverted both VMware and Microsoft VirtualPC [2].

Blue pill

The Blue Pill project took hypervisor exploits a step further by demonstrating a malware payload that was itself a hypervisor that could be installed on-the-fly beneath a natively running Windows operating system. Since this work, a hypervisor that runs beneath and unbeknownst to another hypervisor is referred to generically as a blue pill hypervisor. Platform attestation is required in order to prevent hypervisors from being subverted in this manner. Platform attestation is the process of validating that only known good firmware (e.g., the hypervisor) is booted and controlling the computing platform at all times.

Ormandy

Tavis Ormandy performed an empirical study of hypervisor vulnerabilities. Ormandy's team of researchers generated random I/O activity into the hypervisor, attempting to trigger crashes or other anomalous behavior. The project discovered vulnerabilities in QEMU, VMware Workstation and Server, Bochs, and a pair of unnamed proprietary hypervisor products [3].

Xen owning trilogy

At the 2008 Black Hat conference, security researcher Joanna Rutkowska and her team presented their findings of a brief research project to locate vulnerabilities in Xen. One hypothesis was that Xen would be less likely to have serious vulnerabilities, as compared to VMware and Microsoft Hyper-V due to the fact that Xen is an open-source technology and therefore benefits from the "many-eyes" exposure of the code base.

Rutkowka's team discovered three different, fully exploitable vulnerabilities that the researchers used to commandeer the computer via the hypervisor [4]. Ironically, one of these attacks took advantage of a buffer overflow defect in Xen's Flask layer. Flask is a security framework, the same one used in SELinux, that was added to Xen to improve security. This further underscores an important principle: software that has not been designed for and evaluated to high levels of assurance must be assumed to be subvertible by determined and well-resourced entities.

VMware's security certification

As VMware virtualization deployments in the data center have grown, security experts have voiced concerns about the implications of "VM sprawl" and the ability of virtualization technologies to ensure security.

On June 2, 2008, VMware attempted to allay this concern with its announcement that its hypervisor products had achieved a Common Criteria EAL 4 + security certification. VMware's press release claimed that its virtualization products could now be used "for sensitive, government environments that demand the strictest security".

On June 5, just three days later, severe vulnerabilities in the certified VMware hypervisors were posted to the National Vulnerability Database. Among other pitfalls, the vulnerabilities "allow guest operating system users to execute arbitrary code" [5].

VMware's virtualization products have continued to amass severe vulnerabilities, for example CVE-2009-3732, which enables remote attackers to execute arbitrary code.

Clearly, the risk of an "escape" from the virtual machine layer, exposing all guests, is very real. This is particularly true of hypervisors characterized by monolithic code bases. It is important for developers to understand that use of a hypervisor does not imply highly robust isolation between virtual machines, no more than the use of an operating system with memory protection implies assured process isolation and overall system integrity.

I/O Virtualization

One of the biggest impacts on robustness and efficiency in system virtualization is the approach to managing I/O across virtual machines. This section discusses some of the pitfalls and emerging trends in embedded I/O virtualization. An I/O virtualization architecture can be broken down into two dimensions: a peripheral virtualization architecture and a peripheral sharing architecture. First, we discuss the peripheral hardware virtualization approaches. While this discussion is germane to single-core platforms, the architectural design of I/O virtualization for multicore systems is even more critical due to the frequent concurrent use of shared I/O hardware across CPU cores.

Peripheral virtualization architectures

Emulation

The traditional method of I/O virtualization is emulation: all guest operating system accesses to device I/O resources are intercepted, validated, and translated into hypervisor-initiated operations (Figure 7.11). This method maximizes reliability. The guest operating system can never corrupt the system through the I/O device because all I/O accesses are protected via the trusted hypervisor device driver. A single device can easily be multiplexed across multiple virtual machines, and if one virtual machine fails, the other virtual machines can continue to utilize the same physical I/O device, maximizing system availability. The biggest drawback is efficiency; the emulation layer causes significant overhead on all I/O operations.

In addition, the hypervisor vendor must develop and maintain the device driver independent of the guest operating system drivers. When using emulation, the hypervisor vendor has the option to present virtual hardware that differs from the native physical hardware. As long as the guest operating system has a device driver for the virtual device, the

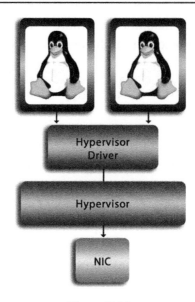

Figure 7.11
I/O virtualization using emulation.

virtual interface may yield improved performance. For example, the virtual device may use larger buffers with less inter-device communication, improving overall throughput and efficiency.

The hypervisor vendor also has the option to create a virtual device that has improved efficiency properties and does not actually exist in any physical implementation. While this paravirtualization approach can yield improved performance, any gains are offset by the requirement to maintain a custom device driver for the guest.

Pass-through

In contrast, a *pass-through* model (Figure 7.12) gives a guest operating system direct access to a physical I/O device. Depending on the CPU, the guest driver can either be used without modification or with minimal paravirtualization. The pass-through model provides improved efficiency but trades off robustness: an improper access by the guest can take down any other guest, application, or the entire system. This model violates a primary goal of system virtualization: isolation of virtual environments for safe coexistence of multiple operating system instances on a single multicore computer.

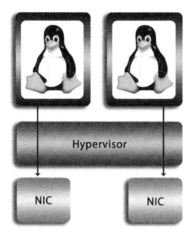

Figure 7.12
I/O virtualization using pass-through.

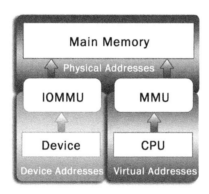

Figure 7.13
IOMMU is used to sandbox device-related memory accesses.

If present, an IOMMU enables a pass-through I/O virtualization model without risking direct memory accesses beyond the virtual machine's allocated memory. As the MMU enables the hypervisor to constrain memory accesses of virtual machines, the IOMMU constrains I/O memory accesses (especially DMA), whether they originate from software running in virtual machines or the external peripherals themselves (Figure 7.13).

IOMMUs are becoming increasingly common in embedded multicore microprocessors, such as Intel Core, Freescale QorIQ, and ARM Cortex A15. Within Intel processors, the IOMMU is referred to as Intel Virtualization Technology for Directed I/O (Intel VT-d). On Freescale's virtualization-enabled QorIQ processors such as the eight-core P4080, the IOMMU is referred to as the Peripheral Access Management Unit (PAMU). On Cortex A15 (and other ARM cores that support the ARM Virtualization Extensions), the IOMMU is not part of the base virtualization specification. Rather, ARM Ltd. has a separate intellectual property offering, called a SystemMMU, which is optionally licensable by ARM semiconductor manufacturers. In addition, the manufacturer may use a custom IOMMU implementation instead of the ARM SystemMMU. In addition, ARM TrustZone provides a form of IOMMU between the normal and secure zones of an ARM processor: normal-zone accesses made by the CPU or by peripherals allocated to the normal zone are protected against accessing memory in the secure zone.

The IOMMU model enables excellent performance efficiency with increased robustness relative to a pass-through model without IOMMU. However, IOMMUs are a relatively new concept.

A number of vulnerabilities (ways to circumvent protections) have been discovered in IOMMUs and must be worked around carefully with the assistance of your systems software/hypervisor supplier. In most vulnerability instances, a faulty or malicious guest is able to circumvent the hypervisor isolation via device, bus, or chipset-level operations other than direct memory access. Researchers at Green Hills Software, for example, have discovered ways for a guest operating system to access memory beyond its virtual machine, deny execution service to other virtual machines, blue pill the system hypervisor, and take down the entire computer — all via IOMMU-protected I/O devices. For high-reliability applications, the IOMMU must be applied in a different way than the traditional pass-through approach where the guest operating system has unfettered access to the I/O device.

A major downside of a pass-through approach (with or without the IOMMU) is that it prevents robust sharing of a single I/O device across multiple virtual machines. The virtual machine that is assigned ownership of the pass-through device has exclusive access, and any other virtual machines must depend upon the owning virtual machine to forward I/O. If the owning virtual machine is compromised, all virtual machines shall be denied servicing for that device.

Mediated pass-through

Mediated pass-through is a special case of the pass-through approach described above. Similar to pure pass-through, the guest operating system device drivers are reused (unmodified). However, the hypervisor traps accesses that are system security/reliability relevant (for example programming of DMA registers). The hypervisor can validate the access and pass through, modify, or reject the access based on policy (e.g., is the DMA accessing a valid virtual machine memory location?). The additional performance overhead of mediated pass-through relative to pure pass-through depends very much upon the specific hardware device and its I/O memory layout. The trade-off in reliability and performance is often acceptable. The hypervisor usually contains only a small amount of device-specific logic used to perform the mediation; a complete hypervisor device driver is not required.

Peripheral sharing architectures

In any embedded system, invariably there is a need for sharing a limited set of physical I/O peripherals across workloads. The embedded operating system provides abstractions, such as layer two, three, and four sockets, for this purpose. Sockets provide, in essence, a virtual interface for each process requiring use of a shared network interface device. Similarly, in a virtualized system, the hypervisor must take on the role of providing a secure virtual interface for accessing a shared physical I/O device. Arguably the most difficult challenge in I/O virtualization is the task of allocating, protecting, sharing, and ensuring the efficiency of I/O across multiple virtual machines and applications.

Hardware multiplexing

This deficiency has led to emerging technologies that provide an ability to share a single I/O device across multiple guest operating systems using the IOMMU and hardware partitioning mechanisms built into the device I/O complex (e.g., chipset plus the peripheral itself). One example of shareable, IOMMU-enabled, pass-through devices is Intel processors equipped with Intel Virtualization Technology for Connectivity (Intel VT-c) coupled with PCI-express Ethernet cards implementing Single-Root I/O Virtualization (SR-IOV), a PCI-SIG standard. With such a system, the hardware takes care of providing independent I/O resources, such as multiple packet buffer rings, and some form of quality of execution service amongst the virtual machines. This mechanism lends itself well to networking devices such as Ethernet, RapidIO, and FibreChannel; however, other approaches are required for robust, independent sharing of peripherals such as graphics cards, keyboards, and serial ports. Nevertheless, it is likely that hardware-enabled, IOMMU-protected, shareable network device technology shall grow in popularity across embedded processors.

Hypervisor software multiplexing

When using an emulated peripheral virtualization approach, it is often natural to also employ the hypervisor to provide sharing of the peripheral across multiple virtual machines. Guest I/O accesses are interposed by the hypervisor and then funneled to/from the physical device based on policy. Hypervisor multiplexing enables the enforcement of quality of service (QoS) and availability that is not possible when a pass-through or even a mediated pass-through approach is used. In the mediated/ unmediated pass-through approaches, one guest must be trusted for availability and quality of service. If that guest fails, then all use of the shared peripheral is lost. Hypervisor multiplexing incurs minimal performance overhead associated with analyzing and moving data appropriately between virtual machines.

Bridging

Regardless of which peripheral virtualization approach is employed, bridging can be used to provide one guest with access to I/O by using a peripheral operated by another guest. For example, a "master" guest that has access to a virtual Ethernet device (virtualized using any of the aforementioned methods) can offer an Ethernet or IP bridge to other "slave" guests. Similar techniques can be used for other stream/block-oriented I/O. Bridging is commonly used when pass-through or mediated pass-through is employed for such devices. Many modern operating systems, including Linux, provide convenient facilities for establishing virtual interface adapters that the hypervisor can use to make sharing seamless in the guest device driver model. The hypervisor must still provide the underlying inter-VM communication facilities. In some cases, this communication will include additional intelligence such as network address translation (NAT) routing for an IP bridge. Bridging can provide a simpler virtualization solution than a hypervisor-multiplexed solution, with the trade-off that a guest must be trusted for system availability and quality of service.

Dom0/console guest

Introduced earlier in this chapter, the console or Dom0 guest virtualization architecture employs a special case of bridging in which the "master" is a guest dedicated for the purpose of providing I/O services to the system's other virtual machines. A master I/O guest approach can provide additional robustness protection relative to bridging since the master is shielded by the hypervisor from the software environments of other guests.

Combinations of I/O virtualization approaches

Various combinations of peripheral virtualization and sharing architectures have advantages and disadvantages. While many of these trade-offs have been discussed in the individual sections above, it is also important to remember the overall system design may require employment of multiple architectures and approaches. For example, while paravirtualization and hypervisor multiplexing may be required when a high-performance CPU is shared between virtual machines, a pass-through or mediated pass-through approach may provide the best performance, cost, and reliability trade-offs for some devices in some systems. Furthermore, the impact on overall system performance of an I/O virtualization approach varies based on many factors, including the applications' use of the peripheral, the CPU (speed, bus speed, hardware virtualization capability, etc.), the type of device (block/streamed, low-latency, etc.), and more. The subject of I/O virtualization is indeed complex, and striking the proper balance for performance, reliability, and maintainability may require advice from your system virtualization supplier.

I/O virtualization within microkernels

As discussed earlier, virtual device drivers are commonly employed by microkernel-style operating systems. Microkernel-based hypervisors

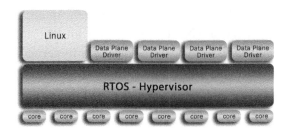

Figure 7.14
Virtual device drivers in microkernel-based system virtualization architecture.

are also well suited to secure I/O virtualization: instead of the typical monolithic approach of placing device drivers into the hypervisor itself or into a special-purpose Linux guest operating system (the dom0 method described earlier), the microkernel-based hypervisor uses small, reduced-privilege, native processes for device drivers, I/O multiplexors, health managers, power managers, and other supervisory functions required in a virtualized environment. Each of these applications is provided only the minimum resources required to achieve its intended function, fostering robust embedded system designs. Figure 7.14 shows the system-level architecture of a microkernel-based hypervisor used in a multicore networking application that must securely manage Linux control-plane functionality alongside high-throughput, low-latency data-plane packet processing within virtual device drivers.

Without virtualization, the above application could be implemented with a dual Linux/RTOS configuration in which the control- and data-plane operating systems are statically bound to a set of independent cores. This is referred to as an asymmetric multiprocessing (AMP) approach. One advantage of virtualization over an AMP division of labor is the flexibility of changing the allocation of control- and data-plane workloads to cores. For example, in a normal mode of operation, the architect may only want to use a single core for control and all other cores for data processing. However, the system can be placed into management mode in which Linux needs four cores (SMP) while the data processing is temporarily limited. The virtualization layer can handle the reallocation of cores seamlessly under the hood, something that a static AMP system cannot support.

Case study: power architecture virtualization and the freescale P4080

Power Architecture has included hardware virtualization support since 2001 in its server-based instruction set architecture (ISA). However, with the advent of Power Architecture ISA 2.06, hardware virtualization is now found in general-purpose embedded processors. The eight-core Freescale™ P4080 is an example of a commercial embedded processor with Power Architecture 2.06 embedded hardware virtualization compatibility. The P4080 was the first microprocessor to provide hardware virtualization capability in a multicore configuration targeted for embedded applications.

Hypervisor performance is highly dependent upon hardware support for virtualization. With the advent of Power Architecture 2.06's ISA Book III-E [6] virtualization hardware support is now available in embedded Power Architecture-based processors such as the Freescale P4080 [7].

Power architecture hardware hypervisor features

Power Architecture ISA 2.06 embedded virtualization capabilities include a new hypervisor mode, guest interrupt injection, guest translation look-aside buffer (TLB) management, and inter-partition communication mechanisms.

The embedded hypervisor mode converts the traditional two-level user/supervisor mode architecture to a three-level guest-user/guest-supervisor/hypervisor mode hierarchy described earlier. The hypervisor level enables hypervisor-mediated access to hardware resources required by guest operating systems. When a guest performs a privileged operation, such as a TLB modification, the processor can be configured to trap this operation to the hypervisor. The hypervisor can then decide whether to allow, disallow, or modify (emulate) the guest operation. The hypervisor mode also provides for an additional guest processor state, enabling guest operating systems to access partition-specific logical registers without trapping to the hypervisor.

Guest interrupt injection enables the hypervisor to control which interrupts and exceptions cause a trap into hypervisor mode for hypervisor handling, and which interrupts can be safely redirected to a guest, bypassing the hypervisor. For example, an I/O device that is shared between multiple partitions may require hypervisor handling, while a peripheral dedicated to a specific partition can be handled directly by the partition's guest operating system. Interrupt injection is critical for achieving real-time response when relying on a guest RTOS for real-time workloads.

TLB and page table management is one of the most performance-critical aspects of hardware virtualization support. The TLB and page tables govern access to physical memory and therefore must be managed by some combination of virtualization and software to enforce partitioning. Without any special hardware support, attempts to manipulate hardware TLB or page tables via supervisor mode instructions are trapped and emulated by the hypervisor. Depending on the dynamic nature of a guest operating system, this mediation can cause performance problems.

This mediation is avoided in microprocessors that support a two-level virtualization-enabled page table system in hardware. In such a system, the guest operating system is able to manage its TLB and page tables directly, without hypervisor assistance. The hypervisor only needs to configure the system to ensure that each guest's page tables reference memory allocated to the guest's virtual machine.

ARM virtualization extensions (ARM VE) and Intel chipsets that support the second-generation VT-x with Extended Page Tables (EPT) are examples of hardware virtualization implementations that avoid the bottleneck of hypervisor page table management. Power 2.06 does not currently specify such a two-level guest-managed page table architecture. Thus, the approach taken by the hypervisor in managing guest page tables may be critical for overall system performance; the approaches are described later, in section 8.2.

Power Architecture ISA 2.06 supports doorbell messages and interrupts that enable guest operating systems to communicate with software outside its virtual machine, without involving the hypervisor. For example, a control-plane guest operating system can efficiently inform data-plane applications of changes that might affect how packets are processed. Besides being a general shoulder-tap mechanism for the system, the hypervisor can use doorbells to reflect asynchronous interrupts to a guest. Asynchronous interrupt management is complicated by the fact that an interrupt cannot be reflected until the guest is ready to accept it. When an asynchronous interrupt occurs, the hypervisor may use the Power Architecture *msgsnd* instruction to send a doorbell interrupt to the appropriate guest. Doorbell exceptions remain pending until the guest has set the appropriate doorbell enable bit. When the original exception occurs, it is first directed to the hypervisor, which can now safely reflect the interrupt using the asynchronous doorbell.

Power Architecture also provides additional facilities for managing interrupts between the hypervisor and guests. For example, if the hypervisor is time-slicing multiple guest operating systems, a scheduling timer may be used directly by the hypervisor. If a guest operating system is given exclusive access to a peripheral, then that peripheral's interrupts can be directed to the guest, avoiding the latency of trapping first to the hypervisor before reflecting the interrupt back up to a guest.

Power Architecture defines a new set of guest registers that mirror prior non-guest versions to enable intelligent interrupt management. These registers include:

- Guest Save/Restore registers (GSRR0/1)
- Guest Interrupt Vector Prefix Register (GIVPR)
- Guest Interrupt Vector Registers (GIVORn)
- Guest Data Exception Address Register (GDEAR)
- Guest Exception Syndrome Register (GESR)
- Guest Special Purpose Registers (GSPRG0..3)
- Guest External Proxy (GEPR)
- Guest Processor ID Register (GPIR)

Most of these registers provide state information important to servicing exceptions and interrupts and duplicate those available in hypervisor mode (or prior Power Architecture® implementations without guest mode). These registers have different offsets from the originals. To ensure legacy operating systems can run unchanged while in guest mode, the processor maps references to the original non-guest versions to the guest versions.

When an external interrupt is directed to a guest and the interrupt occurs while executing in hypervisor mode, the interrupt remains pending until the processor returns to guest mode. The hardware automatically handles this transition.

Power Architecture introduces some additional exceptions and interrupts to support hypervisors. To allow a paravirtualized guest operating system to call the hypervisor, a special system call causes the transition between guest and hypervisor mode. It uses the existing Power Architecture system call instruction but with a different operand.

Some exceptions and interrupts are always handled by the hypervisor. When a guest tries to access a hypervisor privileged resource such as TLB, the exception always triggers execution to hypervisor mode and allows the hypervisor to check protections and virtualize the access when necessary. In the case of cache access, normal cache load and store references proceed normally. A mode exists for guests to perform some direct cache management operations, such as cache locking.

Power architecture virtualized memory management

One important role of hypervisors in virtualization is managing and protecting virtual machines. The hypervisor must ensure that a virtual machine has access only to its allocated resources and must block unauthorized accesses. To perform this service, the hypervisor directly manages the MMU of all cores and, in many implementations, also manages an IOMMU for all I/O devices that master memory transactions within the SoC.

In the case of the MMU, the hypervisor typically adopts one of two primary management strategies. The simplest approach from the hypervisor's perspective is to reflect all memory management interrupts to the guest. The guest will execute its usual TLB-miss handling code, which includes instructions to load the TLB with a new page table entry. These privileged TLB manipulation instructions will, in turn, trap to the hypervisor so that it can ensure the guest loads only valid pages.

Alternatively, the hypervisor can capture all TLB manipulations by the guest (or take equivalent paravirtualized hypercalls from the guest) to emulate them. The hypervisor builds a virtual page table for each virtual machine. When a TLB miss occurs, the hypervisor searches the virtual page table for the relevant entry. If one is found, it performs a TLB fill directly and returns from the TLB miss exception without generating an exception to the guest. This process improves performance by precluding the need for the hypervisor to emulate the TLB manipulation instructions that the guest would execute when handling the TLB miss. These TLB management alternatives are a great example of how hypervisor implementation choices can cause dramatic differences in performance and why it is important to consult your hypervisor vendor to understand the trade-offs.

To allow the physical TLB to simultaneously carry entries for more than one virtual machine, the TLB's virtual addresses are extended with a partition identifier. Furthermore, to allow the hypervisor itself to have entries in the TLB, a further bit of address extension is used for hypervisor differentiation.

In Power Architecture, the TLB's virtual address is extended by an LPID value that is a unique identifier for a logical partition (virtual machine). The address is further extended with the GS bit that designates whether an entry is valid for guest or hypervisor mode. These extensions allow the hypervisor to remap guest logical physical addresses to actual physical addresses. Full virtualization is then practical because guest operating systems often assume that their physical address space starts at 0.

Freescale P4080 IOMMU

As described earlier, an IOMMU mediates all address-based transactions originating from mastering devices. Although IOMMUs are not standardized in Power Architecture as of ISA version 2.06, as previously mentioned, the P4080 incorporates an IOMMU (named the PAMU, Peripheral Access Management Unit, by Freescale). DMA-capable peripherals such as Ethernet or external storage devices can be safely dedicated for use by a virtual machine without violating the system partitioning enforced by the hypervisor. Any attempt to DMA into physical memory not allocated to the virtual machine will fault to the hypervisor. The IOMMU keeps internal state that relates the originator of the transaction with authorized address regions and associated actions. The hypervisor initializes and manages the IOMMU to define and protect logical partitions in the system.

Hardware support for I/O sharing in the P4080

How hardware resources on a multicore SoC are allocated and possibly shared across virtual machines is one of the most important performance attributes of the system. Resources dedicated to a single virtual machine can often execute at full speed, without hypervisor inter-positioning. However, in many systems, at least some of the limited hardware resources must be shared. For example, a single Ethernet device may need to be multiplexed between virtual machines by the hypervisor. The hypervisor can handle the Ethernet interrupts, transmit packets on behalf of the virtual machines, and forward received packets to the appropriate virtual machine. In theory, the hypervisor could even implement firewall or other network services between the virtual machines and the physical network.

Certain applications require shared hardware access but cannot tolerate the overhead of hypervisor mediation. Modern microprocessors are rising to this challenge by providing hardware-mediation capabilities designed specifically for virtualized environments. For example, the P4080 implements encryption and pattern matching engines that support a variety of networking functions. The ability of multiple partitions to share these resources is an important feature of the P4080.

The P4080 provides multiple dedicated hardware portals used to manage buffers and submit and retrieve work via queues. By dedicating a portal to a virtual machine, the guest interacts with the encryption and pattern-matching engines as if the guest completely owned control of the devices. In the P4080, the Ethernet interfaces participate in this same hardware queue and buffer management facility. Inbound packets can be classified and queued to a portal owned by a virtual machine without involving hypervisor software during reception. Similarly, each virtual machine can submit packets for transmission using its private portal.

Virtual machine configuration in power architecture

The Power.org Embedded Power Architecture Platform Requirements (ePAPR) specification [8] standardizes a number of hypervisor management facilities, including a device tree data structure that represents hardware resources. Resources include available memory ranges, interrupt controllers, processing cores, and peripherals. The device tree may represent the virtual hardware resources that a virtual machine will see or the physical resources that a hypervisor sees. In addition, ePAPR defines a boot image format that includes the device tree. ePAPR enables Power Architecture bootloaders to understand and boot hypervisors from multiple vendors and provides a standardized approach for hypervisors to instantiate virtual machines.

Power Architecture's device-tree approach addresses what many believe to be the most significant barrier to adoption of system virtualization technology in multicore embedded systems. Without this form of configuration standardization, hypervisor vendors create proprietary interfaces and mechanisms for managing virtual hardware. That is certainly the case on Intel and ARM architecture-based systems at this time of writing. The lack of standardized hypervisor configuration and management promotes vendor lock-in, something that ultimately hurts rather than helps those same vendors.

The device-tree configuration approach is gaining increased adoption in the Linux community and is at least an optional capability supported for most Linux CPU architectures. The next step is to promulgate this approach to virtualized Linux operating systems and then to other types of operating systems. At the time of writing, support for standardizing virtual machine configuration is being considered in Linaro (which is quickly adopting device trees) as well as in the GENIVI automotive alliance.

Example use cases for system virtualization

The use of virtualization in multicore embedded systems is nascent, yet incredibly promising. This section covers just a few of the emerging applications.

Telecom blade consolidation

Virtualization enables multiple embedded operating systems, such as
Linux® and VxWorks®, to execute on a single multicore processor, such
as the Freescale P4080. In addition, the microkernel-based virtualization
architecture enables hard real-time applications to execute natively.
Thus, control-plane and data-plane applications, typically requiring
multiple blades, can be consolidated. Telecom consolidation provides the
same sorts of size, weight, power, and cost efficiencies that enterprise
servers have enjoyed with VMware.

Electronic flight bag

Electronic flight bag (EFB) is a general-purpose computing platform that
flight crews use to perform flight management tasks, including
calculating take-off parameters and viewing navigational changes, more
easily and efficiently. EFBs replace the stereotypical paper-based flight
bags carried by pilots. There are three classes of EFBs, with Class 3
being a device that interacts directly with the onboard avionics and
requires air-worthiness certification.

Using a safety-rated hypervisor, a Class 3 EFB can provide a typical
Linux desktop environment (e.g., OpenOffice.org™) for pilots while
hosting safety-critical applications that validate parameters before they
are input into the avionics. Virtualization enables Class 3 EFBs to be
deployed in the portable form factor that is critical for a cramped
cockpit.

Intelligent Munitions System

Intelligent Munitions System (IMS) is a next-generation US military net-
centric weapon system. One component of IMS includes the ability to
dynamically alter the state of munitions (for example, mines) to meet the
requirements of an evolving battlescape. Using a safety-rated hypervisor,
the safety-critical function of programming the munitions and providing
a trusted display of weapons state for the soldier is handled by secure
applications running on the safety-certified hypervisor or trusted guest

operating system. A standard Linux graphical interface is enabled with virtualization.

Automotive infotainment

Demand for more advanced infotainment systems is growing rapidly. In addition to theater-quality audio and video and GPS navigation, wireless networking, and other office technologies are making their way into the car. Despite this increasing complexity, passenger expectations for *instant on* and high availability remain. At the same time, automobile systems designers must always struggle to keep cost, weight, power, and component size to a minimum.

Automobile passengers expect the radio and other traditional head-unit components to never fail. In fact, a failure in one of these components is liable to cause an expensive (for the automobile manufacturer) visit to the repair shop. Even worse, a severe design flaw in one of these systems may result in a recall that wipes out the profit on an entire vintage of cars. Exacerbating the reliability problem is a new generation of security threats: bringing Internet into the car exposes it to all the viruses and worms that target general-purpose operating systems.

One currently deployed solution, found on select high-end automobiles, is to divide the infotainment system onto two independent hardware platforms, placing the high-reliability, real-time components onto a computer running a real-time operating system, and the rear-seat infotainment component on a separate computer. This solution is highly unworkable, however, because of the need to tightly constrain component cost, size, power, and weight within the automobile.

System virtualization provides an ideal alternative. Head-unit applications run under the control of a typical infotainment operating system such as GENIVI Linux or Android. In the back seat, each passenger has a private video monitor. One passenger could even reboot without affecting the second passenger's Internet browsing session (e.g., two GENIVI operating systems). Meanwhile, critical services such as rear-view camera, advanced driver assistance systems (ADAS), and 3D clusters can be hosted on the same multicore processor, using a

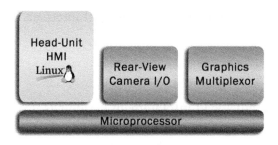

Figure 7.15
Automotive consolidation of center stack/head-unit and critical functions.

hypervisor that supports hard real-time, safety-critical applications, and a virtualized, multiplexed graphics processing unit (GPU), as shown in Figure 7.15.

Medical imaging system

Medical scanners typically consist of three distinct subsystems: an operator console, a real-time data-acquisition component, and a high-end graphical display. Sometimes the system is implemented with three separate computers in order to avoid any resource conflicts and guarantee real-time performance. System virtualization, however, can consolidate these systems onto a single multicore processor. For example, a desktop Linux OS can be used for the operator console while specialized real-time operating systems are used to handle the graphics rendering and data acquisition. This architecture can provide the required isolation and performance guarantees while dramatically reducing size, weight, power, and cost of the system.

Conclusion

Increases in software and system complexity and connectivity are driving the evolution in how embedded systems manage I/O and in the architecture of the operating systems and hypervisors that are responsible for ensuring their security. The combination of reduced-privilege, component-based designs as well as intelligent I/O virtualization to enable secure consolidation without sacrificing efficiency shall remain a focus of systems software suppliers in meeting the flexibility, scalability, and robustness demands of next-generation embedded systems.

References

[1] Whitaker, et al. Denali: Lightweight virtual machines for distributed and networked applications. USENIX Annual Technical Conference, (2002).

[2] S. King, et al. Sub*Virt:* Implementing malware with virtual machines. IEEE Symposium on Security and Privacy, (2006).

[3] T. Ormandy. An empirical study into the security exposure to hosts of hostile virtualized environments. <http://taviso.decsystem.org/virtsec.pdf>, 2006.

[4] J. Rutkowska, A. Tereshkin, R. Wojtczuk. Detecting and preventing the Xen hypervisor subversions; Bluepilling the Xen hypervisor; Subverting the Xen Hypervisor. Black Hat USA, (2008).

[5] CVE-2008-2100, National vulnerability database: <http://web.nvd.nist.gov/view/vuln/detail?vulnId = CVE-2008-2100>.

[6] Available from: http://www.power.org/resources/downloads/PowerISA_V2.06_PUBLIC.pdf.

[7] Available from: http://www.freescale.com/webapp/sps/site/prod_summary.jsp?fastpreview = 1&code = P4080.

[8] Available from: http://www.power.org/resources/downloads/Power_ePAPR_APPROVED_v1.0.pdf.

CHAPTER 8

Communication and Synchronization Libraries

Max Domeika

Technical Project Manager, Intel Corporation, Intel Corporation, Hillsboro, OR, USA

Chapter Outline

Introduction

Effective parallelization of embedded applications requires use of communication and synchronization primitives. In order to increase developer efficiency, several libraries have been developed with the end result of easing application development. These libraries offer higher-level abstractions to the lower-level hardware primitives detailed in a later chapter. This chapter details several communication and synchronization libraries commonly used in embedded development, starting with an

Real World Multicore Embedded Systems.
DOI: http://dx.doi.org/10.1016/B978-0-12-416018-7.00008-0

overview and then moving into details on several common libraries, including Windows Threads, POSIX Threads, C11 Threads, Threading Building Blocks, Boost Threads, OpenMP, and MCAPI.

Library overview and basics

The implementation of communication and synchronization libraries and the APIs that are made available are primarily determined by the memory architecture of the underlying platform. Platforms employing shared memory tend to favor thread-based APIs. Platforms employing distributed memory tend to favor message-passing-based APIs.

Thread APIs

There are several techniques in use today to thread applications. Two broad categories are library-based and compiler-based threading. An example of library-based threading is the POSIX threads API; an example of compiler-based threading is OpenMP. The library-based threading technologies are termed explicit in that the developer is responsible for explicitly creating, controlling, and destroying the threads via function calls. These mechanisms rely upon shared memory to enable access to application data and as such are suseptible to common concurrency issues such as data races and deadlock.

Message-passing APIs

Message-passing APIs employ communications between execution elements to coordinate work. The messages typically contain application data. In addition, the communication would also contain signals serving to coordinate the processing. An example of a message-passing API employed in embedded applications is the Multicore Communications Application Programming Interface (MCAPI).

Explicit threading libraries

Explicit threading libraries require the programmer to control all aspects of threads, from creating threads and associating threads to functions to synchronizing and controlling the interactions between threads and shared resources. The two most prominent threading libraries in use today are POSIX Threads (Pthreads) and Windows Threads by Microsoft. While the syntax is different between the two APIs, most of the functionality in one model will be found in the other. Each model will create and join threads, features synchronization objects to coordinate execution between threads, and controls access to shared resources by multiple threads executing concurrently.

Windows Threads

Windows Threads use the ubiquitous kernel object *HANDLE* type for the handle of a thread. The *CreateThread()* function will return the *HANDLE* of a spawned thread. If the code will be using the C run-time library, it is recommended that the alternate *_beginthreadex()* function be used to create new threads. These two functions have the exact same set of parameters, but the latter is safer to use with regard to initialization of thread resources and more reliable in the reclamation of allocated resources at thread termination.

Making one thread wait for another thread to terminate is accomplished by calling *WaitForSingleObject()*. To be more precise, since any kernel object in a program is referenced through a *HANDLE*, this function will block the calling thread until the *HANDLE* parameter is in the signaled state. If the *HANDLE* is a thread, the object will be signaled when the thread terminates. What it means for a *HANDLE* to be signaled is different for each type of kernel object. Windows also provides the *WaitForMultipleObjects()* function, which can be used to wait until one or all of up to 64 *HANDLE*s are in the signaled state. Thus, with a single function call, a thread can join multiple threads.

Windows Threads provides two basic mutual exclusion synchronization objects: the mutex and the critical section. A mutex is a kernel object accessed and managed through a *HANDLE*. The *CreateMutex()* function will initialize a mutex object. To lock the mutex, *WaitForSingleObject()* is called; when the mutex handle is in the signaled state, the mutex is available, and the wait function will return. If a thread holding a mutex is terminated unexpectedly, the mutex will be considered abandoned. The next thread to wait on the mutex will be able to lock the mutex and the return code from the wait function will indicate that the mutex had been abandoned. Once a thread is finished with the mutex, *ReleaseMutex()* unlocks the object for another thread to gain control. Windows mutexes, like other kernel objects, can be shared between different processes to create mutually exclusive access to shared resources.

Windows *CRITICAL_SECTION* objects function like mutexes, but they are only accessible within the process in which they have been declared. Critical section objects are initialized with the *InitializeCriticalSection()* function before use. The programmer places *EnterCriticalSection()* and *LeaveCriticalSection()* calls around critical regions of code with a reference to an appropriate *CRITICAL_SECTION*. While critical section objects will be unavailable through termination of the program if the thread holding the object is terminated, the overhead of using this method of mutual exclusion will be considerably faster than use of a mutex object.

Windows events are used to send signals from one thread to another in order to coordinate execution. Events are kernel objects and are manipulated by use of a *HANDLE*. Threads use one of the wait functions to pause execution until the event is in the signaled state. The *CreateEvent()* function initializes an event and selects the type of event. There are two types of events: manual reset and auto-reset. The *SetEvent()* function will set either type of event to the signaled state. Any and all threads waiting on a manual reset event, once it has been signaled, will return from the wait function and proceed. Plus, any thread that calls a wait function on that event will be immediately released. No threads will be blocked waiting on the signaled event until a call to *ResetEvent()* has been issued. In the case of auto-reset events, only one thread waiting or the first thread to wait for the event to be signaled will return from the wait function and the event will be automatically reset to the non-signaled state. Unlike condition variables in Pthreads (discussed next), the signals to Windows events will persist until either reset or the required number of threads have waited for the event and been released.

POSIX Threads

POSIX Threads (Pthreads) has a thread container data type of *pthread_t*. This type is the handle by which the thread is referenced. In order to create a thread and associate that thread with a function for execution, the *pthread_create()* function is used. A *pthread_t* is returned through the parameter list. When one thread must wait on the termination of some other thread before proceeding with execution, a call to *pthread_join()* is called with the handle of the thread to be waited upon.

The two synchronization objects most commonly used with Pthreads are the mutex (*pthread_mutex_t*) and the condition variable (*pthread_cond_t*). Instances of these objects must first be initialized before use within a program. Besides providing functions to do this initialization, defined constants are included in the Pthreads library that can be used for default static initialization when objects are declared. Mutex objects can be held by only one thread at a time. Threads request the privilege of holding a mutex by calling *pthread_lock()*. Other threads attempting to gain control of the mutex will be blocked until the thread that is holding the lock calls *pthread_unlock()*.

Condition variables are associated (through programming logic) with an arbitrary conditional expression and are used to signal threads when the status of the conditional under consideration may have changed. Threads block and wait on a condition variable to be signaled when calling *pthread_cond_wait()* on a given condition variable. At some point in the execution, when the status of the conditional may have changed, an executing thread calls *pthread_cond_signal()* on the associated condition variable to wake up a thread that has been blocked. The thread that receives the signal should first check the status of the conditional expression and either return to waiting on the condition variable (conditional is not met) or proceed with execution. Signals to condition variables do not persist. Thus, if there is no thread waiting on a condition variable when it is signaled, that signal is discarded.

Figure 8.1 is a small application employing POSIX Threads to compute an approximation of the value for Pi using numerical integration and the midpoint rectangle rule. The code divides the integration range into *gNumSteps* intervals and computes the functional value of $4.0/(1 + x^2)$ for the midpoint of each interval. The functional values (*partial*) are summed up and multiplied by the width of the intervals in order to

```
#include <stdio.h>
#include <pthread.h>

#define NUM_THREADS 64
double gPi;
int gNumSteps;            //typically >> gNumThreads
int gNumThreads;
double gStep;
pthread_mutex_t gMutex;

void *partialpi (void *arg) {
  int i;
  int myNum = (int)arg;
  double partial, x;

//run the ith step within each block of 64 steps
  for (i = myNum; i < gNumSteps; i+= gNumThreads) {
    x = (i + 0.5) / gNumSteps;
    partial += 4.0 / (1.0 + x * x);
  }
  pthread_mutex_lock(&gMutex);
  gPi += partial * gStep;     //add to the total, safely
  pthread_mutex_unlock(&gMutex);
}

int main()
{
  pthread_t tid[NUM_THREADS];
  int i;
  gPi = 0.0;
  gNumSteps = 100000;
  gNumThreads = NUM_THREADS;
  gStep = 1.0 / gNumSteps;
  pthread_mutex_init(&gMutex, NULL);

//64 threads are created; they handle the first 64
iterations, then the second 64 iterations, etc. So each
thread handles the ith, i+64th, i+(2*64)th steps, etc.
  for (i = 0; i < NUM_THREADS; i++) {
    pthread_create(&tid[i], NULL, partialpi, i);
  }

  for (i = 0; i < NUM_THREADS; i++) {
    pthread_join(tid[i], NULL);
  }
  printf("%12.9f\n", gPi);
}
```

Figure 8.1
Sample POSIX Threads program.

approximate the area under the curve of the function. Two observations are worth pointing out:

1. Use of *pthread_create()* and *pthread_join()* to create a team of threads and await completion of their work.
2. Use of *pthread_mutex_init()* to synchronize access to the global *gPi* value.

C11 *and* C++11 *Threads*

C11 is the informal name for the C programming language standard, ISO/IEC 9899:2011, which was ratified by ISO in late 2011. C++11 is the informal name for the C++ programming language standard, ISO/IEC 14882:2011, ratified in 2011. The C11 standard introduces multi-threading functions that are optional to compiler implementors. The library APIs differ from POSIX Threads functions, which are a standard in their own right and have already been broadly adopted in user applications. Over time, applications based on C11 Threads may increase as applications employing POSIX Threads decrease; however, the expectation is that it will be several years before that occurs. Some early implementations merely employ shim functions that call equivalent POSIX Threads functions for each of the C11 functions and provide additional implementation functionality only where needed.

The C++11 standard describes a thread class that enables explicit thread creation, control, and synchronization primitives.

The following section summarizes several C11 Threads functions.

```
/* create a thread – start executing func(arg), thr = thread id upon
success */
int thrd_create(thrd_t *thr, thrd_start_t func, void *arg);
/* terminate execution of thread */
void thrd_exit(int res);
/* joins thread thr with current thread by blocking until thr has
terminated */
int thrd_join(thrd_t thr, int *res);
/* opposite of join, enables a thread to terminate without awaiting for
another thread */
```

```
int thrd_detach(thrd_t thr);
/* identifies thread that called it */
thrd_t thrd_current(void);
```
/* suspend calling thread until either duration has elapsed or a signal is received. If signal received, place remaining time in *remaining*/
```
int thrd_sleep(const struct timespec *duration, struct timespec *remaining);
```
/* permit other threads to run */
```
void thrd_yield(void);
```
/* determine whether thread *thr0* refers to thread *thr1* */
```
int thrd_equal(thrd_t thr0, thrd_t thr1);
```

Synchronization functions

/* ensure *func* is called exactly once */
```
void call_once(once_flag *flag, void (*func)(void));
```
/* unblocks all threads that are blocked on condition variable *cond* at time of call */
```
int cnd_broadcast(cnd_t *cond);
```
/* unblocks one thread that is blocked on condition variable *cond* at time of call. If more than one thread blocked, then unblocked thread selected by scheduling policy*/
```
int cnd_signal(cnd_t *cond);
```
/* atomically unlocks mutex *mtx* and tries to block until condition variable *cond* is signaled by a call to *cnd_signal* or to *cnd_broadcast* */
```
int cnd_wait(cnd_t *cond, mtx_t *mtx);
```
/* blocks until it locks the mutex *mtx* */
```
int mtx_lock(mtx_t *mtx);
```
/* tries to block until it locks the mutex *mtx* or after the time *ts* */
```
int mtx_timedlock(mtx_t *restrict mtx, const struct timespec *restrict ts);
```
/* tries to lock the mutex *mtx* */
```
int mtx_trylock(mtx_t *mtx);
```
/* unlocks the mutex *mtx* */
```
int mtx_unlock(mtx_t *mtx);
```

Thread-specific (Thread local) storage functions

```
/* creates a thread-specific storage pointer with destructor dtor */
int tss_create(tss_t *key, tss_dtor_t dtor);
/* releases any resources used by the thread-specific storage key */
void tss_delete(tss_t key);
/* returns the value for the current thread held in the thread-specific
storage key */
void *tss_get(tss_t key);
/* sets value for the current thread held in the thread-specific storage
key to val */
int tss_set(tss_t key, void *val);
```

OpenMP

OpenMP [1] is a set of compiler directives, library routines, and environment variables that is used to specify shared memory concurrency in Fortran, C, and C++ programs. The OpenMP Architecture Review Board (ARB), which oversees the specification of OpenMP, is made up of many different commercial and academic institutions. The rationale behind the development of OpenMP was to create a portable and unified standard of shared memory parallelism. OpenMP was first introduced in November 1997 with a specification for Fortran, while in the following year a specification for C/C++ was released. At the time of writing, OpenMP 3.1 is the current revision of the standard.

All major compilers support the OpenMP language. This includes, for Windows, the Microsoft Visual C/C++ .NET and, for Linux, the GNU gcc compiler. The Intel C/C++ compilers, for both Windows and Linux, also support OpenMP.

OpenMP directives demarcate code that can be executed in parallel, called parallel regions, and control how code is assigned to threads. The threads in code containing OpenMP operate under the fork-join model. As the main thread executes the application, when a parallel region is encountered, a team of threads are forked off and begin executing the code within the parallel region. At the end of the parallel region, the threads within the team wait until all other threads in the team have finished running the code in the parallel region before being "joined" and the main thread resumes execution of the code from the statement following the parallel region. That is, an implicit barrier is set up at the end of all parallel regions (and most other regions defined by OpenMP). Of course, due to the high overhead of creating and destroying threads, quality compilers will only create the team of threads when the first parallel region is encountered and then simply put the team to sleep at the join operation and wake the threads for subsequent forks.

For C/C++, OpenMP uses pragmas as directives. All OpenMP pragmas have the same prefix of *#pragma omp*. This is followed by an OpenMP construct and one or more optional clauses to modify the construct. To define a parallel region with a code, use the *parallel* construct:

   #pragma omp parallel

The single statement or block of code enclosed within curly braces that follows this pragma, when encountered during execution, will fork a team of threads, execute all of the statements within the parallel region on each thread, and join the threads after the last statement in the region.

In many applications, a large number of independent operations are found in loops. Rather than have each thread execute the entire set of loop iterations, parallel speedup of loop execution can be accomplished if the loop iterations are split up and assigned to unique threads. This kind of operation is known as worksharing in OpenMP. The *parallel for* construct will initiate a new parallel region around the single for-loop following the pragma:

   #pragma omp parallel for

The iterations of the loop will be divided amongst the threads of the team. Upon completion of the assigned iterations, threads will sit at the implicit barrier that is at the end of all parallel regions. It is possible to split up the combined *parallel for* construct into two pragmas: the *parallel* construct and the *for* construct, which must be lexically contained within a parallel region. This separation would be used when there was parallel work for the thread team other than the iterations of the loop. A set of *schedule* clauses can be attached to the *for* construct to control how iterations are assigned to threads. The *static* schedule will divide iterations into blocks and distribute the blocks among threads before execution of the loop iterations begin execution; round robin scheduling is used if there are more blocks than threads. The *dynamic* schedule will assign one block of iterations per thread in the team; as threads finish the previous set of iterations, a new block is assigned until all blocks have been distributed. There is an optional *chunk* argument for both *static* and *dynamic* scheduling that controls the number of iterations per block. A third scheduling method, *guided*, distributes blocks of iterations like *dynamic*, but the sizes of the blocks decrease for each successive block assigned; the optional *chunk* argument for *guided* is the smallest number of iterations that will ever be assigned in a block.

By default, almost all variables in an OpenMP threaded program are shared between threads. The exceptions to this shared access rule are the loop index variable associated with a worksharing construct (each thread must have its own copy in order to correctly iterate through the assigned set of iterations), variables declared within a parallel region or declared within a function that is called from within a parallel region, and any other variable that is placed on the thread's stack (e.g., function parameters). For C/C++, if nested loops are used within a worksharing construct, only the loop index variable immediately preceding the construct will automatically be made private to each thread. If other variables are needed to be private to threads, such as the loop index variables for nested loops, the *private* clause can be added to many directives with a list of variables that will generate separate copies of the variables in the list for each thread. The initial value of variables that are used within the *private* clause will be undefined and they must be assigned a value before they are read within the region of use. The *firstprivate* clause will create a private copy of the variables listed, but will initialize them with the value of the shared copy; if the value of a private variable is needed outside a region, the *lastprivate* clause will create a private copy (with an undefined value) of the variables listed, but, at the end of the region, will assign to the shared copy of the variables the value that would have been last assigned to the variables in a serial execution (typically, this is the value assigned during the last iteration of a loop). Variables can appear in both a *firstprivate* (to initialize) and *lastprivate* (to carry a final value out of the region) clause on the same region.

In cases where variables must remain shared by all threads, but updates must be performed to those variables in parallel, OpenMP has synchronization constructs that can ensure **mutual exclusion** to the critical regions of code where those shared resources are accessed. The *critical* construct acts like a lock around a critical region. Only one thread may execute within a protected critical region at a time. Other threads wishing to have access to the critical region must wait until the critical region is empty.

OpenMP also has an *atomic* construct to ensure that statements will be executed in an atomic, uninterruptible manner. There is a restriction on what types of statements may be used with the *atomic* construct and, unlike the *critical* construct that can protect a block of statements, the *atomic* construct can only protect a single statement. The *single* and *master* constructs will control execution of statements within a parallel region so that one thread only will execute those statements (as opposed to allowing only one thread at a time). The former will use the first thread that encounters the construct, while the latter will allow only the master thread (the thread that executes outside the parallel regions) to execute the protected code.

A common computation is to summarize or reduce a large collection of data to a single value; for example, the sum of the data items or the maximum or minimum of the data set. To perform such a data reduction in parallel, you must allocate private copies of a temporary reduction variable to each thread, use this local copy to hold the partial results of the subset of operations assigned to the thread, then update the global copy of the reduction variable with each thread's local copy (making sure that only one thread at a time is allowed to update the shared variable). OpenMP provides a clause to denote such reduction operations and handle the details of the parallel reduction. This is accomplished with the *reduction* clause. This clause requires two things: an associative and commutative operation for combining data and a list of reduction variables. Each thread within the parallel team will receive a private copy of the reduction variables to use when executing the assigned computations. These private variables will be initialized with a value that depends on the reduction operation. At the end of the region with the reduction clause, all local copies are combined using the operation noted in the clause and the result is stored in the shared copy of the variable.

Figure 8.2 is a small application that computes an approximation of the value for pi using numerical integration and the midpoint rectangle rule. The code divides the integration range into *num_rect* intervals and computes the functional value of $4.0/(1 + x^2)$ for the midpoint of each interval (*rectangle*). The functional values (*height*) are summed up and multiplied by the width of the intervals in order to approximate the area under the curve of the function.

A pragma to parallelize the computations within the iterations of the for-loop has been added. When compiled with an OpenMP-compliant compiler, code will be inserted to spawn a team of threads, give a private copy of the *mid*, *i*, and *height* variables to each thread, divide up the iterations of the loop between the threads, and finally, when the threads are done with the assigned computations, combine the values stored in all the local copies of *height* into the shared version. This shared copy of *height* will be used to compute the pi approximation when multiplied by the *width* of the intervals.

OpenMP can also handle task parallel threading. This is accomplished using the *parallel sections* worksharing construct. Parallel sections are

```
static long num_rect=100000;
double width, area;
void main()
{   int i;
    double mid, height;
    width = 1.0/(double)num_rect;
#pragma omp parallel for private(mid)
reduction(+:height)
    for (i=0; i< num_rect; i++)
    {
        mid = (i+0.5)*width;
        height += 4.0/(1.0 + mid*mid);
    }
    area = width * height;
    printf("Pi = %f\n",area);
}
```

Figure 8.2
Sample OpenMP program.

independent blocks of code that are non-iterative. Within the sections construct, tasks are delineated by inserting a *section* construct before a block of code. These blocks of code, enclosed within curly braces if there is more than one line in the block, will be assigned to threads within the team. For example, in a three-dimensional computation, the calculation of the X, Y, and Z components could be done by calling three separate functions. To parallelize these three function calls, a *parallel sections* pragma is used to create a parallel region that will divide the enclosed block of code into independent tasks. Above each function call, a *section* construct will set the execution of each function as a task that will be assigned to one of the threads within the team.

The OpenMP specification includes a set of environment variables and API functions in order to give the programmer more control over how the application will execute. Perhaps the most useful environment variable is OMP_NUM_THREADS, which will set the number of threads to be used for the team in each parallel region. The corresponding API function to set the number of threads is *omp_set_num_threads()*. This function takes an integer parameter and will use that number of threads in the team for the next parallel region encountered. If neither of these methods is used to set the number of threads within a team, the default number will be used. This default is implementation-dependent, but is most likely the number of cores available on the system at runtime.

The OpenMP specification contains many more directives, environment variables, and API. Consult the specification document for full details.

Threading Building Blocks

Threading Building Blocks [2] (TBB) is a C++ template-based library for loop-level parallelism that concentrates on defining tasks rather than explicit threads. The components of TBB include generic parallel algorithms, concurrent containers, low-level synchronization primitives, and a task scheduler. TBB has been published as both a commercial version and an open source version [3].

Programmers using TBB can parallelize the execution of loop iterations by treating chunks of iterations as tasks and allowing the TBB Task Scheduler to determine the task sizes, number of threads to use, assignment of tasks to those threads, and how those threads are scheduled for execution. The task scheduler will give precedence to tasks that have been most recently in a core with the idea of making best use of the cache that probably contains the task's data. The Task Scheduler utilizes a task-stealing mechanism to load balance the execution.

The *parallel_for* template is used to parallelize tasks that are contained within a for-loop. The template requires two parameters: a range type over which to iterate and a body type that iterates over the range or a subrange. The range class must define a copy constructor and a destructor, *is_empty()* (returns TRUE if the range is empty) and *is_divisible()* (returns TRUE if the range can be split) methods, and a splitting constructor (to divide the range in half). Besides the lower and upper bounds for the range, the range type also requires a grain size value to determine the smallest number of iterations that should be contained within a task. The TBB library contains two pre-defined range types: *blocked_range* and *blocked_range2D*. These ranges are used for single- and two-dimensional ranges, respectively.

The body class must define a copy constructor and a destructor as well as the *operator()* method. The *operator()* method will contain a copy of the original serial loop that has been modified to run over a subrange of values that come from the range type.

The *parallel_reduce* template will iterate over a range and combine partial results computed by each task into a final (reduction) value. The range type for *parallel_reduce* has the same requirements as the *parallel_for*. The body type needs a splitting constructor and a *join()* method. The splitting constructor in the body is needed to copy read-only data required to run the loop body and to assign the identity element of the reduction operation that initializes the reduction variable. The *join()* method combines partial results of tasks based on the reduction operation being used. Multiple reductions can be computed simultaneously; for example, the minimum, maximum, and average of a data set.

Other generic parallel algorithms included in the TBB library are:

- *parallel_while*, to execute independent loop iterations with unknown or dynamically changing bounds;
- *parallel_scan*, compute the parallel prefix of a data set [4];
- *pipeline*, for data-flow pipeline patterns;
- *parallel_sort*, an iterative version of Quicksort that has been parallelized.

TBB also defines concurrent containers for hash tables, vectors, and queues. The C++ STL containers do not allow concurrent updates. In fact, between the point of testing for the presence of data within such a container and accessing that data it is possible that a thread can be interrupted; upon being restored to the CPU the condition of the container may have been changed by intervening threads in such a way that the pending data access will encounter undefined values and might corrupt the container. The TBB containers are designed for safe use with multiple threads attempting concurrent access to the containers. Not only can these containers be used in conjunction with the TBB parallel algorithms, but they can be used with native Windows or POSIX threads.

Mutex objects, on which a thread can obtain a lock and enforce mutual exclusion on critical code regions, are available within TBB. There are several different types of mutexes, of which the most common is a *spin_mutex*. A *queuing_mutex* variation that is scalable (tends to take the same amount of time regardless of the number of threads) and fair — both properties that a *spin_mutex* does not have — is the other type available. There are also reader-writer lock versions of these two mutex types, as well. None of these mutex types are reentrant. The other type of synchronization supported by TBB is atomic operations. Besides a small set of simple operators, there are atomic methods *fetch_and_store()* (update with given value and return original), *fetch_and_add()* (increment by given value and return original), and *compare_and_swap()* (if current value equals second value, update with first; always return original value).

A good resource covering TBB in more detail is from Reinders [5]. Figure 8.3 is a small application employing TBB that computes an approximation of the value for pi using numerical integration and the midpoint rectangle rule.

```
#include <iostream>
#include "tbb/parallel_reduce.h"
#include "tbb/task_scheduler_init.h"
#include "tbb/blocked_range.h"

using namespace std;
using namespace tbb;

#define NUM_THREADS 4
#define GRAIN 100
int gNumSteps =  100000;

class Pi {
  double *const delta;
public:
  double sum;
  void operator()( const blocked_range<size_t>& r ) {
    double step = *delta;
    double x;
    for (size_t i=r.begin(); i!=r.end(); ++i) {
      x = (i + 0.5)*step;
      sum = sum + 4.0 / (1.0 + x*x);
    }
  }
  Pi( Pi& x, split ) : delta(x.delta), sum(0) {}
  void join( const Pi& y ) {sum += y.sum;}
  Pi(double *const step) : delta(step), sum(0) {}
};

int main()
{
  double Val;
  double Step = 1.0 / (double)gNumSteps;
  Pi chunk((double *const)&Step);
  task_scheduler_init init(task_scheduler_init::deferred);
  init.initialize(NUM_THREADS);
  parallel_reduce(blocked_range<size_t>(0,gNumSteps,GRAIN), chunk);
  Val = chunk.sum*Step;
  cout << "The value of PI is " << Val << endl ;
  return 0;

}
```

Figure 8.3
TBB program.

Boost Threads

The Boost Thread library is a C++ class and set of methods used
for explicit thread management. The library provides a boost::thread
class which can be instantiated to provide a single thread of execution.
The thread library provides for synchronization via locks which can be

either private or shared. Condition variables and barriers are also provided by the API. For further details on Boost, consult online references [6].

MCAPI

The Multicore Association, an industry standards group, is defining a set of multicore processor-focused standards. These are termed Multicore Communication API (MCAPI), Multicore Resource Management API (MRAPI), and Multicore Task Management API (MTAPI). The goal is to create a set of OS-agnostic multicore software building blocks. The first standard, MCAPI, was released in 2008. MRAPI was released in late 2010. At the time of writing, MTAPI was under development.

The communications library, MCAPI, defines three fundamental communication types and operations. The communication types are summarized as:

- Messages — Connectionless data transfers
- Packet channels — Connected, unidirectional, arbitrary sized data transfers
- Scalar channels — Connected, unidirectional, fixed sized data transfers

MCAPI is a message-driven API targeting heterogeneous multicore. In addition, the API is envisioned to be employed in embedded systems where the topology of the cores is known at design time and thus dynamic discovery and partitioning are not a focus. The API strives to have low overhead in terms of both code size and execution performance. An implementation of MCAPI would consist of a C library (and all the requirements of linking in and using) that has been mapped to your specific heterogeneous multicore processor. Companies such as Polycore Software offer an implementation.

Employing MCAPI requires setup of the communications topology using various initialization and creation calls. The function *mcapi_initialize()* starts up the runtime on the host machines and performs necessary setup. The topology of the communications network is organized into domains, nodes, and ports. A domain is the top-level organization of the topology used for routing purposes and could be used to group together multiple cores on a chip or multiple cores on a board. A node is a thread of control and is implementation dependent; however, for an individual domain, the same definition would be assumed throughout. A port is an individual communication conduit on a given node.

The function *mcapi_endpoint_create()* is used to initialize a topology-unique entity. The functions *mcapi_sclchan_connect_i()* and *mcapi_pktchan_connect_i()* connect two endpoints to form a unidirectional commucations channel. If a user wanted bidirectional communication, a second set of endpoints and channel would need to be set up in the opposite direction. Data sending is as specified previously

```
#include <stdio.h>
#include "mcapi.h"

#define DOMAIN 0
#define NODE2 1
#define PORT1 0

mcapi_node_t node = 0;
mcapi_node_attributes_t mcapi_node_attributes;
mcapi_param_t mcapi_parameters;
mcapi_status_t err;
mcapi_sclchan_send_hndl_t chan;


void initialize_comms() {
//point to node attributes stored in memory
  mcapi_node_init_attributes(&mcapi_node_attributes, &err);

//initialize a domain with parameters and other info in memory
  mcapi_initialize(DOMAIN, node, &mcapi_node_attributes,
    &mcapi_parameters, &mcapi_info, &err);

  s_ep = mcapi_endpoint_create(PORT1, &err);
  r_ep = mcapi_endpoint_get(DOMAIN, NODE2, PORT1, 250,
        &err);
  mcapi_sclchan_connect_i(s_ep, r_ep, &req, &err);
  mcapi_sclchan_send_open_i(&chan, s_ep, &req, &err);
}

void send_val(int value)
{
  mcapi_status_t err;
  mcapi_sclchan_send_uint32(chan, value, &err);
}
```

Figure 8.4
Sample MCAPI program.

and can be one of message, packet, or a scalar payload depending on the particular connection setup. Messages which are by definition connectionless only require creation of two endpoints; no explicit connect call is required.

Figure 8.4 shows an MCAPI example where a scalar channel is setup between two endpoints. Once set up, the example's *send_val()* function illustrates how to send data on the channel. This example is simplified in that the receive-side implementation is not shown. Similar commands would be employed to set up the receive endpoint and fully configure the topology.

Conclusion

This chapter detailed synchronization and communication libraries in common use today for creating parallel embedded applications. Thread-based APIs were discussed including POSIX Threads, OpenMP, and Boost. Message-passing-based APIs such as MCAPI were detailed.

References

[1] OpenMP. <www.openmp.org>.
[2] Intel Threading Building Blocks 2.0. <http://www.intel.com/cd/software/products/asmo-na/eng/threading/294797.htm>.
[3] Intel Threading Building Blocks 2.0 for Open Source. <http://www.threadingbuildingblocks.org>.
[4] Prefix Sum. <http://en.wikipedia.org/wiki/Prefix_sum>.
[5] J. Reinders., Intel threading building blocks: Outfitting C++ for Multi-core Processor Parallelism, O'Reilly, 2007.
[6] Boost Thread API Reference. <http://www.boost.org/doc/libs/1_49_0/doc/html/thread.html>.

CHAPTER 9

Programming Languages

Gitu Jain
Software Engineer, Synopsys

Chapter Outline

Programming languages for multicore embedded systems

When writing a software application for an embedded system, the choice of programming language must produce not only an application that executes correctly, but one that does so under the resource and timing

Real World Multicore Embedded Systems.
DOI: http://dx.doi.org/10.1016/B978-0-12-416018-7.00009-2
© 2013 Elsevier Inc. All rights reserved.

constraints imposed by the device on which it runs. This device can be limited in terms of memory, battery power, data transfer bandwidth, or input/output capabilities such as a keyboard or display screen. It can have safety or portability considerations. The application may be limited by the development environment as well.

An embedded system may come with an operating system as elaborate as a regular desktop or it may not have an operating system at all. If the device has no operating system, code must be written to deal with all the low-level details of the device, usually in *assembly language*. Most embedded systems today come with an operating system that has limited or reduced functionality as compared to a desktop or laptop. If an operating system and a suitable development environment are available on the embedded system, you can use a mid- or high-level language such as C or C++. For developing web applications, *Java* or *Python* is suitable. For real-time safety-critical applications such as air traffic control systems, consider *Ada*. A hybrid of two or more languages can also be used: a high-level language for most of the complex code and assembly language for timing-critical portions and for instructions not supported by the high-level language.

In this chapter we will look at the most popular programming languages used for development of multicore embedded systems today. The languages will be presented in order of popularity. The features of the languages that support programming for embedded as well as those that support multi-processing or multi-threading will be illustrated with suitable example code throughout the sections.

C

The two most common programming languages used in embedded systems are C and *assembly language*. Assembly language lets programmers squeeze out the maximum performance from their applications in terms of speed and memory. But recent advancements in compiler technology for languages such as C have enabled compilers to generate code that is comparable to hand-written assembly language code. There is a distinct advantage in using a mid-level programming language such as C in terms of ease of development and maintenance, shorter debug cycles, testability, and portability.

C has emerged as the language of choice for many embedded system programmers because of its ability to access, modify, or update the hardware directly through language features such as *pointers* (Figure 9.1) and *bit manipulation* (Figure 9.2).

C has the ability to declare members of a data structure at the *bit level*, as shown in Figure 9.3. Members of a data structure declared in this

```
unsigned char* aPtr; /* aPtr holds address to a char value */
char varA;            /* a char variable */
aPtr = &varA;         /* load aPtr with address of variable varA */
aPtr = 1;             /* modify varA indirectly using aPtr */
```

Figure 9.1
Pointers.

```
unsigned char varA = 0; /* initial declaration of a variable */
varA ^= 0x01;           /* xor operation - flip bit 0 */
varA &= ~0x01;          /* and(not 1 (0xfe)) - clear bit 0 */
varA << 2;              /* left shift by 2 - multiply by 4 */
varA |= 0x01;           /* or operation - set bit 0 to 1 */
```

Figure 9.2
Bit manipulation.

```
struct thisIsAStruct {
      flagA : 1;        // 1 bit field
      flagB : 1;        // 1 bit field
      char varA ;       // 8 bits
      int varB ;        // 32 bits
}
struct thisIsAStruct aStruct;
aStruct.flagA = 0;
aStruct.flagB = 1;
aStruct.varA = 20;
aStruct.varB = -343;
```

Figure 9.3
Bit fields.

manner, such as *flagA* and *flagB*, can be used and addressed in exactly the same manner as other members of the data structure, such as *varA* and *varB*.

The C language provides a limited set of language features. An embedded device may need certain features that the language does not support (for example bit-wise rotation), which you may need to program in assembly. You can do that in the form of *in-line assembly* embedded in the C program (see Figure 9.8 in the Assembly section).

Dynamic memory allocation is a property of C, and other high-level programming languages, where a program can determine, at run-time, whether it needs a certain amount of memory to store a variable, and gets that memory via a system call, *malloc()*. The memory allocated is placed in the heap. Many embedded systems have limited heap or no heap at all, in which case it may be necessary to disable the dynamic memory management feature of C and do the memory allocation statically in your program. For example, you can change dynamic allocation to *static* allocation for a linked list as shown in Figure 9.4. The program prints '0 1 2'.

Another programming trick for embedded systems is to replace *recursion* by *iteration*, as recursion is very inefficient in terms of space and time due to the added cost of a function call for each recursion. See Figure 9.5 to see how recursion can be replaced with iteration in some cases.

Another rule of thumb when programming for embedded systems is to declare your variable or function argument as a *const* if it is not going to be modified. This is because these values can then be stored in ROM (read-only memory) rather than RAM (random-access memory), ROM being cheaper and more plentiful in embedded systems; see Figure 9.6.

Certain read/write compiler optimizations can cause caching that does not work if the embedded device needs to communicate with I/O peripherals. Use the *volatile* keyword in such cases. For example, in Figure 9.7, suppose a variable *cvar* is not being used anywhere other than the two lines where it is being set to 1 and 2 in the function *ControlFunc()*. The

```
struct node {
    int value;
    struct node * next;
};

/* dynamic memory allocation */
int main() {
    struct node* ll = (struct node*) malloc(sizeof(struct node)*3);
    ll[0].value = 0;   ll[0].next = &ll[1];
    ll[1].value = 1;   ll[1].next = &ll[2];
    ll[2].value = 2;   ll[2].next = NULL;

    struct node* temp = &ll[0];
    while (temp != NULL) {
        printf("%d ", temp->data);
        temp = temp->next;
    };
}

/* static memory allocation */
int main() {
    struct node ll[3];
    ll[0].value = 0; ll[0].next = &ll[1];
    ll[1].value = 1; ll[1].next = &ll[2];
    ll[2].value = 2; ll[2].next = NULL;

    struct node* temp = &ll[0];
    while (temp != NULL) {
        printf("%d ", temp->data);
        temp = temp->next;
    };
}
```

Figure 9.4
Replace dynamic memory allocation by static allocation.

lines *cvar = 1;* and *cvar = 2* are optimized away by the "smart" compiler because it thinks those values are never used and can be removed. But suppose this variable was the control line to an external I/O device and setting the memory location pointed by *cvar* told that device to start some operation and writing 2 told it to stop. Optimizing these two lines will cause this operation to be completely lost. This can be prevented by using the *volatile* keyword as shown in Figure 9.7. Note, however, that the volatile keyword, when used, may also turn off other compiler optimizations such as SIMD, loop unrolling, parallelizing and pipelining. *Use it with care for embedded multicore processors.*

```
/* Recursion */
int RecursiveFactorial(int n) {
     if (n <= 1) return 1;
     return n * RecursiveFactorial(n - 1);
}

/* Iteration */
int IterativeFactorial(int n) {
     int factorial = 1;
     if (n <= 1) return factorial;
     while (n > 1) {
          factorial *= n;
          n--;
     }
     return factorial;
}
```

Figure 9.5
Recursion vs. iteration.

```
void printString(const char *str ); /* contents of str cannot be
modified */

const char* const weekDays[] = {
     "Monday", "Tuesday", "Wednesday",
     "Thursday", "Friday"
};
```

Figure 9.6
Use of *const*.

```
int *cvar;
void ControlFunc( ) {
     *cvar = 1;
        ...
     *cvar = 2;
}

change to:

volatile int *cvar;
```

Figure 9.7
Use of *volatile*.

Multi-threading support in C

C11 is the new C language standard published in 2011, and it has added multi-threading support directly in the language in the form of a library <threads.h>. It defines macros, declares types, enumeration constants, and functions that support multiple threads of execution. For more information, you can find the published standard at [1].

If you are using an older version of C, you can add multi-threading to your programs by using a standard threading library such as POSIX Threads, Win32 Threads or Boost Threads (for portability). Refer to the Synchronization libraries chapter for further details.

Assembly language

Assembly language enables programmers to optimize their applications in terms of speed and memory requirements, which is desirable in embedded systems. Most microcontrollers used to be programmed in assembly language, and it is still used by embedded system programmers today as a standalone language or in conjunction with C. Even though assembly is not portable (it is very architecture specific) and is hard to code, test, debug and maintain, there are several cases in which knowledge and use of assembly comes in handy:

- In some low-cost embedded systems with no support for C compilers, such as small microcontrollers. Assembly is the only programming interface available to you.
- Situations where the high-level language compiler does not provide support for all the hardware features. You will need to write the functionality in assembly, either in-line or as separate functions. For example, in Figure 9.8, you can do bit-wise rotation using assembly, something C does not support.

```
asm("mov %[result],%[value],ror#1":[result]"=r"(y):[value]"r"(x));
```

Figure 9.8
Rotating bits example in assembly.

- Situations where you need to squeeze every last bit of real-time performance out of the device, for example, when writing digital signal processing (DSP) applications.
- For writing device drivers where the timing needs to be strictly controlled.
- Programmers frequently use the assembly output of the C compiler to debug, hand-edit, and hand-optimize the assembly code to maximize the performance in ways the compiler cannot. This lets them go down to architectural details of their code and count instructions to obtain accurate timing characteristics of their application and look at the execution of their code at the machine instruction level.

When none of the above holds true, you should use a higher-level language such as C, as the benefits of using it far outweigh any performance penalties you may incur, which may not be as heavy as you think given the advances in compiler technologies today.

Writing assembly code for embedded systems usually consists of either *inline assembly*, where short assembly routines are embedded directly in C or C++ code and compiled using a common C compiler, or *linked assembly*, where all the assembly routines are isolated in a ".asm" file, assembled using a separate assembler, and linked with other object files, possibly from C/C++ code.

This ability to mix C and assembly can help you write programs where you can do most of your programming in C, and write only the performance or memory-critical portions in assembly if needed.

Multi-threading and assembly

If you need to create threads from assembly code, you should ideally call the C/C++ thread routines supported on your platform to do so. If you are not using a language such as C/C++, then you can create threads using system calls. As far as assembly is concerned, there is no difference between one thread and multiple threads; each has its own register set. The operating system takes care of the scheduling of threads.

C++

C++ is a high-level object-oriented language that offers embedded system programmers certain features not found in the C programming language. However, caution must be exercised when using these features as they sometimes come at a price. C++ programs run more slowly and produce larger executable code than C. Embedded system programmers usually use C++ in a reduced form since it has several features that are bulky, inefficient, or inappropriate for use in the resource-constrained environments normally found on such devices.

Features of C++ that work well for embedded systems

Class definition

Definition of a *class* includes a list of public, private, and protected data members and functions. The C++ compiler is able to use the *public* and *private* keywords to determine which method calls and data accesses are allowed and disallowed at compile time, so there is no run-time penalty. Good object-oriented use of classes can lead to modular designs that lead to well-designed applications.

Constructors and destructors

A *constructor* is called every time an object of a class type is created. A *destructor* is called every time the object goes out of scope. There is a slight penalty associated with the calls, but it eliminates an entire class of C programming errors having to do with uninitialized data structures and memory leaks. This feature also hides awkward initialization sequences that are associated with complex classes. In Figure 9.9 *String* is a class declaration with two constructors and a destructor.

Function and operator overloading

With *function overloading,* functions having the same names but different parameters are each assigned unique names during the compilation process. The compiler alters the function name each time it appears in your program, and the linker matches them up appropriately. This does not affect performance (Figure 9.10).

```
class String
{
public:
  String() {            //constructor with no arguments
      str = NULL;
      size = 0;
  }
  String(int len) {     //constructor with one argument
      str = new char[len];
      size = size;
  }
  ~String() {           //destructor
      delete [] str;
  }
private:
      char *str;
      int size;
};
```

Figure 9.9
String class.

```
#include <iostream>
using namespace std;

void print(int i) {
      cout << i  << " is an integer." << endl;
}
void print(double f) {
      cout << f << " is a float." << endl;
}

int main() {
      print(8);
      print(3.14);
}

The following is the output:

8 is an integer.
3.14 is a float.
```

Figure 9.10
Function overloading.

Similarly, with *operator overloading*, whenever the compiler sees
an operator declaration such as the ' + ' operator in the example
in Figure 9.11, it simply replaces it with the appropriate function call.

```
#include <iostream>

class CVector {
public:
        CVector () {}
        CVector (int a, int b) : x(a), y(b) {}
        CVector operator+(CVector vec);
private:
        int x,y;
};

CVector CVector::operator+(CVector vec) {
        CVector temp;
        temp.x = x + vec.x;
        temp.y = y + vec.y;
        return (temp);
}

int main () {
        CVector a (3,1);
        CVector b (1,2);
        CVector c = a + b;
        cout << c.x << "," << c.y;
        return 0;
}

The following is the output:
4,3
```

Figure 9.11
Operator overloading.

Virtual functions (Polymorphism)

C++ would not be a true object-oriented language without virtual functions and polymorphism. Polymorphism is the characteristic of being able to assign a different meaning or usage to something in different contexts — specifically, to allow an entity such as a variable, a function, or an object to have more than one form. In C++, virtual functions are defined in base classes with the same interface, and are over-ridden in derived classes with different implementations.

Objects with virtual functions have an additional pointer called a *vptr* (virtual table pointer). The *vptr* points to a virtual table of pointers called a *vtbl*. Each pointer points to a virtual function in the class. When a virtual function is invoked on an object, the actual function called at run-time is determined by following the object's *vptr* to its *vtbl* and looking up the appropriate function pointer.

For example, in Figure 9.12, the call to *poly[0]->area()* will access the *vtbl* of the *Rectangle* class as poly[0] is an object of type *Rectangle*, and call the appropriate virtual function *area()* defined in that class. Virtual functions have a reasonable cost/benefit ratio for large classes since they need only one additional memory lookup before being called.

In addition to these, the techniques previously outlined for C apply equally well to C++. These include *pointers* (Figure 9.1) and *bit manipulation* (Figure 9.2) to access hardware directly; *in-line assembly* for hardware features not supported by the language (Figure 9.8); *static* instead of *dynamic* memory allocation (Figure 9.4); replacing *recursion* with *iteration* (Figure 9.5); and *const* (Figure 9.6) or *volatile* (Figure 9.7) declarations where appropriate.

Features of C++ that do not work well for embedded systems

The features of C++ that are too expensive for embedded systems are generally the ones that cause a dramatic increase in run-time or have negative impact on code size.

Templates

Templates in C++ allow a single class to handle several different data types. For example, you can use function templates to apply the same algorithm to different data types. Another good use is for container classes such as vectors, lists, or queues (Figure 9.13). Templates are implemented at compile time by making a complete copy of the code for each data type inserted for the template type. It is not hard to see how this can lead to code size explosion, which could overflow the memory of a smaller embedded system.

You can use templates if your compiler does a good job of compiling them efficiently as they reduce the effort of coding, testing, and debugging for different data types to a single set of code.

```
// base class for Polygon
#include <iostream>
using namespace std;

class Polygon {
protected:
     int width;                    // width of polygon
     int height;                   // height of polygon
public:
     Polygon(int w, int h) : width (w), height(h) {} //constructor
     virtual int area () = 0;      // pure virtual function to
calculate area
  };

// Derived Class for Rectangle
class Rectangle: public Polygon {
public:
     Rectangle(int w, int h) : Polygon(w, h);
     virtual int area () {
          return (width * height);
     }
  };

// Derived class for Triangle
class Triangle: public Polygon {
public:
     Triangle(int w, int h) : Polygon(w, h);
     virtual int area () {
          return ((width * height) / 2);
     }
  };

int main () {
     Polygon *poly[2];
     poly[0] = new Rectangle(4,5);
     poly[1] = new Triangle(4,5);

     cout << poly[0]->area() << endl;    // will call the area()
function in Rectangle
     cout << poly[1]->area() << endl;    // will call the area()
function in Triangle
     return 0;
}

The following is the output:
20
10
```

Figure 9.12
Polymorphism and virtual functions.

```
// Class Queue using templates
template <class T>
class MyQueue
{
public:
        void Add(T const &d);
            ...
private:
        std::vector<T> data;
};

// defining member functions
template <class T>
void MyQueue<T> ::Add(T const &d)
{
        data.push_back(d);
}

// C++ class templates usage
void main()
{
        MyQueue<int> qint;
        MyQueue<float> qfloat;
        MyQueue<String> qstr;
        qint.Add(1);
        qfloat.Add(3.14);
        qstr.Add("thisisaString");
            ...
}
```

Figure 9.13
Templates.

Exceptions

In C++, *exception handling* separates the error handling code from the code written to handle the tasks of the program. Doing so makes reading and writing the code easier. Exception handling propagates the exceptions up the call stack. An exception is *thrown* at the place where some error or abnormal condition is detected. This will cause the normal program flow to be aborted. In handled exceptions, execution of the program will resume at a designated block of code, called a *catch* block (Figure 9.14). Exception handling in C++ is not efficient in terms of both space and run-time, so it's best to avoid using this feature in embedded systems.

```
#include <iostream>
using namespace std;

int main ()
{
      try {
            throw 20;
      }
      catch (int exception_no) {
            cout << "An exception occurred - " << exception_no << endl;
      }
      return 0;
}

The following is the output:

An exception occurred - 20
```

Figure 9.14
Exceptions.

Run-time type identification

Run-time type identification (RTTI) lets you find the exact type of an object when you have only a pointer or reference to the base type. The *dynamic_cast* < > operation and *typeid* operator in C++ are part of RTTI. With dynamic_cast < >, the program converts the base class pointer to a derived class pointer and allows the derived class members to be called. The typeid operator is used to determine the class of an object at run-time. There is a space and runtime cost for using this feature and it is recommended that, for embedded systems, you should disable RTTI for the sake of more efficient code.

Also be aware that C++ does not allow you to declare bit fields in your data structures like C does (Figure 9.3), so memory alignment can be an issue and can result in an increase in the footprint of your code. Locality of reference in the data cache will also be affected, as some memory in cache is never referenced after being fetched.

Multi-threading support in C++

C++11 is the new C++ language standard published in 2011, and has added multi-threading support directly into the language in the form of a < *thread* > library. It provides functions for managing threads, mutual exclusion (mutex) management, generic locking algorithms, call once functions, and condition variables. For more information, you can refer to the published standard at [2].

If you are using an older version of C++, you can add multi-threading to your programs by using a standard threading library such as POSIX Threads, Win32 Threads or Boost Threads (for portability). Refer to the Synchronization libraries chapter for further details.

Java

Java is a high-level language with which you can write programs that can execute on a variety of platforms. Java was designed to produce code that is simpler to write and easier to maintain than C or C++. Java applications are typically compiled into *bytecode* (files with a *.class* extension) that can run on any computer architecture with the help of a Java interpreter and run-time environment called a Java Virtual Machine (JVM). Bytecode can also be converted directly into machine language instructions by a just-in-time compiler (JIT).

Java is one of the most popular programming languages in use today, and has become popular in high-end embedded systems such as smartphones, PDAs, and gaming consoles. It is well suited for web applications.

Java is an object-oriented language with features very similar to C++, such as class structure, polymorphism, and inheritance. There are some differences and improvements as outlined below.

- Java does not support C/C++ style pointers; it uses references instead. This prevents errors caused by using pointers to trick the compiler into storing any type of data at an address.
- It provides a technique known as "native methods", where C/C++ or assembly code can be called directly from Java to manipulate the hardware registers and memory directly using pointers. This can sometimes be useful in embedded systems.
- It has automatic garbage collection that simplifies dynamic memory management and eliminates memory leaks.
- Multiple class inheritance in C++ is replaced with *interfaces* in Java.

- It has support for automatic bounds-checking that prevents writing or reading past the end of an array.
- It has a fixed size for primitive data types; for example, an *int* in Java is always 32 bits, unlike C or C++ where it can be 32 or 64 depending upon the compiler.
- All test conditions must return either true or false. For example, the error *if (x = 3)*, which should correctly be *if(x = = 3)*, will be detected at compile-time, while it would be allowed by a C or C++ compiler as unorthodox but acceptable.
- It has built-in support for strings and string manipulation that allows statements like *"Hello"* + *"World"*.
- It has built-in multi-threading support that makes applications portable by providing consistent thread and synchronization APIs across all operating systems.
- It has many useful standardized libraries for graphics, networking, math, containers, internationalization, and other specific domains.

Multi-threading support in Java

Java defines two ways in which you can create a new thread of execution — you can provide a *Runnable* object (Figure 9.15); or you can extend the *Thread* class (Figure 9.16).

Java provides thread communication and synchronization mechanisms through the use of *monitors*. Java associates a monitor with each object. The monitor enforces mutually exclusive access to *synchronized* methods defined in the object such as shown in Figure 9.17. When a thread calls a

```
public class IamRunnable Implements Runnable {
      public void run() {
            System.out.println("Hello I am a Runnable Thread.");
      }
      public static void main(String args[]) {
            (new Thread(new IamRunnable())).start();
      }
};
```

Figure 9.15
Implement *Runnable* interface.

```
public class IamAThread extends Thread {
      public void run() {
            System.out.println("I am a Thread.");
      }
      public static void main(String args[]) {
            (new IamAThread()).start();
      }
};
```

Figure 9.16
Extend class *Thread.*

```
class Counter
{
      private int count = 0;
      public void synchronized Increment() {
            int n = count;
            count = n+1;
      }
}
```

Figure 9.17
Synchronized method in Java.

synchronized method, the JVM checks the monitor for that object, and, if the monitor is free, ownership is assigned to the calling thread, which proceeds with execution. If not, the calling thread has to wait until the monitor is freed by the thread currently owning it. Note that Java monitors are not like traditional critical sections, as they are associated with objects and not blocks of code. For example, two threads may execute the method *Increment()* concurrently if they are invoking it on different objects.

Some of the classes and containers in the **java.util.concurrent** package provide atomic methods that do not rely on synchronization and are still thread-safe.

The following methods can be used for *inter-thread communication.*

- **wait( ):** this method tells the calling thread to give up the monitor and go to sleep until some other thread enters the same monitor and calls notify( ).
- **notify( ):** this method wakes up the first thread that called wait( ) on the same object.

- **notifyAll( ):** this method wakes up all the threads that called wait( ) on the same object. The highest priority thread will run first.

See the producer-consumer example in Figure 9.18. The consumer will keep getting values in *itemCount* as long as the producer keeps putting it there. Notice the use of the *synchronized* keyword to control access to the shared *Buffer* object and the use of *wait()* and *notify()* for inter-thread communication.

In Java, a semaphore is created using the **java.util.concurrent. Semaphore** class; see Figure 9.19. You can find out more about semaphores in the Synchronization chapter.

Running Java programs requires a JVM, which can take up significant resources and slow down performance. Because of this "managed run-time", Java is considered a costly language and is not suitable for most embedded devices which have limited resources such as memory and battery power. It is also not suitable for real-time and safety-critical applications. For this reason, Java is only popular in high-end embedded devices such as mobile phones, where portability and the ability to browse the internet are needed.

There have been attempts to tailor Java for the embedded developers' community called Embedded Java, where Java can be run in three ways:

- Run using JVM with a JIT compiler − JIT compilers use too much memory.
- Run using a special-purpose JVM and core libraries − stripped-down version.
- Run compiled Java (instead of interpreted) − best run-time behavior but not portable.

Many vendors offer run-time interpreters, environments and compilers for Embedded Java with stripped-down versions of the core Java libraries and run-time environment.

Python

Python is a general-purpose, highly flexible, high-level programming language that is gaining popularity due to its ease of use and ability to

```java
class Buffer {
  int itemCount=0;
  boolean value = false;
  synchronized int get() {
    if (value==false)
      try {
        wait();
      }
      catch (InterruptedException e) {
        System.out.println("InterruptedException caught");
      }
      System.out.println("Getting " + itemCount);
      value=false;
      notify();
      return itemCount;
  }
  synchronized void put(int itemCount) {
    if (value==true)
      try {
        wait();
      }
      catch (InterruptedException e) {
        System.out.println("InterruptedException caught");
      }
      this.itemCount=itemCount;
      System.out.println("Putting " + itemCount);
      value=true;
      notify();
  }
}

class Producer extends Thread {
  Buffer buf;
  Producer(Buffer buf) {
    this.buf=buf;
    this.start();
  }
  public void run() {
    int i=0;
    buf.put(++i);
  }
}

class Consumer extends Thread {
  Buffer buf;
  Consumer(Buffer buf) {
    this.buf=buf;
    this.start();
  }
  public void run() {
    buf.get();
  }
}

public class ProducerConsumer {
  public static void main(String[] args)
  {
    Buffer buf=new Buffer();
    new Producer(buf);
    new Consumer(buf);
  }
}
```

Figure 9.18

Inter-thread communication using producer-consumer example.

```
static Semaphore sVar = new Semaphore(2); // initialized to 2
...
sVar.acquire();
criticalCode();                           // critical section
sVar.release();
...
```

Figure 9.19
Use of semaphore in Java.

create custom code quickly. It supports multiple programming paradigms, from object-oriented styles to use as a scripting language for web applications. Python is much slower than programming languages such as C, so it should not be used for timing-critical applications. Python also requires a lot of memory or disk space to run, so it cannot be your language of choice for smaller embedded systems.

Python can be used to create applications for high-end embedded systems such as smartphones, which host many web applications.

Python is used by embedded system developers to create custom prototype code quickly, which is one of the strong points of the language. Python is very easy to learn for people of various programming backgrounds, such as Java, C or Perl. Python leads to the creation of highly readable, compact, and well-structured code. Python code, compiled and running on PCs or emulators, can be used to test applications.

Multi-threading support in Python

The *threading.py* library provides high level threading interfaces on top of the lower-level *_thread* module, which is based upon *POSIX Threads*. This module defines the following functions and objects, modeled after Java's **Thread** class and **Runnable** interface:

- Thread — this class represents an activity that is run in a separate thread of control. There are two ways to create a thread: by implementing the *Runnable* interface, or by overriding the run() method by deriving a subclass from the *Thread* class.
- Lock — a *primitive lock* is the lowest synchronization primitive available in Python. It has two basic methods, acquire() and release().

- RLock — a *reentrant lock* is a synchronization primitive that may be acquired multiple times by the same thread. A thread calls acquire() to lock and release() to unlock. Acquire()/release() call pairs may be nested.
- Condition — a condition variable has acquire() and release() methods that call the corresponding methods of the associated lock. It also provides wait(), notify(), and notifyAll() methods. The wait() method releases the lock, and then blocks until it is awakened by a notify() or notifyAll() call for the same condition variable in another thread. Once awakened, it re-acquires the lock and returns.
- Semaphore — a semaphore manages an internal counter which is decremented by each acquire() call and incremented by each release() call. The counter can never go below zero; when acquire() finds that it is zero, it blocks, waiting until some other thread calls release().
- Event — This is one of the simplest mechanisms for communication between threads: one thread signals an event and another thread waits for it.

Ada

Ada is a high-level, object-oriented programming language that has built-in support for parallelism. In the late 1980s, the US Department of Defense (DoD) mandated the use of Ada for all its software projects. Today, Ada is still used for the majority of DoD's projects, although the mandate has been lifted and many more languages are in use, such as C, Fortran, and C++. Ada is also used in many embedded and real-time safety-critical commercial systems such as air traffic control, railway transport, and banking systems.

Ada's features include exception handling, concurrency (tasks), modularization (packages), hierarchical namespaces, object-oriented programming, and generic templates. Unlike other programming languages such as C++ or Java, dynamic allocation for data such as arrays and records is not performed unless explicitly requested by the programmer; this is a useful characteristic for embedded systems.

Ada provides a large number of compile-time and run-time checks that help produce high-quality, maintainable software applications. For example, you can specify a range of values for a scalar variable in Ada, where an attempt to assign an out-of-range value will be detected. Ada compilers also help detect potential deadlocks, a software bug characteristic of multi-threaded applications. Because of all these run-time checks the performance of Ada applications can suffer, making this a poor choice for performance-critical applications (for which assembly or C might be better suited). The performance can, however, be improved by turning off some of these checks.

Concurrency support in Ada

The unit of concurrency in Ada is the *task*, and tasks generally interact with each other through encapsulated data (*protected objects*) or via direct communication (*rendezvous*). A good comparison of the real-time features of Ada and Java are presented in the paper in [3]. Section 7 compares the mutual exclusion mechanisms of Ada and Java and section 8 compares the task/thread synchronization and communication controls provided by the two languages. Ada has predictable and portable thread scheduling that works well for real-time applications.

- Task is the basic unit of concurrency in Ada; it is equivalent to a thread. It has a declaration and a body. Tasks begin as soon as they are created. They can pass messages and can share memory. Two or more tasks communicate by sending messages using *entry* and *accept* methods.
- Rendezvous represents synchronization via message passing where sending and receiving tasks have to wait.
- Protected Objects encapsulate data and operations. The operations then have automatic mutual exclusion. Guards can be placed on these operations for conditional synchronization.

Summary

There are a wide variety of programming languages available to developers today. Not all of them are suitable for developing applications for embedded systems and, in particular, for *multicore* embedded systems. There may be resource, run-time, and safety constraints that dictate the choice of language. In addition, you have to take into account the ease of development and the familiarity of the programmer with the language of choice. If you are developing for a multicore embedded system, you have to choose a language that supports efficient development for these systems, especially in terms of memory synchronization and debugging issues.

In this chapter we presented the top choices for programming languages for multicore embedded systems with suitable examples illustrating how you can apply them to your needs. Some languages were designed with inherent concurrency built into the language, such as Ada and Java. Others have concurrency support added late, such as C and C++, even though these languages have been used to do multi-threaded programming for several decades using thread libraries. Some offer function-level locking, such as C or C++, while others offer object-level locking, as in Java. Some offer reliable scheduling of threads, such as Ada. In the end, all the mid- to high-level languages presented here — C, C++, Java, and Ada — provide support for developing applications on multicore embedded systems.

You will need to consider all these factors carefully when choosing a programming language. In addition, the language features which affect performance, memory requirements, safety, and ability to access and modify hardware directly, ease of use, and popularity will also be critical when making a decision.

References

[1] C11 the C language standard, <http://www.open-std.org/jtc1/sc22/wg14/www/docs/n1570.pdf>, 2011.
[2] C++11 the C++ language standard, <http://en.cppreference.com/w/cpp/thread>, 2011.
[3] B.M. Brosgol, A comparison of the concurrency and real-time features of Ada 95 and Java. Ada UK '98 Conference, October 1998. <ftp://ftp.cs.kuleuven.ac.be/pub/Ada-Belgium/mirrors/gnu-ada/jgnat/papers/ada-java-concurrency-comparison.pdf>.

CHAPTER 10

Tools

Kenn Luecke

Boeing Test & Evaluation in the Vehicle Management Subsystem Labs, Peters, MO, USA

Chapter Outline

Real World Multicore Embedded Systems.
DOI: http://dx.doi.org/10.1016/B978-0-12-416018-7.00010-9
© 2013 Elsevier Inc. All rights reserved.

Introduction

Finding adequate software tools continues to be one of the biggest challenges for software developers working with multicore microprocessors. Too often, software tools aimed at the multicore microprocessor market are developed by a hardware vendor to support only their products. In addition, some software vendors' tools only support certain languages. C, C++, Java, and Fortran are supported with the most tools. In addition, there is very little support for functional programming languages that some developers feel are better suited for multicore microprocessors than traditional modular and object-oriented languages.

Many software tools provide limited results or require greater manual labor than what is expected when compared to single-core microprocessors. Unfortunately, it is even more difficult stringing together a tool suite for an entire software development lifecycle starting with requirements generation through architecture development, code generation, and ending with system integration for an entire application.

For example, often the output from one tool cannot be directly fed as input into another tool. Hence, one tool's output may require labor-intensive manual reformatting prior to inputting the data into the next tool, thus forcing a software developer to spend more time performing repetitive formatting tasks as opposed to more important software engineering tasks. Also, some tool vendors repackaged their multi-microprocessor software tools with a few modifications to handle inter-core processing as tools for multicore software development. In theory, this is a good approach, but often the execution does not succeed due to underestimating the true effort along with the actual differences between supporting multiprocessing environments and multicore processing environments.

However, the good news is that a few software development and debug tools have entered the market that are mature, are focused on products from multiple vendors, and provide a greater automation so the developer can spend more time on engineering intensive tasks. The rest of this chapter will discuss several software development, communication, and debugging tools for multicore processing. Most of the information below comes from tool investigations that the author has performed, demonstrations that the author has witnessed, or the tool vendors themselves.

Real-Time operating systems (RTOS)

Functions performed by an RTOS become more layered and more complex in a multicore environment. An RTOS that supports multicore processing must schedule tasks dynamically and transparently between microprocessors to efficiently balance workloads between the available microprocessors in a symmetric multiprocessing (SMP) environment. For RTOSes the task is even more difficult for preserving determinism and real-time latency. The RTOS design has to adapt in the areas of scheduling, interrupt handling, synchronization and load balancing.

Many RTOS vendors have developed RTOSes that support both asymmetric multiprocessing (AMP) and SMP with tools for developing and debugging applications to run on multicore microprocessors. Most of these tools assist in parallelizing software, developing thread safe code, and debugging issues with multicore microprocessors. Each of these tools allows the developer to "see under the covers" of their application while it is executing.

DEOS by DDC-I

DDC-I develops the DEOS Time and Space Partitioned (TSP) RTOS and software application development tools. DEOS 7.2 and later RTOSes have the following embedded tools for multicore debugging:

- Memory usage – assists developers in pinpointing common memory errors.
- Synchronization – provides assistance with synchronizing a software application's tasks to ensure the overall application's correctness.
- System profiling – assists developers by showing "what is taking place underneath the hood" of a software application. Provides insights into your application that would be difficult if not impossible without system profiling.
- Communication debugging mechanisms – isolate applications from changes in I/O format and bus hardware/source.

- Integration tool – provides insight for the developer into what physical resources are required at execution to prevent resource contention.

Unlike most RTOS vendors, DDC-I does not emphasize virtualization support for their RTOS products. DDC-I also provides the HeartOS microkernel RTOS with multicore support for Digital Signal Processors (DSP). For more information, see http://www.ddci.com.

Enea OSE

OSE implements a microkernel and supports multicore embedded processors. OSE provides the following support for multicore processors:

- Black-box recorder – traces all events that fired leading up to an application fault. This tool can be used during application development and release.
- Log analyzer – contains advanced logging and post-processing capabilities. It provides assistance for collecting and refining data, searching and filtering logs, and managing multiple data logs for improved analysis and debugging for software applications.

For more information, see http://www.enea.com/software/products/rtos/ose.

Express Logic ThreadX

ThreadX implements an extremely small kernel for supporting multicore embedded processors. ThreadX provides the following support for multicore processors:

- Execution profiling – provides the Execution Profile Kit (EPK) to dynamically track system threads, idle conditions, and interrupt service routines (ISR). It is useful for optimizing the application performance and debugging threading problems.
- Run-time performance metrics – generates optional metrics for threaded objects. Metrics information is provided for thread resumptions, suspensions, preemptions, inversions, and time outs. This information is required for debugging threading errors along with improving overall application performance.

- Memory utilization — moves stacks into fast memory areas for high-priority threads.
- System trace — stores event information for analyzing the system's events. It is used in conjunction with the GUI-based TraceX tool for the developer's use.

For more information, see http://www.rtos.com.

Green Hills Integrity

Green Hills Integrity 10.0 and later RTOSes have the following embedded tools for multicore debugging:

- TimeMachine — allows developers to explore program execution history. TimeMachine captures historical data such as register and memory values, finds specific data points for additional visibility into your applications, and allows developers to examine data structures. TimeMachine allows developers to debug backward and forward in time.
- Path analyzer — provides a view of an application's call stack over time, assists developers in identifying where an application diverts from the expected execution path.
- Event analyzer — displays a view of an operating system's events over time, assists developers in tracking down deadlocks between task and other undesirable system behavior with a GUI interface.
- Performance profiler — contains code coverage tools that use TimeMachine data to determine which code segments in your application have not been exercised in your testing.
- Probe — an advanced hardware debug device that connects to the onboard debug ports present on most modern microprocessors. The Probe allows developers to use all Green Hills debugging tools on the microprocessor itself.
- TraceEdge — collects trace data from nearly all 32-bit and 64-bit microprocessors, even those without integrated trace hardware. Developers can use this trace history with Green Hills Software's MULTI TimeMachine tools to visualize and replay their software's execution to assist with debugging.

For more information, see http://www.ghs.com.

LynuxWorks

LynuxWorks LynxOS 5.0 and later RTOSes have the following embedded tools for multicore debugging:

- SpyKer — provides support for trace history, path and event analysis and performance profiling. SpyKer has special features to support SMP. Spyker provides the developer a view into program execution and timing of events and assists the developer with fine-tuning system performance.
- Luminosity — includes the Cross Process Viewer that can monitor memory utilization and microprocessor utilization per process.
- LynxInsure++ — pinpoints memory leaks and allocation errors resulting from programming errors.
- TotalView — allows software developers to debug multi-process, multi-threaded, and multi-processor systems in the same manner as debugging a single computer.

For more information, see http://www.lynuxworks.com.

Mentor Graphics Nucleus

Nucleus implements a kernel and supports multicore embedded processors. Nucleus, through its incorporation of Sourcery Codebench tools, provides the following support for multicore processors:

- Sourcery analyzer — captures and displays system analysis views for the software application under test. It captures various multicore-related issues such as race conditions, deadlocks, non-deterministic behavior, and under-utilization of individual cores within a multicore system.
- Sourcery probe — assists developers with analyzing memory usage and processor registers dynamically. Allows developers to dynamically analyze the application's execution control.

For more information, see http://www.mentor.com/embedded-software/nucleus.

MontaVista

MontaVista develops embedded Linux system software and software application development tools. MontaVista Carrier Grade Edition (CGE) 5.0 and later RTOSes have the following embedded tools for multicore debugging:

- Flight Recorder — provides a detailed view including a scheduler history, to provide a crash dump snapshot. Flight Recorder provides developers a system post-crash analysis by gathering data prior and during the system crash.
- Performance profiling — a system measurement tool that provides a detailed view of all system activity including operating system, drivers, application, file systems, and stacks.
- Memory usage — a system measurement tool that assists developers in visualizing program memory usage.
- Execution tracing — a system measurement tool that provides a history of the application's execution.
- System profiler — provides a detailed view of all system activity including operating system, drivers, application, file systems, and stacks.
- Memory analysis — assists developers in visualizing program memory usage. It can quickly detect buffer overruns, invalid de-allocations, memory leaks, and other common memory errors.

For more information, see http://www.mvista.com.

QNX

QNX Neutrino RTOS incorporates a microkernel and the Momentics Development Suite. The Neutrino RTOSes which support embedded multicore processors have the following embedded tools for multicore debugging:

- System profiler — provides a detailed view of all system activity including operating system, drivers, application, file systems, and stacks. Also, generates time stamp interactions between the various processing elements.

- Memory analysis — assists developers in visualizing program memory usage. It can quickly detect buffer overruns, invalid deallocations, memory leaks, and other common memory errors via a GUI interface. It also provides a dynamic history of memory usage for identifying memory trends.
- Application profiler — analyzes microprocessor usage for multiple processes and targets. This tool can also be used to assist developers in finding portions of code that are good candidates for parallelization. In addition, performs post-execution profiling and analysis accessing/loading historical statistical files.
- Event analyzer — displays a view of an operating system's events over time, assists developers in tracking down deadlocks between task and other undesirable system behavior with a GUI interface.
- Thread analyzer — determines current thread states and why specific threads are executing.

For more information, see http://www.qnx.com.

Wind River VxWorks

VxWorks was modified for the 6.X releases to provide a multicore API. The 6.X releases support many multicore Intel x86 cores and multicore Freescale PPC microprocessors. The following tools are embedded in VxWorks 6.5 and later RTOSes:

- System viewer — a graphic user interface that shows the dynamic operation of a software system including event analysis.
- Performance profiler — specifies the individual routines within the application that are consuming CPU cycles. Helps with detecting performance bottlenecks. Formerly known as ProfileScope.
- Memory analyzer — assists developers in visualizing program memory usage. It can detect buffer overruns, invalid de-allocations, memory leaks, and other common memory errors via a GUI interface.
- Data monitor — used to examine variables, data structures, and memory locations in your application. Allows developers to catch

any out-of-range values for a variable. Formerly known as Stethoscope.

- Code coverage analyzer – analyzes a software applications code base to determine which code blocks have not been exercised in your testing. Formerly known as CoverageScope.
- Function tracer – traces the code execution in real time by monitoring function call sequences as your code executes. Formerly known as TraceScope.

For more information, see http://www.windriver.com.

Communication tools

PolyCore Software

PolyCore Software provides a multicore software communications framework. The PolyCore software tool suite, which is Eclipse based, consists of:

- Poly-Mapper, which provides a GUI that allows developers to map software communications across multiple cores using XML commands.
- Poly-Generator converts the Poly-Mapper XML commands to C source code files.
- Poly-Messenger contains a software communications library for distributing processing on multiple cores.
- Poly-Inspector allows developers to inspect and analyze applications for communication "hot spots". A "hot spot" occurs where one or more cores have an increased amount of processing activity while other cores are idle.

See Figure 10.1 for a visual depiction of how the PolyCore software tools integrate together.

The PolyCore Software suite of tools is used for generating C code files that can be ported to multiple OSs. The code files execute on microprocessor chips such as x86, PPC, ARM, and MIPS. For more information, see http://polycoresoftware.com.

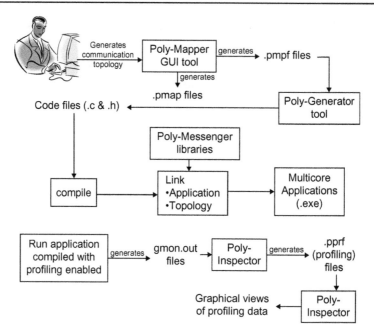

Figure 10.1
PolyCore Software tool chain.

Enea Linx

Enea's Linx is an interprocess communications (IPC) service for multicore DSPs, microcontrollers, and embedded system microprocessors. Linx allows for both homogeneous and heterogeneous cores on an embedded system to communicate with common device drivers, device handlers, and other external devices. Linx is open-source software that is not dependent on the underlying hardware, interconnect, or operating system. It supports both data- and control-plane software applications over reliable and unreliable media. Linx incorporates a gateway server which allows existing software applications to use the gateway client API to communicate with a Linx cluster of applications without rewriting your applications. Linx uses a layered architecture that supports the separation of software applications from the underlying hardware, thereby enabling the software application to be more scalable. For more information, see http://www.enea.com/linx.

Parallelizing serial software tools

There are many varied toolsets for parallelizing serial software. In general, the tool suites that parallelize serial software have the most robust capabilities of any category of tools that support software development for multicore microprocessors.

CriticalBlue Prism

The objective of CriticalBlue's Prism tool is to provide analysis and an exploration and verification environment for embedded software development using multicore architectures. Prism implements a GUI to assist with a developer's multi-threaded software development. The GUI provides multiple views for a look "under the covers" from multiple angles.

The multiple graphical views depict the interactions between threads and dependencies in your multicore system, and display cache analysis along with the microprocessor pipeline, and any data races that may be inherent in your system. All of these views provide a detailed analysis of your application.

The tool works for many microprocessor chips including x86, PowerPC (PPC), MIPS, ARM, and the Android operating system. The author has tested this product and found it to be effective for migrating an application from a single core to a multicore microprocessor. Prism is discussed in more detail in the Partitioning chapter. For more information, see http://www.criticalblue.com.

Vector Fabrics

Vector Fabrics provides the Pareon tool for parallelizing serial software for Intel x86 and the ARM Cortex-A multicore embedded platforms. The Pareon tool:

• analyzes software applications on a model of your multicore hardware and provides feedback to the application developer about loops and data dependencies that can be parallelized, processing and memory bottlenecks;

- identifies code constructs and designs that inhibit parallelism;
- computes performance estimates for parallelizing software applications;
- provides recipes and strategies for parallelizing serial code from the previous analysis tasks. The recipes guide the developer in implementing the parallelized code while avoiding bugs and rework.

The Vector Fabric tools use a GUI for easier use by the software developer. These tools have automated the tedious parts of determining which serial code sections are good candidates for parallelization while leaving decisions and implementation to the programmer. Hence, the developer is still in control over an application's parallelization effort. These tools also have the capability to optimize your parallel software for faster performance. Pareon is discussed in more detail in the Partitioning chapter. For more information, see http://www.vectorfabrics.com.

Open multiprocessing (MP)

OpenMP provides compiler directives, library functions, and environment variables that can be used to parallelize serial applications. OpenMP Fortran provides an extension to the Fortran programming language, while OpenMP C/C++ provides an extension to the C/C++ programming languages.

OpenMP is relatively simple to use since developers can parallelize code by simply inserting preprocessor directives around code blocks to be parallelized. As the serial application is executing, once an OpenMP preprocessor directive is reached, the master thread forks a specified number of slave threads and the task is divided among the threads. These slave threads execute concurrently, and, once the parallelized threads complete, they join back together into the master thread, and execution continues as a single thread. One advantage of OpenMP is that the preprocessor directives are treated as comments when serial compilers are used for compiling the application software. One disadvantage with OpenMP is that it does not have any tools to assist with race condition and synchronization bugs. For more information, see http://www.opnemp.org.

Clean C

Clean C overcomes some single core to multicore conversion problems. IMEC has developed the Clean C utility as an Eclipse plug-in.

First, the developer runs the analysis tool to determine whether the software application is compliant with Clean C's 29 programming rules. Code that is not compliant with the programming rules is flagged and the developer manually updates the application software to meet these rules.

Once this update is complete, the developer then executes the utility that automatically converts C code from a single core microprocessor to a multicore microprocessor. This transformation utility is correct in roughly 99% of all cases.

The developer needs to go back and review the transformed software for correctness and check for the 1% of the code that was not appropriately updated. If the Clean C utility is applied without implementing the 29 programming rules to the C code base, the result will likely be non-operational. The Clean C utility can only be applied to C language software code bases. Clean C does not properly convert C++ based applications. The author has not tested this product. For more information, see http://www.imec.be/CleanC.

Software development and debug tools

Intel Parallel Studio

Intel's Parallel Studio is a C/C++ multicore tool suite that integrates with Microsoft's Visual Studio 2005, 2008, and 2010 Integrated Development Environments (IDE). The integration of Parallel Studio and Visual Studio was enhanced by both Intel and Microsoft adopting a common runtime called the Microsoft Concurrency Runtime. Parallel Studio is composed of:

- Intel Parallel Advisor, which models an application and analyzes the best locations to implement parallelism within your application. This tool finds and resolves race conditions and locks before the developer begins parallelizing serial software. It also provides a recipe for developers to implement task-oriented parallelism.

- Intel Parallel Composer, which contains a C++ compiler, performance libraries, software thread building blocks, and parallel debugging extensions for improved performance on multicore microprocessors. It incorporates Intel Threading Building Blocks, which is a C++ template library for implementing parallel software. Overall, Parallel Composer improves a parallel software application's performance and speeds up the development of parallel software applications.
- Intel Parallel Inspector, which automatically identifies memory and threading errors for the software developer. Parallel Inspector uncovers memory corruption, memory errors, and threading errors which may cause your software applications to crash. This tool identifies the errors, and it is the developer's responsibility to fix the errors.
- Intel Parallel Amplifier, which analyzes processing hot spots, concurrency, and lock and waits with the goal of optimizing performance and scalability. This tool is GUI-based to provide an easier way for the software developer to see how an application's execution can be optimized.

Some Cilk++ components have been embedded into the Intel Parallel Composer product. Cilk++, Cilk, and jCilk products were purchased by Intel. The Cilk++, Cilk, and jCilk products used three simple keywords to parallelize serial applications for the C++, C, and Java programming languages, respectively. For more information, see http://software.intel.com/en-us/articles/intel-parallel-studio-home.

Benchmarking tools

The purpose of benchmarks is to provide metrics on different microprocessor and architecture capabilities. Multicore benchmarking is important because of its recent advancements and high demand. Benchmarks provide developers a comparison between different microprocessors and architectures to assist in choosing the right microprocessor and architecture for the system being developed.

Embedded Microprocessor Benchmark Consortium (EEMBC)

EEMBC provides one of the best multicore industry standard benchmarking environments for embedded software and hardware developers to use in comparing different multicore processors used in embedded systems. The EEMBC goal is to provide an industry standard for evaluating the capabilities of microprocessors, compilers, and various embedded system implementations. They currently have benchmarks for software, microprocessors, and systems. They provide benchmarks for the following embedded industries:

- Telecom
- Office automation
- Energy consumption
- Automotive
- Consumer
- Digital entertainment
- Networking
- Network storage
- Virtual machines

EEMBC manages new benchmarks and certifies their existing benchmarking results. For more information, visit http://www.eembc.org/home.php.

Multibench

Multibench is an EEMBC benchmark that focuses on parallelization and scalability impacts to computationally intensive functions. Multibench provides benchmarking analysis for task scheduling, task synchronization, memory usage, and software architectures. Multibench's end goal is to provide analysis that system and software developers can use to improve their own software applications running on a multicore platform. For more information, visit http://www.eembc.org/benchmark/multi_sl.php.

CoreMark

CoreMark is an EEMBC benchmark that focuses on an embedded microprocessor's cores. CoreMark's benchmarking involves executing several standard algorithms such as sorting, state machines, cyclic redundancy checks (CRC) used in finding data errors in digital networks, and matrix operations. Each benchmark algorithm is developed in C code.

Current benchmarks are available for the following hardware vendors' microprocessors: Advanced Micro Devices (AMD), ARM, Intel, Freescale Semiconductor, Hewlett Packard, IBM, Nvidia, and Texas Instruments. CoreMark does not provide benchmarks for a microprocessor's memory or cache. It is expected to replace the Dhrystone benchmark.

CoreMark produces a single value score which allows for quick microprocessor comparisons. For a more meaningful single score, CoreMark sets specific rules about executing code and report results to eliminate inconsistencies between microprocessors. For more information, visit http://www.coremark.org.

Standard Performance Evaluation Corporation (SPEC) CPU2600

SPEC's goal is to provide an industry standard for computer performance benchmarks. The SPEC benchmarks are usually written in C, Java, or Fortran. The SPEC CPU2600 provides benchmarks for the high-performance computing (HPC) market and analyzes CPU, memory, and compiler performance. SPEC CPU2006 is composed of two performance benchmarks:

- CINT2006 – compares computation-intensive integer-based algorithms.
- CFP2006 – compares computation-intensive floating-point-based algorithms.

SPEC CPU2000, which has been retired, targeted the workstation and server markets and focused its benchmarks on integer and floating-point arithmetic. For more information, see http://www.spec.org.

Conclusion

This chapter has discussed several tools for developing software for multicore microprocessors. Most of these tools are mature and provide excellent support for a software developer. Very few tools are direct competitors with another tool. Currently most tools attempt to solve one small piece of the software developer's task in writing software for a multicore environment. In general, when choosing software development tools for multicore microprocessors, keep in mind that many tools, unlike the tools discussed in this chapter, are still immature and are usually programming language, microprocessor, and/or vendor specific. Make sure you have a thorough understanding of the application you are developing or migrating and the development needs for the application you are developing. Ask very detailed, pointed questions of the tool vendors to make sure you understand what their tool can and cannot perform at the time of purchase or use.

What is really needed are a series of software tools for multicore development that integrate well with other software development tools for both single and multicore microprocessors. For maximum benefit the output from one tool should directly feed as input into the next tool in the tool chain. This would eliminate or reduce labor-intensive manual reformatting prior to inputting the data into the next tool, thus enabling a software developer to spend more time performing software engineering tasks as opposed to repetitive formatting tasks. There have been some successes such as Poly Core Software, CriticalBlue Prism, Vector Fabrics, and Intel Parallel Studio, but additional integrated tool suites are needed.

Acknowledgments

The author, Kenn Luecke, would like to thank Andrea Egan, Shawn Rahmani, and Wayne Mitchell for assisting him with his initial research into software development for multicore microprocessors for the Boeing Company. He would also like to thank fellow Boeing engineers Homa Ziai-Cook, Tom Dickens, David Caccia, Heidi Ziegler, Jon Hotra, Ron Koontz, and Don Turner for their assistance with multicore-related issues.

CHAPTER 11

Partitioning Programs for Multicore Systems

Bryon Moyer[a], Paul Stravers[b]

[a]Technology Writer and Editor, EE Journal
[b]Chief Architect and Co-founder of Vector Fabrics

Chapter Outline

Real World Multicore Embedded Systems.
DOI: http://dx.doi.org/10.1016/B978-0-12-416018-7.00011-0

Introduction

A successful multicore system is the result of numerous technologies contributing overall to a computing platform that can handle concurrency efficiently and effectively. But a multicore system only serves to exploit any concurrency available in the program(s) to be executed. Run a sequential (or serial — there appears to be some debate as to which term is correct, but both are used) program in a multicore system, and you'll use one core and waste all of the concurrency-handling capabilities — because the program exhibited no concurrency.

So an essential component of any successful embedded multicore implementation must include creating a program that makes concurrency available for the system to exploit. When writing a new program, code can be written in a way that's inherently more concurrent. More difficult is taking an existing legacy program and re-engineering it to run well on a multicore system.

In fact, of all the challenges inherent in multicore system design, none is more difficult than this: it's often referred to as "the hard part" of multicore, and numerous attempts at completely automating the conversion of a sequential program to a parallel program have failed — to the point that the mere mention of such a tool is likely to end a conversation simply because such a tool has no credibility. There is no magic button that can be pushed to effect that transformation.

There is also no "best" way to split a program up. The obvious trade-offs of performance against engineering work apply, but it's also true that optimizing has to be done with the target platform in mind. Change the number of cores, and you might decide to use a very different partitioning scheme. Even just changing from one vendor's four-core processor (for example) to another's might change your approach, since the details of the hardware platform — caching, bus bandwidth, and other characteristics — do matter.

This chapter is therefore dedicated to the partitioning of programs for multicore. The focus is on the partitioning of existing code because the steps are concrete. The architecting of a new program must adhere to the same principles as the dividing up of an existing program, but it happens at a more abstract level, and is fundamentally harder to illustrate in a book like this. In some regards, it's also easier to do because you can design your program from scratch rather than trying to pull apart legacy code that you likely didn't write.

We will discuss both fully manual and tool-assisted techniques. Lest this appear to contradict the unviability of a parallelization tool, note that there are tools that can help enormously in identifying opportunities and barriers to parallelism, but they aren't push-button solutions. It is even possible that they might be able to do some code refactoring automatically, but programmers generally don't like tools to write code for which they will be held accountable. They want to write and understand their own code. So even that portion of the process tends to be manual — despite the real potential for automation.

But the part that simply can't be automated is the making of partitioning decisions. Tools are very effective at providing good information to inform the decisions, but you, as the programmer, retain control over the conclusions to which the information leads you.

One critical concept that dominates all decisions is that of the semantics of the program. The ideal situation is the ability to create a parallel version of a sequential program that maintains all aspects of the semantics of the original program. Said more plainly, it works in every way like the original program, so that if you ran a set of inputs on both programs and monitored all available outputs and effects and externally visible artifacts, they would appear to be identical.

That can, in some cases, be a high standard to meet. So we will include discussion of situations where it might be acceptable to alter the semantics of a program in ways that still accomplish its overall goals.

Finally, the result of partitioning may be a single process with multiple threads (for an SMP implementation, for example) or it may be multiple processes (for an AMP implementation). It could, in theory, even be a combination. Multi-threaded solutions tend to be more common than multi-process ones, and so many of the examples as well as the focus of the tools will be on multi-threaded scenarios. But the fundamental concepts themselves apply to either case; the only difference comes down to the mechanics of synchronization and the details of how the resulting program is constructed.

What level of parallelism?

Concurrency can exist at different levels, or "granularity", as it is sometimes referred to. At a very low level, there is instruction-level parallelism. By contrast, threads or processes that run in parallel represent much larger "chunks" of concurrency. We must also make a distinction between the concurrency inherent in an abstract algorithm and that provided by a specific implementation of that algorithm.

Threads of control

A sequential program reflects a single program flow with one thread of control, and its progress reflects an explicit series of states. The resources used to implement the state are:

- the program counter (PC);
- the internal processor state;

- the local registers; and
- the memory.

A multicore system replicates three out of these four elements: each core gets a PC, a processor, and a register set. But, with respect to the semantics of the memory, the original program assumes a single memory space that can be accessed from anywhere in the program, and, in order to preserve program semantics, the parallel version needs to appear to have a similar structure.

The actual underlying memory structure in a multicore system may or may not consist of a single memory shared across all of the cores. The partitioned program may use different logical spaces if multiple processes are created (and those may be from the same or different physical memories), or it may use a single logical space if you're building a multi-threaded solution.

For this reason, much of the challenge of partitioning boils down to focusing on memory accesses to ensure that a parallel implementation still appears as if it had a shared memory. Synchronization is critical to preserving this behavior. So this is the level at which we will focus in this chapter.

Much farther down in granularity is instruction-level parallelism. This can occur largely for two reasons:

- the low-level details associated with program execution – address calculation, fetches, etc. – can be parallelized, and
- different operations that happen to be near each other in a program may use independent operands, for example, and so can run independently of each other.

These kinds of optimizations can be performed if the low-level hardware permits – the few very-long-instruction-word (VLIW) architectures, for example, can accommodate multiple operations in a single instruction. But such parallelism is related not only to the program being run, but also to the capabilities of the underlying platform. The same program run on different systems may execute very differently due to low-level hardware and compiler differences. The compiler and/or the processor must make the decisions as to how to make this work, and the range of optimization tends to be relatively small. Operations near the beginning and end of the program, for example, that might possibly be run in parallel will not generally be discovered by such systems because they're too far apart.

More fundamentally, four decades of processor architecture engineering have pretty much squeezed all of the possible performance out of pipelining and dynamic dispatching; the only way to improve performance from here is through threading and vectorization. It is for this reason that this chapter will not address instruction-level parallelism, but will focus on establishing an effective implementation containing multiple higher-level threads of control.

There is one more conceptual difference between the implementation of thread-level and instruction-level concurrency. The use of instruction-level parallelism is done in a way that hides it from the programmer; by contrast, the purpose of thread-level concurrency is to expose the concurrency to the programmer for exploitation.

Solutions, algorithms, and implementations

Concurrency is exposed or suppressed at different stages of the creation of a program. At the very top level is the fundamental, abstract solution to a problem. If the problem is the determination as to whether a driver has had a bit too much to drink, then one possible solution is to have that person recite the alphabet backwards. That solution inherently has no concurrency: the letters must be recited in a given order or the solution is incorrect.

If the solution were to have the individual recite all of the letters of the alphabet, regardless of their order, as long as each letter were stated exactly once, then the sequential nature of the solution has been removed, and, in principle, each letter can be stated in parallel with each other letter as long as there is a way of ensuring that there are no duplicates. While this is probably much harder for a human to do, it would be more effective in a multicore system.

More sophisticated solutions to harder problems have another level to consider: the algorithm used. When resizing a photo in Photoshop, you have a choice of different algorithms. Each has its strengths and limitations. The high-level problem is that the picture is the wrong size; the solution is to resize it. The choice of algorithm now adds another degree of freedom to the decision process. Some algorithms will have more inherent concurrency than others.

Finally, once an algorithm is selected, someone must implement that algorithm in a real program. This introduces yet another degree of freedom as lower-level details like data structures and programming style are considered.

For example, linked lists are popular in many programs; they implement a structure that is easy for a programmer to envision, and they're a natural solution for some problems. But they are, by definition, sequential structures. You can only find a specific list element by navigating from the beginning (or end) of the list until the desired element is found. By contrast, using an array to represent the data may, at first blush, appear non-intuitive and unnatural for the problem being solved. But any element of an array can be arbitrarily accessed in one step without regard to any other elements; thus it is much more amenable to a concurrent implementation.

Object-oriented programming has become more popular within embedded systems as efficiency and available resources have reduced the cost of any overhead. There are concrete benefits to an object-oriented approach, but objects can pose a challenge for parallel programming, especially when partitioning an existing program.

By intent, the classes defined in an object-oriented architecture hide implementation details. They encapsulate and render opaque the subtleties of how state is maintained between high-level property accesses or method assertions. Those details may hinder partitioning, forcing a lot of work to re-design the classes to be more parallel-friendly. It may be easier to create a new object-oriented program with concurrency in mind, but even then the design decisions need to be well documented so that future programmers don't inadvertently make changes that compromise those decisions.

By contrast, data-flow languages — obscure though most might be — are based upon the very data relationships that affect partitioning, and, therefore, programs written using them can be examined for concurrency much more easily. Unfortunately, that is mostly of academic value for embedded code.

The basic cost of partitioning

Depending on how a partitioned program is implemented, there will be associated costs. At the most fundamental level, a program can be broken into multiple "independent" programs for execution in an AMP system, or it can be implemented as a single program with multiple threads. Designing for AMP systems is generally harder, but the overhead of program execution is minimal, if any. Each process can start up on its core (coordinated by a master core, as discussed in the OS chapter), and there is effectively no further overhead except that associated with synchronization (discussed below).

In a multi-threading implementation, however, you have the cost of threading – and that cost is not trivial. Done poorly, a partitioned multicore program can actually run more slowly than its sequential original.

One of the central costs is that of asking the operating system to create and destroy threads. For this reason, it is generally more effective to employ what are called "worker threads": these threads are created at start-up and are then assigned to different tasks as needed while the program is running. It is almost always faster to reassign an existing thread than it is to create a new thread (although the benefit varies by system).

Worker threads are typically implemented with FIFOs that store the tasks that the thread is to perform. Each entry in the FIFO would contain a function pointer, indicating the code to be run, and, in some cases, a data pointer if the data location is abstracted away from the code. In the simplest implementation, an unused thread sleeps until a signal from the FIFO wakes it up; once it has its "marching orders" it joins all of the other active threads vying for time on the processors.

The reason why using worker threads may or may not have a dramatic benefit over creating threads is the very fact that unused threads "sleep".

The process of getting them into the scheduling stream can take time: the OS has to schedule the thread on some core, and that core's cache has to be refilled to service the thread. How long all of this takes varies widely by system.

One way of avoiding this delay is through "busy wait" threads. Rather than the thread going to sleep, it's kept awake doing nothing (Figure 11.1). This, of course, consumes cycles, so it only works on deeply embedded systems where a core is dedicated to some function, where no other code is waiting to use the cycles. In other words, this works best when the wasted cycles would be wasted anyway.

Busy wait threads are also best used when the "duty cycle" is high — that is, for example, when the thread is actually doing real work 80% of the time and idling busily only 20% of the time.

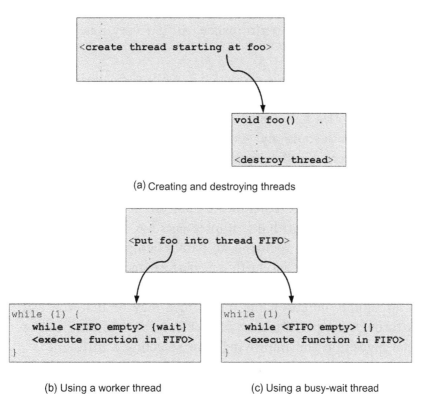

(a) Creating and destroying threads

(b) Using a worker thread   (c) Using a busy-wait thread

Figure 11.1
Different threading options.

A high-level partitioning algorithm

At a high level, partitioning a program can be summarized in the following steps:

1. Choose a partitioning strategy
2. Identify the dependencies that the strategy breaks
3. Determine whether the broken dependencies can be handled
4. Estimate the performance benefit of the strategy
5. Repeat iteratively, looking for better or additional partitions; make your decisions
6. Implement the chosen partitioning strategy(-ies)

In practice, it's extremely unlikely that you will be successful with the first selection you make. At worst, you'll find that there's a problem that makes a given strategy infeasible. At best, what you select will work, but it won't be enough; you'll need to do further partitioning to get more performance. So it's very much an iterative process.

To investigate these steps in more detail, we'll work though the concepts critical for performing them. We'll initially discuss manual approaches to these steps – that helps to clarify the concepts as well as to illustrate why this process is so thorny. As mentioned above, tools have come onto the market that can greatly simplify some of the harder parts of the problem; we'll discuss those later.

One important thing to remember is that the best program can't be parallelized if the algorithm it implements has little intrinsic concurrency. We'll be focusing on parallelizing actual code, but it may be that you can't get sufficient performance from those efforts even if you do the best job possible. That could be because your goals are simply unreasonable, but it also may be that you can break through the performance barrier only by returning to the original algorithm and improving it or selecting a different algorithm. At that point, you may either be writing a parallel program from scratch, or, if the algorithm is complex, you may code and test it as a new sequential program and then come back to partitioning that program later.

The central role of dependencies

The concept of partitioning was touched on in the OS chapter, introducing the abstract notion of cutsets when breaking a program apart. In the ideal case, you would like to find a way to break your program into sections that are completely independent of each other, meaning that they have no interactions. It's fair to say that no serious programs have this characteristic. So the partitioning exercise amounts to getting as close to such a division as possible.

The reason that different portions of a program rely on each other comes down to dependencies: one part of the program consumes data that another part produces. These concepts were introduced at a high level in the Concurrency chapter; we now look at the practical implications of dependencies.

Breaking dependencies

A useful way of thinking about partitioning a program is to think in terms of breaking and repairing dependencies. If you divide your program into two sections, section A and section B, and if data is produced in section A that is needed in section B, then a dependency exists between the two sections. Considered graphically, the partition cuts through the dependency (Figure 11.2).

If you simply divide the program and then carry on, that dependency has been broken: the data that section B relies on is not provided. What's particularly insidious about this is that section B will not necessarily "know" that it isn't getting the right data. It's not like it sits there

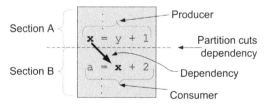

Figure 11.2
Breaking a dependency between a producer and a consumer.

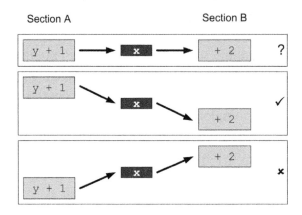

Figure 11.3
If the dependency is left broken, it's impossible to tell whether the producer will save the data before the consumer grabs it.

waiting for data — that would be an instant indication that something was wrong and could be readily addressed.

Figure 11.3 illustrates what can actually happen. The first portion shows the uncertainty as to whether the new value of x will be saved in Section A before it is used in Section B. The middle scenario is one option, in which case things will proceed correctly. But the third option is also possible: section B may grab what it thinks is the data — only it isn't. So the program executes merrily along as if nothing were wrong. If the dependency affects the normal, high-level functioning of the program, then, at the very least, you'll be able to see that something is wrong (although identifying the root cause can be a challenge). But if the dependency relates to a "corner case", some rarely-utilized section of the code, or if the effects of the broken dependency are subtle, then you may not even notice that there's a problem.

To make matters worse, even if things work on one occasion, they may not work the next time, depending on all of the system vagaries that determine actual execution timing.

So the key to effective partitioning is that dependencies will be broken; they must then be repaired so that the data communication in the original sequential program is restored in the partitioned program. The communication of this data over the repaired dependency is more commonly referred to as synchronization: the data in the two (or more) sections of the program is updated so that the sections are in synch. The hard part is identifying all the dependencies, understanding which ones can be repaired and the costs of repairing them. The actual process of doing this manually in a real program is even harder.

Types of dependencies

There are various ways of referring to different dependencies, and it can be a bit confusing to figure out which ones are which. So first we'll review the different kinds of dependencies; we can then discuss their implications.

Canonical dependencies

Dependencies usually arise out of a relationship between the writing of data by a producer and the reading of data by a consumer and the order in which those happen. The best way to approach this question manually is to ask the question, if I split the code at some point and the two sides of the split could run independently, what could possibly go wrong?

Read-after-write

Here, if the read and write are no longer sequential, but run at the same time, then the value read for x in the read operation will be wrong if it occurs before the write operation (Figure 11.4). This is the same scenario as was illustrated in Figure 11.3.

This is also referred to as a "true" dependency or a "flow" dependency. This means that a given read must happen after a given write, and this conforms to what is probably most easily understood as a dependency. If one calculation depends on a value from a prior calculation, then you can't read that value until the prior calculation has written it.

Figure 11.4
A "true" or "flow" dependency.

Write-after-read

This (Figure 11.5) is exactly the reverse of the prior situation. In this case, the read operation is expecting to find a value in x that was set by some prior write operation. Now if what was originally the subsequent write happens before the read operation, then the read operation won't get the value it should get — the desired value has already been overwritten.

This is called an "anti-dependency". It means that a memory location can't be written until some other read has happened. In practical terms, if a producer creates a value that will be consumed, and if it or some other producer will then replace that value with a new value, that new value can't be written until all consumers have had a chance to read the prior value.

Write-after-write

Here the read operation expects the value set by the first write, but if the second write could happen independently of the first write, then the two writes compete, and the value that is read will be unpredictable (Figure 11.6).

This is also called an "output" dependency. It means that two successive writes need to occur successively, not independently of each other. You might think of it as the combination of the read-after-write dependency created by the first two lines and the write-after-read dependency created by the last two lines. The subtle difference is that an anti-dependency is

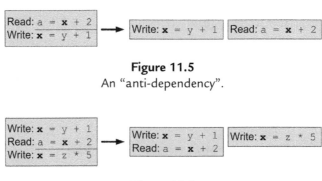

Figure 11.5
An "anti-dependency".

Figure 11.6
An "output" dependency.

more about maintaining one value before writing a new one. With an output dependency, you may in fact have only one persistent value, and you don't know which one it will be.

Note that if there is truly one write operation that immediately follows another, with no intervening read of the written data, then, with one caveat, the first write operation could be (and, by many compilers, will be) optimized away as not being useful. That leaves only the second write operation, and the output dependency no longer exists, and the system state is now predictable.

The one caveat to this point applies particularly to embedded systems: the write operation might deliver a value to a memory-mapped peripheral like a hardware accelerator, so, while the program itself doesn't use the value, the write operation still has an impact on the system — it's just that that impact is outside the scope of the program. In fact, this is where the `volatile` keyword can be useful to ensure that compilers don't optimize away such a line of code. But this consideration is not a defining aspect of the output dependency.

Read-after-read

Finally, there is a read-after-read dependency (Figure 11.7) — in theory. It has the name "input dependency", but, in fact, it's not a real dependency in the sense that the order of operations matters. We include it here just for completeness.

Ordering dependencies

A further type of dependency may be referred to as an "ordering" or "control" dependency, and it simply means that the order of seemingly unrelated statements must be preserved.

Figure 11.7
An "input dependency".

Out-of-scope dependencies

One frequent way this can happen is when one end of a dependency is outside the scope of the program. An example might be sending data to a printer: the program sees a data "write", but the data is never read within the program. That's because the printer "reads" the data outside the program scope. The "caveat" described above in the discussion of output dependencies is another such example. Variable initialization as a program starts up, before actual execution starts, is another out-of-scope dependency.

As to whether this is a problem, such dependencies must be manually reviewed. A printer example provides an easy intuitive sense of the issue. If, during each pass through a loop, a line is to be printed indicating that the pass was made, then there may be an issue if the loop is parallelized. The original sequential program will cause the lines to be printed in a specific order; a parallel version will violate those semantics.

Non-obvious in-scope ordering dependencies

Another example of an ordering dependency reflects the use of cheap-and-dirty data synchronization. Let's say we want the value value_a to be written and then read by some other thread. In order to let the other thread know that the value is ready, we add a second variable flag_a. After writing value_a, we then set flag_a to true; when the other thread sees that flag_a is true, it knows that it can read value_a (Figure 11.8).

Through this mechanism, we have created an ordering dependency between the writing of value_a and flag_a: we cannot re-order those statements without breaking the program semantics. Yet no automated process can intuit the relationship between value_a and flag_a; this dependency must be recognized by a human. If such dependencies exist

Figure 11.8
The code on the left must execute in that order in order for the code on the right to read the correct value of value_a.

in an old legacy program with poor documentation, they may be extraordinarily difficult to identify.

Compute and memory dependencies

There is another categorization of these dependencies that is orthogonal to what we've just described:

- What we can call "compute" dependencies are straightforward dependencies that involve values stored in registers (or anywhere that isn't memory).
- Values produced into and consumed from memory can be referred to as "memory" dependencies.

The reason for distinguishing between non-memory values and memory values can be summed up in one word: pointers. For calculations occurring with register values, simple variable names are used to access those values – specifically because the specific register locations are assigned by the compiler and are not visible to the programmer.

But values in memory are stored in specific locations: a named variable or pointer refers not just to an abstract value-holder like a register; it may refer to a specific memory location that may have a well-defined relationship to the locations of other variables. For example, it is generally expected that, in an array, elements are contiguous in memory, in the order defined by the array. (Yes, that could be virtualized in some manner to make non-contiguous segments appear contiguous, but the overhead would be enormous.)

If memory is dynamically allocated, then those memory locations are not known by the programmer, and they may be different each time the program executes.

For purposes of identifying dependencies, there are two specific reasons why memory dependencies need to be treated differently:

- The C language, which predominates in the embedded world, allows the undisciplined use of pointers. A value can be stored in some arbitrary location in memory whose address may have been derived from the value of some pointer, possibly with pointer arithmetic. That location could therefore be anywhere the memory manager allows it to be, and that location may not have an associated variable name or may not reside within a named structure. Even if it does reside within a named structure, dynamic memory allocation will prevent you from determining the relationships by examining the program. For example, you may be writing some value to a location that may or may not be inside an array. And, if you are, that may be what you intended, or it might be a mistake — you might be corrupting an array.

"Pointer aliasing" affects programs written not just in C, but also in any languages providing "objects" or the ability to refer to some complex structure by a simple name. References to such objects aren't really to the objects themselves, but to their locations: the variable name is really just a pointer. Many languages don't allow pointer arithmetic, meaning that the prior issue goes away, but aliasing can still occur. Aliasing was described in more detail in the Concurrency chapter earlier on.

The practical implications of this distinction between compute and memory dependencies is that, with compute dependencies, it's relatively straightforward (in concept) to trace variables by their names to see what depends on what simply by examining the source code. But that's not possible with memory dependencies: you have to know specifically where the writes and reads are happening in memory regardless of the names of the entities doing the writing and reading.

Locating dependencies

Manually locating dependencies of any kind is difficult work, so, in general, you will want to limit the scope of the dependencies you try to track down. In practice, this would mean that, if you want to break up your program at a particular point, then you want to identify the dependencies that "cross" the break you're creating.

In the discussion that follows, let's assume a sequential program being split into two; the concepts scale for further divisions. Because of the sequential nature of the original, we can characterize one section as the "early" section and the other as the "late" section (corresponding to "Section A" and "Section B" from the prior figures, respectively): all of the early code is executed before the late code. In practical fact, this is an oversimplification for the following reasons:

- Some subroutines might get called on both sides of the divide, so that code is both early and late. The "early" and "late" characterizations really refer to *executed* code — the actual invocation of a function, not simply the source code as written. With that in mind, such a subroutine actually becomes two subroutines: one executed early, the other executed late.
- Loops allow "backwards" flow. If the break comes in the middle of the loop, then the "early" code comes earlier than the "late" code of one loop iteration, but the "late" code of one iteration is earlier than the "early" code of a subsequent iteration.

So there are times when a given section of code might have to be treated as both early and late, and the concept of loop distance, as introduced in the Concurrency chapter, impacts the breaking of loops.

Locating compute dependencies

Relatively speaking, <u>compute</u> dependencies are much easier to trace. A high-level algorithm would look something like this (illustrated in Figure 11.9):

1. Identify each variable that gets written (left-hand side) in the early section.
2. Identify the latest write instance for each of those variables in the early section.
3. Search the late code for read (right-hand side) references to those variables.

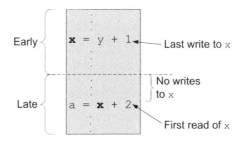

Figure 11.9
Locating a true dependency broken by a partition.

4. Of the read instances, find those that occur before any writes have been executed.

The read instances identified in Step 4 are relying on the last writes identified in Step 2. Each of these is a <u>true dependency</u> that has been broken by the partition. The implications of that will be considered below.

Next, repeat those steps, but now exchange writes and reads (Figure 11.10). You're identifying instances in the late section where data is written to variables that are read in the early section: these are <u>anti-dependencies</u> that have been broken by the partition.

Finally, repeat these steps looking for the last write to a variable in the early section and the first write to that same variable in the late section (Figure 11.11). This lets you identify <u>output dependencies</u> that your strategy breaks.

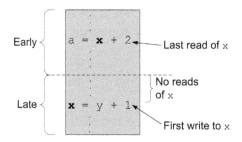

Figure 11.10
Locating an anti-dependency broken by a partition.

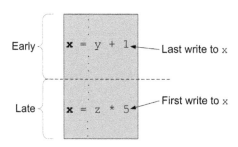

Figure 11.11
Locating an output dependency broken by a partition.

There is no need to locate input dependencies.

While this is a tractable algorithm, you can well imagine that, in practice, there are many variables, and tracing them all down this way can be a lot of work — especially considering loops, subroutines, and partitioning into more than two sections.

It becomes even more difficult with legacy code — particularly poorly structured code. The key to doing a good job of this lies in understanding the program in detail, and everything that works against that knowledge makes the problem harder.

Locating memory dependencies

As mentioned, identifying <u>memory</u> dependencies can't be done by inspection (except in the unlikely case of dependencies involving only global variables that never have their addresses taken explicitly, such as through the "&" operator, or implicitly, as when passing a global array as a parameter). There is, in fact, no really good way to do this manually. So between the difficulty of identifying compute dependencies and the impossibility of identifying memory dependencies, the job ends up relying on the knowledge and expertise of the programmer.

There are some approaches that can help manage the process:

- Instead of looking for dependencies in the program itself (the implementation), look at a higher level at the algorithm that the program implements. You may be able to identify high-level dependencies at this level. The good thing about this approach is that you don't get bogged down with any dependencies that are not intrinsic to the algorithm and that might have been introduced in the implementation. The bad news is that you end up rewriting the algorithm implementation.
- Significantly, as mentioned, tools have started to appear that simplify the process of identifying dependencies. We'll take a look at these tools shortly.

Handling broken dependencies

We've identified different kinds of dependencies and ways to identify them; what does this mean for partitioning purposes? Broken dependencies must be "repaired" somehow to ensure that the original program semantics are preserved.

Repairing broken canonical dependencies

True dependencies obey intuitive producer/consumer behavior. So the key lies in ensuring that produced data reaches the intended consumer, and that the consumer doesn't consume anything until it knows that the right data is in place. This is the standard role of data synchronization (Figure 11.12). There are different ways of synchronizing data, and the best ways depend on the kind of concurrency being exploited. We'll look at data synchronization in more detail below.

Repairing a broken true dependency leaves the dependency in place, but ensures that the dependency is honored in the multi-threaded version.

Both anti-dependencies (Figure 11.13) and output dependencies (Figure 11.14) hinge on one write operation not clobbering a prior write operation. This can be managed by replicating the storage — that is, defining two separate variables with different names — so that each write affects a different location in memory, guaranteeing that they won't interfere with each other.

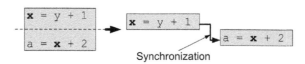

Figure 11.12
Repairing a broken true dependency.

Figure 11.13
Repairing a broken anti-dependency.

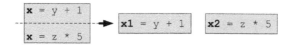

Figure 11.14
Repairing a broken output dependency.

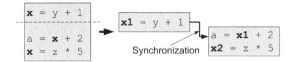

Figure 11.15
An output dependency repair coupled with a true dependency repair.

This storage replication effectively removes the anti-dependency or output dependency from the program. Subsequent read operations must, of course, refer to the correct written value, meaning that you must be sure that each read refers to the correct new variable.

If any reads of the first write end up in the second thread, then a true dependency exists with respect to the first write, requiring synchronization. But that synchronization must then be to a different memory location than that used by the second write. Note, however, that the very synchronization means that the output dependency is no longer an issue: the fact that the first read in the late code has to wait for the last write of the early code guarantees that the two writes cannot collide. So this example shows, conceptually, how different dependencies can interact (Figure 11.15).

If the resource that needs to be duplicated is expensive — say, a very large matrix — and it's not used very often, then, instead of duplicating the resource, you can instead create a critical section that forces the order of execution, as discussed below. This is the only way to solve the problem if the resource can't be replicated — say for the case where an embedded hardware resource is mapped as a variable. A critical section would be the only way to enforce correct behavior.

Special cases

There are dependencies that qualify as one of the canonical types, but have specific structures that can be handled differently. Because we know something about what's going on with them, even if we don't know anything about the underlying algorithm being computed, we can re-design the way the computation is done to break the dependencies. These are examples of very specific patterns that can be detected and managed more effectively than if we were simply to deal with the dependencies in a brute-force manner. Two examples follow.

Induction expressions

Many loops use an *iterator*: this is the variable that is used to count the iterations in a for loop. Let's call this variable i. We do two things with i: increment it and compare it to some limit. The incrementing involves a statement like i = i + 1. This is called an *induction expression*, and it involves a loop-carried dependency because the value of i for the current iteration of the loop depends on the value of i for the prior iteration. Even if the body of the loop itself creates no loop-carried dependencies — even if it were, in theory, possible to execute all of the iterations exactly in parallel — the induction expression now messes up what otherwise would look clean.

Obviously, if you ran each instance on a separate core, there would be no more looping, so that's something of a degenerate case. But if you take, say, a 16-iteration loop and distribute it amongst four cores, now you have four 4-iteration loops. Figure 11.16 illustrates this scenario, with the "for" loop syntax recast into the individual initialize/increment/compare sequence, which makes the dependency more obvious.

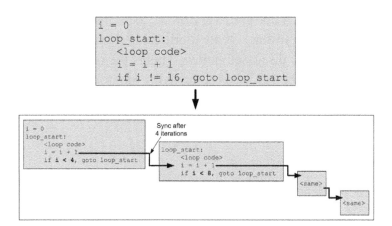

Figure 11.16
An induction expression creates a loop-carried dependency. (The third and fourth boxes, marked <same>, are just like the second box, with the exception of the loop test value; abbreviated for purposes of space.)

As expected, this implementation provides no improvement over the sequential version because each core has to wait for the prior one to complete before starting, which is specifically the problem the loop iterator introduces. Assuming there are no other dependencies in the loop body, we can parallelize this by giving each loop its own iterator; now they don't depend on each other anymore.

Based on our knowledge of how iterators work, we can break that dependency by re-computing the loop starting values independently for each core when dividing up which iterations get handled by which cores. As shown in Figure 11.17, i can be set to run from 0 to 3, 4 to 7, 8 to 11, and 12 to 15, respectively. No core has to check any other core in order to compute its iterator value.

Reduction

It's very common for calculations to be performed on a wide data vector, and for those calculations then to be summed together into a single scalar result. This is referred to as *reduction*. A simple case might involve simply moving through an array to add all of the values in the array. That process makes the sum at any particular cell dependent on the cell before it, forcing sequential execution (Figure 11.18).

This can easily be re-designed by taking advantage of commutativity and associativity: it doesn't matter which order or grouping the values are summed in. So the array can be partitioned, with different cores summing different portions, and then the sums for each of the cores can themselves be added together to get the final result (Figure 11.19). Note that there's still a dependency, but, rather than the entire second half of the loop having to wait to get started until the first half finishes,

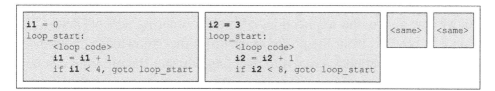

Figure 11.17
Recalculating the loop start values independently for each core eliminates the dependencies.

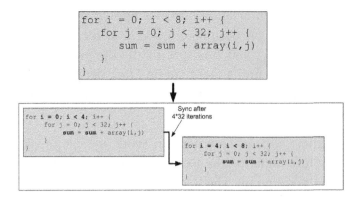

Figure 11.18
Reduction creates loop-carried dependencies.

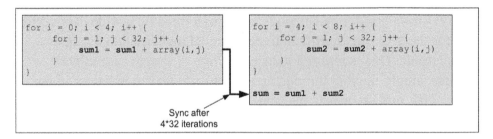

Figure 11.19
Reduction re-designed to calculate partial sums.

both halves can run in parallel, and only the final sum has to wait until the partial sums are ready.

Repairing broken ordering dependencies

As we saw above, the implications of broken ordering dependencies won't be obvious from the program itself. Further knowledge about the intent of the program is critical to deciding whether the ordering can be broken. It's important to be clear here: there is no simple way to repair a broken ordering dependency. It becomes a question of whether a semantic change is acceptable or whether it is possible to "fake" the old behavior with additional new code.

Out-of-scope dependencies

We'll continue here with the printing example we used above to illustrate the decision process. The original program prints out the lines in a specific order (Figure 11.20).

When parallelized, the ordering will be different — and it will likely not be entirely predictable (Figure 11.21). This represents a blatant violation of the original program semantics. So the key to this decision lies in understanding whether the original program semantics represent a strict requirement or merely the artifact of the original implementation. If the

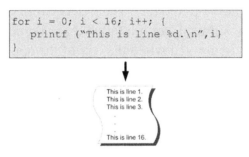

Figure 11.20
Original ordering of printed lines.

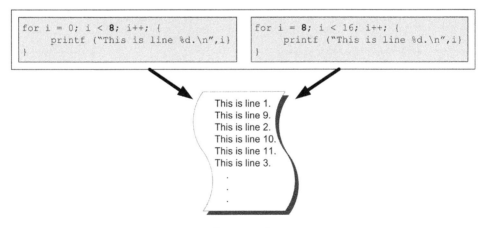

Figure 11.21
Parallelizing changes the ordering of the printed lines. The order may or may not be as shown, and it may change each time the program is run.

intent of the program is that lines be printed in a specific order, then parallelizing will violate that intent, and it will be a problem.

On the other hand, if the intent is simply that each line be recorded, with no need for any specific ordering, then, even though the original semantics are broken, the intent is not so strict, and a parallel implementation can still meet the application requirements.

If performance hinges on parallelizing such a program but the semantics require ordered printing, then it may be possible, for example, to send all the print messages, in whatever order they occur, to some temporary location if they can include an ordering number of some sort (like a proxy for the original loop iterator). Then, once all of the threads have completed the loop, a thread can be started that re-orders the messages and causes them to be printed while the rest of the application proceeds (Figure 11.22).

Such a solution is, of course, highly application-dependent; each such situation must be evaluated by hand. The decisions will also rely on access to the original program requirements (which can be difficult to come by for legacy programs) and, harder yet, knowledge about whether any subsequent work has relied on that ordering (much like a

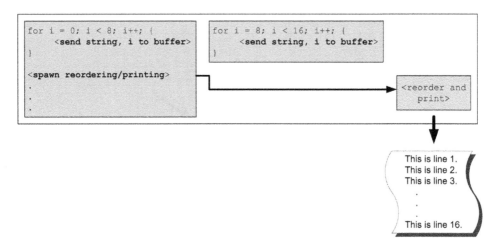

Figure 11.22
Using a buffer to re-order the printed lines can re-establish the original print behavior. Details of the reordering buffer omitted as straightforward.

microprocessor bug has to be replicated in later versions because software has been written assuming it exists).

Non-obvious in-scope ordering dependencies

There is a solution available for ordering dependencies that remain within scope. Assuming a simple dependency between two statements, insertion of a barrier (also referred to as a "fence") between the two statements will ensure that the first statement will be completed before the second takes place.

When we discussed this kind of dependency above (Figure 11.8), we used a "ready" flag as an example. If the flag ended up in a different thread, then it might be set before the value of value_a was written. To solve that situation, first we need to ensure that the writing of value_a and its flag occur in the same thread. Then we place a fence after the flag is written to ensure that the reading of that flag in another thread happens only after value_a has been written (Figure 11.23). It's also useful to include a fence on the other core to ensure that the compiler doesn't re-order the reading of value_a ahead of the testing of flag_a, since, as far as the compiler is concerned, there is no relationship between those statements.

The degree to which such fences need to be added depends on the memory consistency model of the processor core. Some processors tend to require the addition of more fences; some less.

For ordering dependencies involving a longer sequence of statements, this is more effectively handled by declaring the set of statements to be a critical section; we deal with those next.

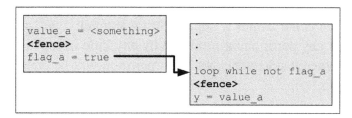

Figure 11.23
Using a barrier (or fence) to force ordering.

Critical sections

Certain sequences of operations in a program can be particularly problematic in multicore systems. It may be essential that a series of steps be done in a manner that looks atomic. It's an all-or-nothing operation: either all of the statements succeed or the operation as a whole fails because the system might be left in an inconsistent state.

The simple transferring of funds between two bank accounts, for instance, is a classic example. Money is removed from one account and added to the other. If the withdrawal succeeds but the deposit fails, then you have an unhappy bank customer. If the deposit succeeds but the withdrawal fails, you have a happy bank customer, but an unhappy bank.

Even if both the withdrawal and deposit succeed, but, between them, some other program comes in to read the balances to create a statement, then the statement will be wrong because it will show one account after the withdrawal, but the other before the deposit. Now you have a panicked bank customer.

While this example is more appropriate to the concept of database transactions, it has its analog in programming as well. For this reason, such sequences are marked as "critical sections" (or "critical regions") so that, during execution of that section, only one thread has access to the resources. Critical sections are a separate topic on their own, and are addressed in other chapters of the book. But, for partitioning purposes, their implications are simple: you should not partition a program anywhere within a critical region.

Note that, as mentioned before, long sequences of code with required ordering can be protected by declaring them as a critical section. In this case, the issue is not so much implementing atomicity, but simply ensuring that any partitioning proposed by any kind of tool will not disrupt the ordered statements, since no partition will be created within a critical region.

Note that when critical sections are used for atomicity, it is OK for forks and joins to occur entirely within a critical region as long as all forked threads are rejoined before the end of the critical section. If, on the other hand, the critical section is used to maintain ordering, then such forks and joins may cause the ordering to be broken, and they should be approached cautiously.

By definition, critical sections create a serialized execution bottleneck, so they can't be used willy-nilly. Their use should be evaluated against other approaches or changes to the program design to maintain acceptable program performance.

Synchronizing data

As we saw above, broken true dependencies can be "repaired" by rebuilding the producer/consumer behavior through data synchronization – that is, finding some way to deliver the produced data to the consumer in a manner that ensures that the consumer won't consume the data until the data is valid.

There are different ways to do this, and we'll illustrate two different ones that relate to data and functional parallelism, respectively.

Using counting semaphores

With data parallelism, by definition, data is broken into chunks that are handled by different cores. So it may be possible to keep all of the data, for example, in a single array, with different operations happening on different parts. Where data produced in one portion affects data consumed in another portion, semaphores can be used to indicate whether the produced data is ready for consumption.

In the likely chance that loops are involved, any loop distance issues (as discussed in the Concurrency chapter) can be handled with counting semaphores. Ordinary binary semaphores simply indicate locked/unlocked. But counting semaphores have "post" and "wait" operations that increment and decrement a counter, respectively. Each time the producer writes data, a "post" is issued; when the consumer uses it, a "wait" is issued.

```
<init semaphore to -127>             for i = 4; i < 8; i++ {
for i = 0; i < 4; i++ {                  for j = 1; j < 32; j++ {
    for j = 1; j < 32; j++ {                 sum2 = sum2 + array(i,j)
        sum1 = sum1 + array(i,j)         }
        <issue "post">               }
    }
}                                    <loop while semaphore <=0  >
                                     sum = sum1 + sum2
```

Figure 11.24
Using a semaphore to delay synchronization by 128 iterations.

Such semaphores can be used to issue a number of "tokens" or simultaneous reader locks, so a positive number typically indicates permission to proceed. In this example, however, we're not issuing permission for some number of operations; we're looking for a downstream operation to wait some number of iterations. We can do this by initializing the semaphore with a negative number. So, for example, with a loop distance of 4*32 = 128 as shown in Figure 11.19 – meaning that the consumer has to lag behind the producer by 128 cycles – the semaphore can be initialized to -127 so that, after 128 posts, the semaphore will be in positive territory, and the consumer can start reading (Figure 11.24).

Using FIFOs

For functional parallelism, data is typically delivered "downstream" to a later process rather than "across" to a concurrent process. A FIFO is a more typical way of doing that. FIFOs are more expensive to implement than semaphores, although it is possible to create lock-free FIFOs with better performance (see reference [2]).

There is a distinct difference between using a FIFO and using a semaphore. With semaphores, multiples cores are accessing the same memory locations; the semaphore is merely coordinating the timing. With FIFOs, you're literally delivering data to a different location. This is common with streaming applications, where some incoming stream of data (like a packet or a video frame) is passed through a series of operations. Instead of those operations taking turns accessing the same memory, they have their own regions and move the data around. The FIFO becomes the means of this delivery.

There are two critical considerations when using FIFOs: the FIFO size and the frequency with which synchronization happens. The FIFO size depends on four characteristics: the amount of data produced in one "delivery" by the producer; the loop distance; the need to avoid deadlock; and the amount of elastic clearance you wish to provide.

The size and frequency of synchronization

The size of each delivery — we'll call it a "chunk" for lack of a better technical term — can vary, and is typically traded off against synchronization frequency. Let's look at the simple example of Figure 11.25. Here we have two nested loops that traverse an array with i scanning rows and j scanning columns. There are two steps in the loop body. The first writes a value and is hence a producer. The second reads that value, using it to sum up all the array values. In a parallelized version, that loop has been distributed over two cores, with one handling the writing and the other handling the summing.

Clearly, the first iteration of the second loop can't happen before the first iteration of the first loop. But if we want to minimize latency, then, as soon as a value is written by the first core, it should be synchronized with the second core so that it can get started. Here we're delivering a small chunk of data that is communicated after each write — that is, with high frequency (Figure 11.26). (Note that, for this example, the actual

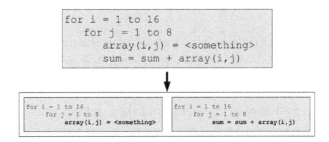

Figure 11.25
A loop distributed over two cores so that one core executes the first part of the loop body and the other does the second part.

```
for i = 1 to 16
    for j = 1 to 8
        array(i,j) = <something>
        <array(i,j) → FIFO>
```

```
for  i = 1 to 8*16
    while FIFO empty, wait
    sum = sum + <read FIFO>
```

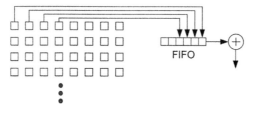

Figure 11.26
Synchronizing one array cell at a time. Each sync is small, but happens for each cell.
The summer can read from the FIFO directly because the content is scalar.

```
for i = 1 to 16
    for j = 1 to 8
        array(i,j) = <something>
    <array(i) → FIFO>
```

```
for i = 1 to 16
    while FIFO empty, wait
    for j = 1 to 8
        sum = sum + <FIFO(j)>
```

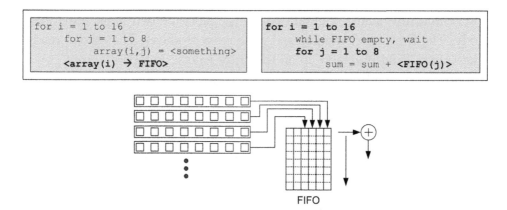

Figure 11.27
Here an entire row is synchronized, delaying the start of the summing operation, and
increasing the size of each FIFO entry.

depth of the FIFO is arbitrary since we're concerning ourselves now with the size of each entry. We'll deal with the number of entries shortly.)

On the other hand, you could, for example, have the first core complete its work on the first row of the array and then deliver that entire row in one chunk to the second core. That second core would have to wait until the first core was done with the entire first row, sacrificing some latency. On the other hand, synchronization is much less frequent. As you can see in Figure 11.27, the size of each FIFO entry now has to be big enough to contain an entire row's worth of data, and the summer must scan that array in order to get the total.

Between these two cases, the total amount of data transferred is the same. The difference is that, in the first case, a small FIFO cell is required while frequent interconnect utilization is needed; in the second case, the interconnect is burdened less frequently, although perhaps for longer (depending on how the size of the chunk being communicated aligns with the interconnect structure), and more memory is needed for the FIFO in order to handle the larger chunk. These are system decisions that have to be made by the programmer; there is no one right answer.

Delaying delivery loop distance

Any concurrency delay due to a reduction in synchronization frequency is "voluntary"; that is, it's a design decision as opposed to being mandatory. Loop distance, however, requires a delay in concurrency in order to wait for the produced data to be ready for consumption. If the synchronization granularity — that is, the chunk size — is smaller than the loop distance, then several syncs have to happen before consumption can start. For example, if the chunk is simply an individual array cell and the loop distance is four, then four syncs have to occur before the consumer can start going. This means that the FIFO has to be at least four "chunks" deep.

This means that the minimum FIFO size can be calculated as follows:

- Divide the loop distance by the number of cells per sync. For example, if the loop distance is 8 and 2 cells are written per synch, then it takes four syncs to traverse the loop distance, and the FIFO depth must be at least 4.
- Multiply the FIFO depth by the chunk size, and you have the minimum memory needed for the FIFO.

Note that if one dependency has a loop distance that requires delay of synchronization, then other dependencies within that loop body — including ones with no loop distance issues — must also be delayed so that all are delivered at the same time.

Avoiding deadlock cycles

In order to manage data traffic, the actual communication of synchronized data can be designed to occur in bursts. So, for example, the producer may load up a FIFO, but the consumer may not read until the FIFO has some number of "chunks" populated just to bundle the reads together for better efficiency.

But such a design may involve a feedback path — a kind of side-channel indicating that the consumer is ready for more data. This can be managed on a "token" basis, with the consumer issuing a certain number of tokens against which the producer writes data.

We can easily construct a simple example [figure] of where deadlock can occur. If, for example, the consumer is awaiting 10 writes before reading, but the FIFO is only sized to hold 5 writes, then the consumer will never read the data and the system is deadlocked. Likewise on the return path: if the consumer successfully reads all the data and sends tokens back, but the producer reads the tokens in blocks of 10 and the FIFO has only 5 slots, then the system locks up waiting to send more data across.

These may sound like trivial scenarios, and, indeed, these specific examples are. The problem is that real-world cycles are typically not this simple, and they can be very hard to detect. The best way of avoiding the problem is to make sure that partitions don't create any cycles, that they're all "feed forward". As with the other partitioning considerations, this can be extraordinarily difficult to do by hand, but tools can help.

Providing elasticity

Some elastic space is also an important consideration. If the cores are doing more than one thing, then the consumer may not be consuming in lock-step right behind the producer, and the producer might be interrupted in its duties. If only the minimum FIFO size is used, then as soon as there is any producer interruption, the consumer is blocked. Likewise, if the consumer doesn't keep consuming, the FIFO will fill up and the producer will be blocked.

Blocked cores mean waste. But, even worse, they may mean that a blocked thread gets scheduled out — and that can result in a much more serious delay because, as we've seen, context switching is expensive.

For this reason, it's prudent to make the FIFO longer than the minimum. This gives some extra space for the producer to add more chunks if the consumer gets distracted for a moment. And, if the FIFO is more than minimally full, it means that, even if the producer gets distracted, the consumer will still have work to do.

Deciding how much elastic storage to build into the FIFO is not a trivial matter, and it cannot realistically be analyzed by hand. The interactions between the threads and the scheduling algorithms, coupled with possible bursty data arrivals or congestion between the producer and consumer, are extraordinarily complex. For data-flow languages, there are tools that can help determine how often a thread might be swapped out, but for the languages in common use, there is no such assistance available. It's entirely possible that the amount of elastic built into the FIFOs has to be adjusted once a real system is running real scenarios on real data.

Implementing a partitioning strategy

Once you've decided how you want to partition your program, you now have to make changes to the code to do that. This is a "re-factoring" exercise that, depending on the scope and number of the partitions, can be tedious and error-prone. You must break apart the selected tasks at the positions you chose and stitch things back together in a way that:

- creates new threads (SMP) or processes (AMP);
- replicates any storage to accommodate anti- and output dependencies;
- creates any critical sections if you chose that as an approach; and
- implements synchronization to repair broken true dependencies.

There's no way of sugar-coating this: forgetting any small detail can result in difficult-to-debug problems. A methodical approach is essential to ensuring that everything gets done properly.

Using tools to simplify partitioning

So far we have described what needs to be done in order for a program to be successfully partitioned. This is useful in concept, but, in practice, it's very hard to do manually. It tends to be the domain of very experienced specialists, since there's a high learning curve, and it's easy to get things wrong.

Over the last few years, tools have emerged that can help with some of the harder jobs associated with partitioning a program. Let's start by reprising the simplistic algorithm we proposed earlier in this chapter:

1. Choose a partitioning strategy
2. Identify the dependencies that the strategy breaks
3. Determine whether the broken dependencies can be handled
4. Estimate the performance benefit of the strategy
5. Repeat iteratively, either looking for better or additional partitions; make your decisions
6. Implement the chosen partitioning strategy

We can divide up these steps into two categories:

• Taking some kind of action or decision
• Getting information that supports the action or decision.

In particular, steps 2, 3, and 4 are of the second type: they are what tells you whether the partition you've chosen is viable or optimal. And they are particularly hard to do by hand. In fact, they really constitute the "hard part" of this whole business. Given the information gleaned in those steps, it's not so hard to decide on a set of partitions to implement. Implementing them is tedious, but can be a deterministic process that leads to a correct-by-construction result. In fact, it could, in theory, be automated itself. Except for the fact that most programmers want to write and understand their own code — they're not fond of some tool generating code for which they will be held accountable.

There are tools that can take a program that you have instrumented with parallelization instructions through pragmas or some other mechanism and do some of the mechanics of splitting it apart, but, in order for that to work, you must already know how you want to parallelize it. That's the hard part. So we won't consider such tools.

The tools we will look at help with the information-gathering and what-if evaluation of different ways of partitioning a plain, un-instrumented, un-annotated program.

Vector Fabrics's Pareon

This is the latest edition of a tool that has evolved over the last few years, starting with a more limited analysis program called vfAnalyst and progressing to its fuller-functioned incarnation as Pareon. Pareon can analyze programs written in C or C++.

Static and dynamic analysis test coverage

Pareon starts by using static analysis to identify compute dependencies, which can be found "by inspection". It then instruments the program in order to track all of the loads and stores that occur so that memory dependencies can be identified using dynamic analysis. Of course, this means that test coverage is important: if there are sections of code that aren't exercised, then any memory dependencies associated with that code will not be identified.

This is an important consideration: if you make partitioning decisions without full knowledge of all the dependencies, then you may create a parallel version that works fine until you encounter a flow that exercises those un-analyzed corners. When that happens, things can go wrong in potentially hard-to-debug ways.

So it's critical that you provide the program with a test set that provides as close to 100% coverage as possible. Pareon will provide a measure of how much of each subroutine was tested so that you can see if you need to improve the test set.

Dependency analysis

The analysis identifies all dependencies encountered, and it characterizes them by type and by whether or not it's possible to break and repair them. These dependencies can be viewed in a unique visualization approach that is illustrated in Figure 11.28.

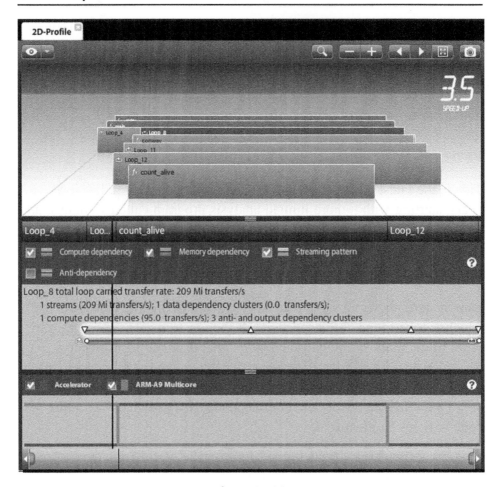

Figure 11.28
A view of the call stack and dependencies.

Each bar in the top represents a subroutine, and the length of the bar reflects the amount of time each routine contributes to the overall computation time. Longer bars become targets for parallelization, since that's where time can be saved. You can dive in layer-by-layer to see which routines represent the most work; the "3-D" view maintains visibility into this call stack.

Underneath those bars are lines that represent dependencies. The coloring of the line indicates the type of dependency, and clicking on a dependency provides information on the nature and cause of the dependency, including the ability to cross-probe into the source code to see where the dependency arises.

Parallelizing

While it may be interesting to explore the dependencies, that's only really necessary if one or more are getting in the way of parallelization. If you're looking to re-design the program in a way that eliminates the dependency, then this can be useful.

But one would typically first try to identify where parallelization is possible in the program as it exists. Pareon can take a routine and identify where it's possible to partition it in a way that's guaranteed not to cause issues like deadlocks or data-flow cycles (discussed above). By selecting a routine and telling the tool to partition it, the tool will come back with a series of options for partitioning (Figure 11.29), including:

- the implications of the partition in terms of the number of threads created
- the overhead introduced due to thread management, data synchronization, and increased pressure on the memory subsystem
- the factor by which the program can be sped up.

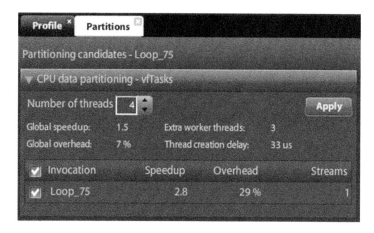

Figure 11.29
A display of partitioning candidates, including the cost and benefit of one of them.

By selecting one of them and "implementing" it, the display will now show the program with the selected routine parallelized and with the overall program speedup indicated in the upper right (Figure 11.30).

A "properties" view provides information regarding the cost of synchronization (Figure 11.31).

There is also a schedule view that shows the execution schedule, along with the dependencies that determine that schedule (Figure 11.32). This is particularly useful if the task you selected for partitioning yielded no useful partitions; the schedule will typically show where the blocking dependencies are. Such an inability to partition is not common, but might occur in certain cases involving complex recursion or broken ordering dependencies (section 5.4.3).

If you want to explore other options or change your mind for any other reason, you can "undo" any of the task parallelizations.

Of course, you can't really estimate performance accurately without considering the architecture on which the code will run. There's a mapping capability (Figure 11.33) that identifies which parts of the code get implemented on which cores. From this information, the cost of threading and the implications on cache and bus activity for that specific architecture can be estimated, providing a more accurate reflection of the performance implications of the partitioning strategy.

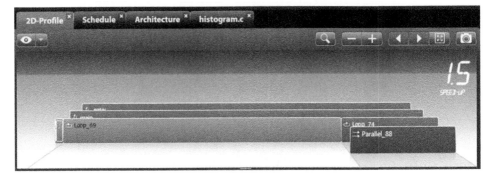

Figure 11.30
The program after the partition in Figure 11.29 has been applied; the overall program is sped up by a factor of 1.5.

When calculating the overhead of threading, worker threads are assumed (rather than creating and destroying threads). The architecture view allows you to describe whether or not a core is "lost" if not used (in other words, whether or not some other function can be scheduled on it). Based on that, the tool may select busy-wait threads or sleep-wake threads according to the overhead implications.

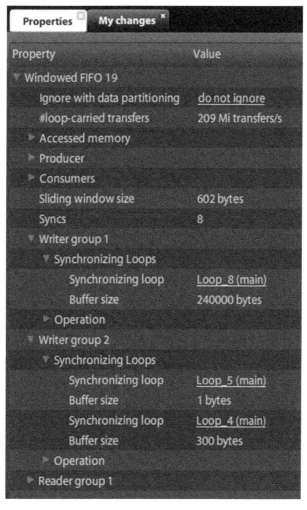

Figure 11.31
Dependency properties.

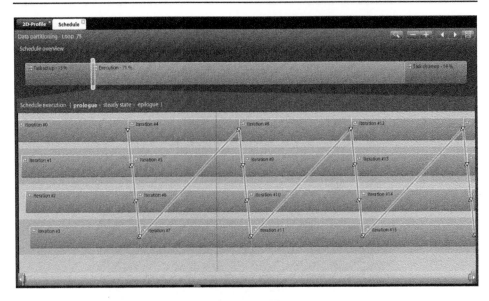

Figure 11.32
Schedule view, showing dependencies that determine the schedule.

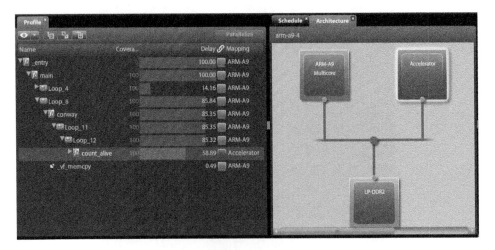

Figure 11.33
Mapping the tasks to a specific architecture.

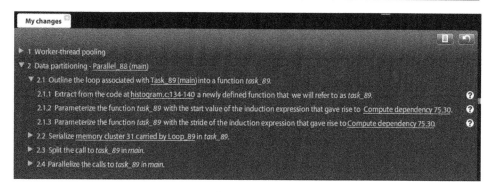

Figure 11.34
Re-factoring steps.

Re-factoring

Once the partitioning decisions have been made, it's time to implement them. The Pareon tool provides a set of re-factoring steps, down to a very specific detailed level, leading to implementation of the partitioning strategy. The steps can be done piecewise, allowing re-testing after each step to ensure that no errors have been introduced.

The tool can provide instructions on implementing:

- threading;
- storage replication;
- synchronization; and
- handling of special patterns like induction expressions and reduction.

It does not help with implementation of critical regions. It also assumes an SMP implementation. If an AMP version is desired, then the threading steps can give insight into the creation of separate processes, but the specifics are left to you (Figure 11.34).

The overall process of partitioning using Pareon is summarized in Figure 11.35.

CriticalBlue's Prism

Prism is another tool that helps with parallelization of code. At its core, it operates on the same principles as Pareon, but its working assumptions — and therefore how the user uses the tool — are different.

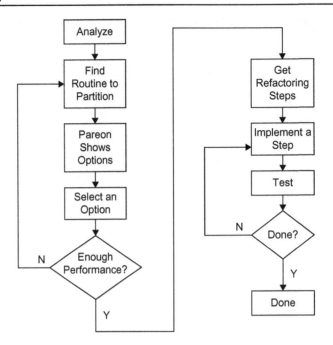

Figure 11.35
Partitioning process using Vector Fabrics's Pareon tool.

While Pareon focuses on finding places for partitioning a program, Prism works under the assumption that the user already has a good idea of what would be nice or good as places to partition, but doesn't know for sure what the impact would be — or if it's even feasible.

Users of Prism tend to fall into one of two camps: programmers new to multicore who are trying to figure out how to implement multi-threading efficiently, and experienced users who have already done their partitioning and multi-threading by hand and are now looking to optimize the performance.

Analysis

Prism starts by instrumenting and analyzing the original program, which can be either sequential or already multi-threaded. This gives you information on how your program is running as is. Some of the information resembles standard profiling information — which routines are running the longest, for instance, but also how often different routines are called, how memory is used, and how things are scheduled

Figure 11.36

The function profile for a jpeg encoder example. The macroblock encodes
(`mb_encode`) are mostly running in parallel.

(Figure 11.36). This also gives a good view of program data-flow, which can help a programmer understand what's going on as the program executes.

"What-if"

At this point, you can experiment with different ways of threading your program.

Forcing threads

You can start by "assigning" or forcing functions to new threads. You can use "generic" threads, which aren't bound to any particular processor (and are therefore less accurate in timing), or you can create a specific architecture and bind different threads to specific cores.

When you assign or "force" a function onto a thread, Prism simulates the scheduling that would occur. This is a "what-if" scenario: you're asking, what would happen if I put this function in a different thread? If the function then is scheduled perfectly in parallel with other functions, you get the best benefit you can from that particular assignment. More often, however, there's an underlying dependency that blocks full concurrency.

Prism shows you the blocking dependencies and allows you to see the source code that creates the dependency. By examining the dependency, you can decide whether it's possible to "remove" it from your program.

Forcing dependencies

If you decide that you can remove the dependency, then you need to see what the performance impact of that would be — is it worth the effort? Is there a different dependency that will still be in the way? So you can "force" the dependency off, essentially telling Prism to assume it's gone and simulate the new schedule.

Figure 11.37 shows an example of a situation where what could be parallel is completely serialized due to an anti-dependency.

By simulating the removal of (i.e., ignoring) the anti-dependency, much more concurrency is shown to be available (Figure 11.38), although full parallelization is still not possible due to another true dependency that became visible once the anti-dependency was ignored. The Huffman encoding calls used in the algorithm run serially (the smaller pieces cascading down the right side of each mb_encode), while the other portions of the mb_encodes run in parallel (the overlapping left side pieces).

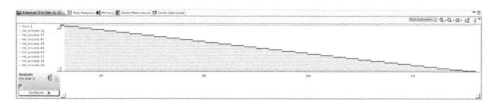

Figure 11.37
Jpeg encoding with macroblock parallelism serialized by an anti-dependency.

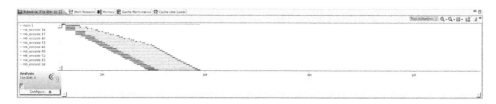

Figure 11.38
The anti-dependency is ignored (using what-if), and now the parallelism is much better, only limited by some true data dependencies. Note that the timescale shown here is the same as that of Figure 11.37.

In this manner, you can walk through the program, making successive assignments to threads and deciding whether or not each of those steps is feasible and makes a useful improvement in performance. The schedule simulates the overhead of threading (and can model new or worker threads), but not the overhead of simulated synchronization. If a pre-threaded program has synchronization explicitly built in already, then the synchronization overhead is measured as part of the overall analysis.

Other issues and optimizations

Prism also lets you visualize any data races, which are a result of incorrect synchronization, since the order in which the producer and consumer will access the data is not well defined and may happen out of

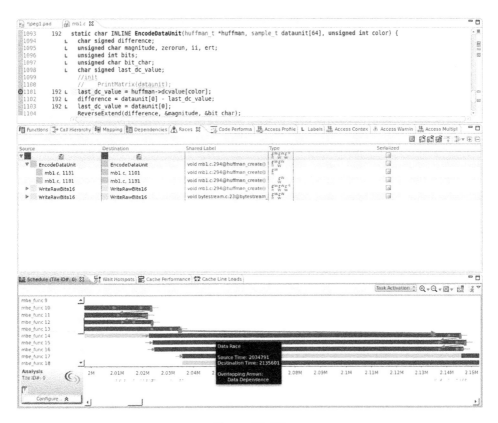

Figure 11.39
Data races are shown in red. Note that some races point forwards (races luckily won), while others point backwards (races lost, in this case, reading values before they were written).

order. Figure 11.39 shows an example where data races are visible as red arrows; those pointing forward were won, those pointing backward were lost. But in different runs under different conditions, the direction of any of those arrows might change.

You can force the scheduler to order the races correctly (Figure 11.40), allowing you to see if simply synchronizing properly will still give you good performance, or whether you end up losing performance because what seemed like useful parallelism turned out to rely on lost data races (as if parts of the program were "jumping the gun").

You can also see a summary of all protected accesses, including synchronization and critical regions.

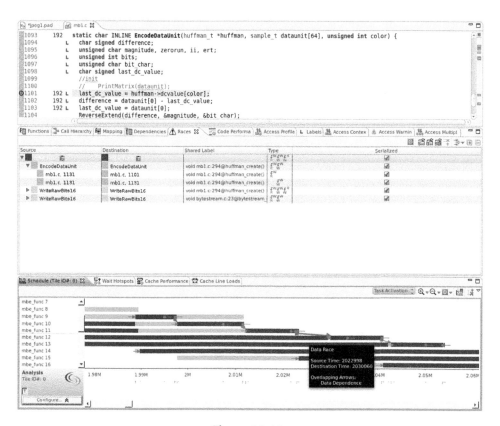

Figure 11.40
This shows "what-if" these races were fixed (serialized). The source code is unmodified, but the scheduler forces all races to be properly ordered — useful to see if the races could be fixed without spoiling the parallelism.

Experienced programmers tend to use Prism for figuring out why their multi-threaded code isn't running as quickly as it should be. All of the above elements can play into that analysis, but one of the biggest culprits is the cache. Prism can do cache hit and miss analysis (most profiling tools focus only on misses, not on hits) to help optimize how data is laid out.

An example of the kind of cache analysis done is shown in Figure 11.41. This table shows, for a particular cache line, how much of it was loaded but sat idle before it got swapped out. That is, of course, wasted cache space, and indicates that data structures aren't well-mapped to the cache.

While data locality is generally thought of as a good thing, with a single cache line serving up multiple pieces of data, that only applies within a single core. With multiple cores accessing adjacent data, they end up forcing each other to refresh their cache lines, slowing things down. This very often leads to code improvements like splitting structures into two, with one containing the "hot", or frequently-accessed, fields and the other containing the "cold", or infrequently accessed, fields. Prism helps with the analysis of which pieces of memory need to be segregated and which should remain local.

In fact, threading — even hardware threading — may not be the main factor limiting performance. Figure 11.42 shows three different implementations of the jpeg encoding algorithm. The second exploits

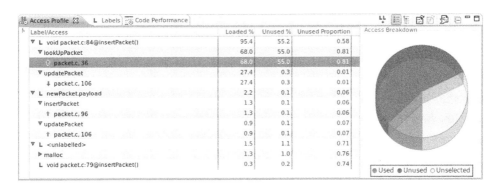

Figure 11.41

Rather than the usual cache miss percentage, this table shows how much of a cache line is loaded but not used before it is evicted — this points to poor data structure layout and/or data access patterns which can be optimized.

- Single core
- No hardware threads
- 8k/16k caches

- Single core with 2 hw threads
- **2k**/16k caches
- Cache misses limit speedup

- **Quad core** with 8 hw threads
- **32k/256k/8M** caches
- Dataset fits in cache

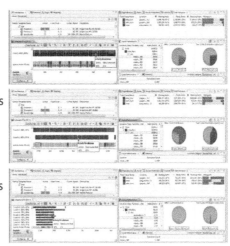

Figure 11.42

This compares how three different architectures will perform a jpeg encode (using the same time scale). Moving to two hardware threads but reducing the cache size yields no increase in performance; the cache is the limitation.

two hardware threads, but, with a small cache, yields a negligible performance improvement. The third option has more threads, but, in particular, has a cache big enough to hold an entire dataset, dramatically reducing the time spent interacting with memory.

Implementation

Once you have identified the threading approach you wish to take, then you can modify your program to implement the threading. Note that you could implement each improvement as you go, meaning that you first simulate the improvement of one change, then implement it, then re-analyze to get a more accurate measure before proceeding, or you can simulate an entire series and then implement them at once (although making small sets of changes and re-verifying the code at each step is highly preferable to coding all the changes at once and then trying to verify it all together; these code changes are error-prone and must be done right).

It's important to perform functional and performance verification as you proceed. Eliminating data races, for example, should be done before trying to further optimize threads. Ensuring that each optimization works correctly before proceeding with more changes ensures that you have a stable performance estimate from which to improve things further; it also

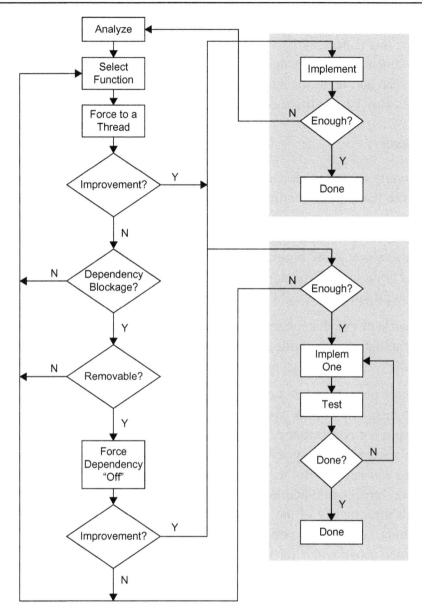

Figure 11.43
Partitioning process using CriticalBlue's Prism tool.

keeps you from accumulating bugs through several revisions that must then be untangled later.

A simplified summary of the process is illustrated in Figure 11.43. It shows both the "implement and try some more" and "simulate and implement all at the end" approaches (the gray boxes). As with the prior process, this is necessarily oversimplified. In particular, a given improvement on its own may seem inconsequential, but, combined with others, can add up to meaningful improvement. This is why several changes may be identified before any of them is implemented.

Summary

Partitioning a program is the single most difficult part of moving into a multicore paradigm. Identifying both available concurrency as well as blockers to that concurrency on a large scale is, practically speaking, impossible to do manually. The best way of approaching it is typically to isolate sections of the program that *seem* to offer the most promise and spend efforts there. Studying the algorithms on which a program is based can also provide clues as to where parallelization is possible.

This has kept this discipline to a few elite teams of programmers that have, through dint of battle, learned how to manage partitioning manually.

Tools that provide more systematic analysis of parallelization opportunities, as well as where improvements in a program design may open up more concurrency, can make the problem more tractable. This has only recently been possible, and the use of such tools has not yet become widespread. But as more embedded programmers are forced into multicore processors, and as the amount of parallel programming exceeds the capacity of the cadres of elite programmers, a combination of training, moderate experience, and tools can equip the everyday embedded programmer for multicore.

References

[1]   A. Edward, Lee. The Problem with Threads. UC Berkeley, Technical Report No. UCB/EECS-2006-1, January 10, 2006. <http://www.eecs.berkeley.edu/Pubs/TechRpts/2006/EECS-2006-1.html>.
[2]   A. Nieuwland et al. *C-Heap: A Heterogeneous Multiprocessor Architecture Template and Scalable and Flexible Protocol for the Design of Embedded Signal processing Systems*. Kluwer, (2002).

CHAPTER 12

Software Synchronization

Tom Dickens
Boeing Associate Technical Fellow at Boeing, Renton, WA, USA

Chapter Outline

Real World Multicore Embedded Systems.
DOI: http://dx.doi.org/10.1016/B978-0-12-416018-7.00012-2

Introduction

An application that uses concurrent code requires a means to synchronize the multiple concurrently executing algorithms. We will explore why and when synchronization is required, how to achieve synchronization, what can happen without synchronization or with incorrect synchronization, and the side-effects of good synchronization. We will also visit language-specific details along with design considerations to achieve the proper level of synchronization within an application. During this discussion we will explore tips and techniques to help you and your team implement a good thread-safe multicore strategy and hopefully provide real-world advice to help you avoid common pitfalls.

The focus here is limited to what an application developer can control and have visibility of from a high-level language such as C, C++, or Java.

With the language discussions and examples I will attempt to be succinct and brief while also thorough, but many times there are advanced and subtle nuances, not all of which can be included in a simple chapter on the subject. I encourage you to research the fine points of specific languages and language implementations to fully understand the design trade-offs and implications.

Disclaimer: *my experience comes from 20 + years of college/university teaching while also working as a Boeing engineer and Technical Fellow building desktop engineering applications for 25 + years. I have no knowledge or experience with Boeing-released software in any Boeing products; my comments, experiences, stories, and opinions herein by no means can reflect on any existing or future Boeing products.*

Why is synchronization required?

Synchronization is required in multi-threaded applications to ensure the correctness of the algorithm, to ensure the validity of the data being used, to properly allocate limited resources, and to provide good response to the user for user-driven GUI-based applications.

It is good to remind ourselves why we are embracing multicore for the masses now — simply because the majority of our computing platforms are now multicore. Historically most systems were not multicore, so most applications did not need to be (could not be) multicore friendly, so most programmers did not need to know multicore programming. Indeed, multicore programming and the needed coordination were inherently difficult and complex. In a famous quote by Seymour "father of supercomputing" Cray, the question was asked, "If you were plowing a field, which would you rather use? Two strong oxen or 1024 chickens?" (Seen as a quote in several books, but without specific citation.) This directly acknowledged the fact that coordinating multiple cores was indeed quite difficult; most people could visualize managing one or two large cores but could not picture managing a high number of cores. Unfortunately that golden era of computing is over [1] as hardware manufacturers are forced to scale into multiple cores to offer more performance, thus making programming for multicore systems the new norm. Understanding how to do this is now every programmer's mission and duty.

Given that multicore systems are now commonplace, let's explore the specific reasons for synchronization.

Data integrity

Data in an application can be classified with many different attributes with respect to synchronization needs (Table 12.1). With proper classification you can determine which data could require synchronization.

Table 12.1: Classification of Data Attributes

Data classification	Description/example	Synchronization requirement
Immutable	Data that cannot be changed. For primitive data types they can be declared as const in C/C++ or final in Java. *Caution: Care must be taken with pointers; the object the pointer is pointing to may be changed even if the pointer variable is declared const or final. This is a common mistake or oversight to watch out for.* *Additionally, a const declaration can be circumvented in C/C++ using the following technique:** `const ten = 10 ;` `*(int*)&ten = 25 ; // defeats const` In object-oriented languages such as C++ and Java, objects can be defined that have private data and only provide getter-methods. Objects, like String, take on a value when created and cannot be changed. `Example: Java: final int ten = 10 ;` `Example: C/C++: const int ten = 10 ;`	None
Mutable but local scope	Local variables are defined within a method and each call to that method creates fresh versions of the data in memory. If the method is called at the same time from two different threads, each thread will see their own private version of that data. *Caution: The use of static in C/C++ causes the local variable to take on a global scope, defeating the benefit of local scope.*	None
Mutable and more global or shared scope	This can include variables in any non-local scope, mutable variables in objects, variables mapped to shared memory location, etc.	Yes

*You would not design code like this, but others, when pressed to "fix" something, will do whatever it takes.

Another data issue requiring synchronization is stale data — the data is out-of-date with the rest of the application but is not retrieved from or pushed to the authoritative source of the data every time it is accessed. This can be caused by compiler/virtual-machine optimizations, caching of the data, etc. In these cases, synchronization is used to guarantee that the data is not out-of-date.

Atomicity

Atomicity is the ability or expectation that an operation, or a set of operations, take place as a single unit without being interrupted; this is also known as an atomic transaction. Obvious cases are when a set of operations needs to be done without interruption and without any other part of the application accessing any of the pieces involved in the transaction before the entire transaction is complete.

Less obvious cases are when the resulting processor instructions generated by a seemingly atomic statement in a high-level language are examined; one may find multiple operations where only a single operation was expected. In cases such as this, synchronization must be used to ensure that the set of operations is done from start to finish without allowing conflicting operations to be executed (Table 12.2).

Examples in actual code

Table 12.3 shows a simple example of two threads setting a 64-bit double value in parallel; thread 1 setting it to 3.0 and thread 2 setting it to 1/3. You would think that a read of that double would either return 3.0 or 1/3, but in a simple test case two other numbers were also returned. Note by looking at the hex values of the numbers, the 32-bit halves of the two numbers can become garbled as the first 32-bits of one number and the last 32-bits of the other creating two different numbers (not to mention the wild possibilities when C unions are used).

There are other cases in which multiple fields of data must be changed in an atomic operation or transaction, or else a read of that data could be a garbled mix or two valid multi-field data values. For example, when changing an address in a business contact you would need the address, country, state, and postal-code to change all at once since you would not

Table 12.2: Descriptions and Examples of Atomic Operations and the Need for Synchronization to Ensure their Atomicity

"Atomic" code	Description/example	Synchronization requirement
numValues++ ; values.add(val) ; mean = sum ()/numValues ;	Here we have three statements that modify three elements (an int, a list, and a double) when we add a new value to the set of values. Unless we synchronize these 3 lines of code to change all 3 numbers as an atomic operation, one thread could change 1 or 2 of them and another thread could get all 3 values and they would not be consistent. This is a typical example of the need for an atomic transaction.	Yes
variable++ ;	While seemingly atomic, not synchronizing a single statement like this* is a common synchronization mistake since the CPU-level operations are get-data, increment-data, and then store-data. Two threads could both be executing these operations and an increment could be lost.	Yes
if(val < min) min = val ;	This is a check-then-do pattern, and if not synchronized then 2 threads could do the checks simultaneously and then both try to change min to their different values. Not good.	Yes
if(p != null) p.doSomething () ;	This is also a check-then-do pattern, but worth mentioning as a common synchronization problem — another thread could set p to null thus causing a null-pointer-exception after the check in your thread. Instead of missing minimums as in the previous case, this causes run-time exceptions.	Yes
intVar = 4 ;	If the integer size is larger than the native data size of the processor or virtual machine, then it is not guaranteed to be atomic. Definitely a Yes for long types.	Maybe
realVar = 1.2 ;	If the floating-point size is larger than the native data size of the processor or virtual machine, then it is not guaranteed to be atomic. Definitely a Yes for double types.	Maybe
Object pointers: p = q ;	Like integer sizes, if the pointer sizes are larger than the native processor or virtual	Maybe

(Continued)

Table 12.2: (Continued)

"Atomic" code	Description/example	Synchronization requirement
	machine, then it is not guaranteed to be atomic. This is a common problem using 64-bit pointers on a 32-bit machine; a pointer-assignment actually results in two operations, which must be atomic.	
Using char: c = 'a' ;	Assigning a value to a char variable.	No**

*And all similar variations, i.e. ++i, i-, -i, i+ = 3, i * = 5, and so on.
**Synchronization is not required in my experience, but there may be some cases on some hardware/compiler that this may result in multiple operations— I personally would not trust it without making 100% sure.

Table 12.3: 64-bit Double Example (also Displaying Hex Bytes) Requiring Synchronization to be Set and Read Atomically

Thread 1 double (64-bit) value	Thread 2 double (64-bit) value	Garbled double values
3.0 $0 \times 4008000000000000$	0.3333333333333333 $0 \times 3fd5555555555555$	3.0000006357828775 $0 \times 4008000055555555$ and 0.33333325386047363 $0 \times 3fd5555500000000$

Table 12.4: Multiple Related Fields Requiring Synchronization to be Updated and Read Atomically

Before address	After address	Read of partial change
P. O. Box 3707 Seattle, Washington 98124 USA	Level 10, Exchange House 10 Bridge St Sydney NSW 2000 AUSTRALIA	P. O. Box 3707 Seattle, NSW 2000 AUSTRALIA

want a read of the information to be done with part of the change complete (Table 12.4).

Please note that the garbled read could occur even if the write started after the read had started, so both the read and the write of the address data must be synchronized to avoid garbled multi-field data.

Do these problems really happen? Yes, I have seen very subtle synchronization errors that crash applications, or worse, subtly change the answer in an equation or change the state of an application. Even if the chance of a synchronization error is 1 in a billion, with processors running typically at 2–3 GHz, that is 2–3 errors per second. If a synchronization error caused one pixel on the screen to be bad for one frame, that error would probably not be considered to be severe. However, if a synchronization error caused a financial transaction to not complete correctly, or caused a spreadsheet to calculate incorrectly, the consequences could be severe. The obvious synchronization errors are the best ones to have; they are highly visible and you know to fix them. The hard problems are the non-obvious synchronization errors that cause subtle errors in calculations that might not be noticed but will cause inaccuracies. (A highly publicized subtle software error, while not a multicore synchronization error but a very interesting study in machine precision and cascading inaccuracies, was the Patriot Missile Software Problem, which resulted in extremely severe consequences.)

It is also preferable if synchronization errors are detected in your code, since you can fix it and test it. In other cases, synchronization errors are found in others' code (open-source, vendor-provided, operating-system provided, etc.) as well as bugs in compilers, operating systems, and even in the hardware and CPUs themselves. Dealing with synchronization errors that are out of your control to fix directly can be much more difficult.

Sequence of processing

Programmers tend to write their algorithms with the operations in a very specific order, and they assume that the program will use the order they specified. It may surprise you to know that this is not guaranteed without synchronization, as compilers, virtual machines, and even processors have the ability to reorder operations as long as the reordering does not affect the outcome of <u>the logic of that thread</u> (Table 12.5). When considering multiple threads, the order that you specify may be changed and could result in errors. It has been widely known, when understanding

Table 12.5: Pseudo-Code in a C/Java Like Style Showing the Reordering of Operations Problem

Version	Thread 1	Thread 2	Result from thread 2
As the code was written.	waitForMe = true ; start_thread_2() ; value = 123 ; waitForMe = false ;	while(waitForMe) sleep(WAIT_TIME) ; print("value = " + value) ;	value = 123
A possible reordering	start_thread_2() ; waitForMe = true ; value = 123 ; waitForMe = false ;	-same-	value = 0 (assuming value was initialized to 0, or else a random value)
A possible optimization	start_thread_2() ; value = 123 ;	-same-	Loops forever since waitForMe was not used in Thread 1 so changing it was optimized away.

the detailed specification of the Java Virtual Machine and the Java Memory Model, that the contract for optimization was to maintain logic correctness "as long as the reordering is not detectable within that thread". (Similar conventions hold for C/C++ compilers and libraries, but they are not standard.) These reordering can occur by the compiler and also by the run-time JIT optimizer. Additionally, instruction reordering can also occur inside the CPU! This fact took me completely by surprise as I learned it from a fellow student, an Intel engineer, during a class we were taking in multicore techniques. The instructions in the on-chip queue can be reordered, plus there are assembly-language fence instructions, which the on-chip reordering cannot jump over. (These fence instructions cannot be inserted by the programmer; they are used by the compiler to ensure correctness.)

If your algorithm depends on multi-threaded logic and the order of operations being executed in one thread is critical to the correct logic in another thread, then the code must be synchronized to prevent reordering from occurring by any of the possible down-stream actors, causing grief in your code.

The C/C++ standard does not address operation ordering/reordering and the effects on multi-threaded applications, thus different compilers will implement this differently.

Java does specify the following effect: "The semantics of the Java programming language allow compilers and microprocessors to perform optimizations that can interact with incorrectly synchronized code in ways that can produce behaviors that seem paradoxical" [2].

The take-away lesson here is to not assume that the order of what you wrote will be the order of what gets executed unless you take great care with synchronization.

Access to limited resources

Limited resources in an application can include files, databases, sensors, I/O ports, etc. Care must be taken to ensure that two or more threads do not try to open or access a limited resource at the same time, as either access errors will occur, or, if the system allows multiple threads to open and write to the same resource (file for instance), then chaos can result. If synchronization is used with all access to limited resources, then a thread will wait for the resource to be free before being able to use it.

Critical timing for real-time

A concern with real-time systems is that certain sections of code run with expected execution time. With multiple threads in your application and the thread scheduling left up to the underlying operating system, the dedication of code to a processor cannot be guaranteed. Your code may be swapped in and out of a processor multiple times while it is running and the overall time of execution of that code can vary wildly.

Problems with not synchronizing (or synchronizing badly)

From experience, someone new to object-oriented programming will tend to throw everything into an object definition, or will continue to write sequential code and just wrap a class around huge methods and have all of their data global. Similarly, someone introduced to design patterns will implement too many patterns everywhere in a disjoined manner or will try to force their entire application into just one pattern. When new to synchronization one tends to either (1) synchronize everything without an architecture (since, of course, "more is better"), or

(2) miss many critical areas that require synchronization. (Even when one thinks they are synchronizing everything, without careful analysis many things can be missed.) The consequences of synchronizing badly can include poor performance, logic errors, and more power consumption.

Slower throughput

The implementation of locks typically takes dozens to hundreds of processor cycles depending on the instructions used, the processor-to-processor coordination required, and the required reads/writes on the memory. The cost of fine-grained synchronization (i.e., synchronizing a value + + operation) can easily be hundreds to thousands of times more costly than the cost of the operation when the interaction between threads and lock-contention is taken into account. Sections of code that were proven to not be thread-safe can be made 100% thread-safe but can also result in degraded performance of 2 to 3 orders of magnitude. (Yes, you can turn unsynchronized code that executes in 1 second into correctly-synchronized code that now takes many minutes, even more than an hour, to run. Correctness has a cost.) Such results can lead to the (erroneous) conclusion that synchronization is bad; it could be the specific synchronization implementation was not optimal. The general goal with a multi-threaded algorithm is to use the available hardware to get better performance; if the best performance is achieved by only using one of the available cores and running things strictly sequentially, then that may be the preferred solution.

Execution speed can also be greatly affected by naively breaking up a sequential problem into parallel threads, synchronizing correctly, but having each thread trying to read/write the same data. The result is these threads are falling over each other with cache misses and moving the same data from processor-cache to processor-cache as it is kept in sync across multiple threads. (The term "cache churning" describes this effect.) This can turn a simple efficient sequential algorithm into a parallel implementation that has much slower throughput. This does not need to be the case, but care must be taken to ensure faster throughput when parallelizing.

Errors in synchronization logic

Properly synchronizing your code is not easy. A classic example (and the source of humorous but sad examples and arguments on the net) is how to initialize a singleton (to ensure a class only has one instance, and provide a global point of access to it) in a thread-safe manner. A singleton with lazy initialization is even more difficult to correctly implement. Implementations may use the infamous double-checked lock (also known as an evil antipattern; http://www.javamex.com/tutorials/double_checked_locking.shtml.), the use of which is almost always an indication that an easier and better way exists. Double-check locking was also used to try to minimize the performance cost of synchronization in the past, but now most synchronization implementations are more optimized, or highly optimized preferred alternatives are available (a result of multi core being pushed to the masses on commodity devices).

Other problems, detailed in a following section, include deadlock, live-lock, race conditions, and so on.

Consumes more power

Power consumption is a key constraint and design requirement for battery-operated devices. With synchronization, a thread waiting for the lock may constantly re-check and thus keep burning through power on the device. Synchronization with low power consumption may be a goal for your application and new and device/system-specific synchronization techniques may be available to minimize power consumption that the developer should be aware of.

Testing for proper synchronization

In addition to developers needing to thoroughly understand synchronization and the issues involved, the testers of the application also need to understand the thread-safe issues and how to test for correctness. A difficulty with testing code for proper synchronization is that the execution and interaction of multiple threads is not deterministic and can be affected by many external factors. Unfortunately, testers, by definition, will execute the same tests over and over, especially when automated testing is used, so they tend to see the same behavior time after time. (I am a strong proponent of extensive automated testing and also in TDD — test-driven development.) Users, on the other hand, will interact with the application in subtly different ways, or wildly unexpected different ways, which can greatly affect the application's behavior and can expose thread-safe problems which were never noticed in testing. In addition, the user (or tester) may only rarely witness a thread-safe problem, which they cannot reproduce due to the non-deterministic nature of multi-threaded programming. If they cannot easily reproduce a problem it may be ignored or otherwise be attributed to be user error and not considered a flaw in the application.

It is out of scope of this modest chapter on software issues with synchronization to dive deeply into issues and solutions (or best practices) with testing synchronized applications. Suffice it to say that the success of implementing synchronization in an application heavily relies on informed, thorough, and inspired testing.

How is synchronization achieved?

Thus far we have discussed the need for synchronization and problems that can occur with it when adding synchronization to an application. We will now dive into specific details on what techniques and offerings are available to the application developer to implement synchronization.

In this section, we will use object-oriented pseudo-code to describe the concepts in a simple manner. The specific language implementation may or may not be object-oriented and will involve a lot of other details. At this point in the discussion we are still at a notional and conceptual level and anticipate the reader to apply these examples to their specific environments.

Since synchronization can be complex and unfamiliar to many on the team, it is strongly suggested to clearly adopt a simple set of synchronization techniques and patterns for your application/project/ team, and to use these in the same manner consistently. Otherwise you will find that the synchronization of the application is not obvious, causing confusion in maintenance and causing the application to be, or become, fragile to modify, even for a development team of one. (I know from experience when looking at complex code I wrote in the past.)

Exclusion

Exclusion is the high-level concept in multi-threaded applications of guaranteeing that only N different threads are allowed to do something. If N is 1, then the exclusion is referred to as MUTual-EXclusion (MUTEX, also commonly written as "mutex"), guaranteeing that two different threads of execution cannot both be executing a piece of critical code at the same time. A critical section of code can require exclusion due to many reasons, such as treating that section of code as an atomic transaction, managing shared resources, maintaining data integrity, managing available processor loading, minimizing cache collisions, or guaranteeing required operation ordering, to name just a few.

There are many synchronization primitives commonly used to achieve exclusion. In addition to being used to implement exclusion, many of these synchronization primitives can be used for additional synchronization needs, such as a thread blocking until another thread's work is complete before it continues, coordinating support threads or children threads, and so on. Given the set of synchronization primitives available in your development environment, as defined by your language, available support libraries, and operating system offerings, you can map out your strategies to use the available synchronization primitives to implement all of these needs.

Semaphores

A semaphore has the ability to be initialized with a set number of "permits", and is then used to "acquire" a permit (either given the permit or rejected), and is used to "release" the previously acquired permit, allowing the semaphore to give it to someone else. A semaphore that only allows one permit is sometimes called a "binary semaphore"; one that allows more than one permit may also be called a "counting semaphore."

Table 12.6 should generate some questions, such as, "How can both threads see s10?"; "What work is being done?"; "Can you ask the semaphore how many permits it was initialized with and how many are

Table 12.6: The API for a Simple Semaphore Object

Semaphore API	Example use

class Semaphore {
    init(int numPermits)
    acquire()
    release()
  }
	Thread 1:
	    s10 = Semaphore.init(10) ;
	    s10.acquire() ; // block until acquired
	    doWork() ;
	    s10.release() ;
	Thread 2:
	    s10.acquire() ; // block until acquired
	    doWorkMore() ;
	    s10.release() ;

Table 12.7: Creation of 20 Threads, But Limited to Running 10 at a Time Using a Semaphore

Initialization code	Thread work code

s10 = Semaphore.init(10) ;
for(i = 0; i<20; i++) {
  t = thread.create(doWork(i))
;
  t.start() ;
}
	doWork(int i) {
	  s10.acquire() ; // blocks
	  // do something using i
	  s10.release() ;
	}

currently available?"; etc. The scope of s10 would need to be such that the multiple threads could see it. The offerings in the semaphore API are specific to the implementation and may include many other methods (notionally availablePermits(), numPermits(), numWaitingThreads(), setNumPermits(num), tryAcquire(), etc.), but the fundamental definition of a semaphore is these three abilities.

Table 12.7 is another example using a semaphore to limit a large number of threads to only run N at a time.

In the initialization code, the semaphore is created specifying 10 permits, and then 20 threads are created, each of them running the doWork() method. When the initialization is done, there are 20 threads running doWork(), but 10 of them are blocked at the s10.acquire() operation. When one of the lucky 10 threads that got the s10 permit is ready to leave doWork(), it releases the s10 permit. At this time, one of the 10 threads waiting in the s10.acquire() call will return from that call and can start the work in doWork(), and so on until all 20 threads have finished. Not shown here is the initialization thread waiting for all 20 of the "children" threads to complete their work, but, depending on the specific semaphore API being used, it could wait until all permits have been returned; it could track and wait for all the children tasks; it could use a countdown latch that is decremented by each thread finishing doWork(); and so on.

If a semaphore is initialized with only one permit, the semaphore can then be used to only allow one thread at a time into a critical section. This is the definition of a lock and is a key synchronization primitive for the mutex.

Locks

A lock is a mechanism for controlling access to a critical section of code (i.e., a shared resource) by multiple threads. Usually a lock provides exclusive access to only one thread at a time. However, some locks may allow coordinated access to a shared resource concurrently by multiple threads, such as a read-write lock that permits multiple readers but requires that a writer obtain exclusive use and will cause all new readers to block until the write is complete (Table 12.8).

This is the fundamental pattern for synchronization. Yet in this example, there is actually no guarantee that the doWork() method will only be called when the correct lock is acquired. (What, you can't believe the comments in the code? What if some other thread, running some other code (that someone else wrote), just calls doWork() without acquiring the lock? Synchronization broken!) A "better" architectural design to

Table 12.8: The API for a Simple Lock Object

Lock API	Example use

`class Lock {`
    `acquire()`
    `release()`
`}`
	Thread 1:
	    `lock.acquire();`
	    `doWork(); // exclusive use guaranteed`
	    `lock.release();`
	Thread 2:
	    `lock.acquire(); // block until acquired`
	    `doWork(); // exclusive use guaranteed`
	    `lock.release();`

Table 12.9: Using a Lock to Make a Method Exclusive

doWork method	Example use

doWork() {
  lock.acquire();
  // exclusive use guaranteed
  // do work here
  lock.release();
}
	Thread 1:
	  doWork();
	Thread 2:
	  doWork ();

**Table 12.10: Method-Scope Locking Syntax in
C# and Java**

```
C# method MUTEX syntax
    [MethodImpl(MethodImplOptions.Synchronized)]
    public void doWork() {
      // do work here
    }
Java method MUTEX syntax
    public synchronized void doWork() {
      // do work here
    }
```

guarantee the exclusive execution of the doWork() logic is to place the lock acquire and release in the doWork method (Table 12.9).

Notice that this code is much simpler, as each thread does not need to have code to manage the locks but the lock code is pushed into the doWork() method itself. We still need to deal with the initialization and scope of the lock, which is implementation specific. Most importantly, any other code that calls doWork() will also use the lock and will thus be properly synchronized.

The next step in simplification is for languages to have built-in syntax for using locks. There are language constructs in C# and Java, for example (Table 12.10), which simplify the synchronization to the method level (as detailed in the Languages chapter). C++ does not currently have built-in language constructs for synchronization; however, the proposed C++11 standard has a thread class, synchronization objects called promises and futures, a thread_local storage type giving thread-specific data, and the async() function template for creating concurrent tasks.

Recursive locks

The simple lock will only allow a single acquire() to return; all additional calls to acquire() will block until the code that acquired the lock subsequently release the lock. In perfectly architected and perfectly implemented code, this works well. However, if the thread that has acquired a lock tried to acquire the same lock again, that thread will wait for itself to release that lock, but that thread is now blocking in the acquire () method. Not good! This is deadlock but can be caused by the actions in just a single thread. Of course you would not write code with nested locks on the same lock, but if doWork() manually implements locking in the method, and doWork() calls another method that also implements locking on the same lock, we have deadlock. Debugging this is not easy, and when applications are written in large teams, the coordination of lock details can become confused and out of sync with each other.

The solution to this problem is to use a recursive lock, also known as a re-entrant lock. Instead of the lock just tracking the state (locked, unlocked), it also tracks which thread currently holds the lock as well as tracking a lock-counter. (In all but the simplest lock implementations, the thread is tracked to ensure the same thread releases the lock.) When a thread acquires the lock, that thread is remembered and the lock-counter is set to 1. If that same thread acquires that lock again, the lock-counter is incremented while other threads trying to acquire the lock will block on the acquire() method. When the thread that has the lock call release(), the lock-counter is decremented, and, if it reaches zero, the lock is free to be acquired again. If the lock count is still greater than zero, the thread would need to release the lock additional times to properly free the lock. This requires all acquire() and release() methods to be carefully balanced so a lock will not be left locked when the thread believes it has fully released the lock. (This is especially tricky in exception-handling code where an error occurs; making sure locks are released is a key use of the finally clause in try/catch blocks, along with closing files or releasing other resources.)

Shared-exclusive (Read-write) locks

As mentioned above, some locks may allow coordinated access to a shared resource concurrently by multiple threads. A common one is the read-write lock that permits multiple readers to acquire the lock but requires that a writer obtain exclusive use and will cause all new readers to block until the write is complete. Specific implementations of a ReadWriteLock vary and may be configurable when created. Here is a list of some of the various details a ReadWriteLock would have to handle.

- Re-entrant – is the lock re-entrant, both for the reading and writing?
- Acquisition order – when locked for reading and acquire requests for both reading and writing are waiting, how is it decided which request to grant when the lock is released?
  - Based on the order of requesting the lock?
  - Write requests first?
  - Read requests first?
  - A random choice from all pending requests?
  - Allow a priority for the acquisition requests?
  - Other based on the application logic?
- Downgrading – can a thread that currently owns the lock for writing change their request status to reading without the need to block?
- Upgrading – can a thread that is currently allowed to read upgrade their lock to a write?

While quite useful, a ReadWriteLock can be tricky to fully understand and complicated to use in a clear and intended manner as there may be subtle unaccounted cases that yield unanticipated behavior.

Spin locks

While much of the implementation details of how synchronization is accomplished (handled by the OS, the kernel, on the processor) is hidden from the application developer, it is key for application developers to understand the effects of specific synchronization techniques in order to choose the best synchronization techniques for their requirements.

Blocking methods used for synchronization will usually put the thread to sleep when waiting, which incurs the overhead of context switching. When the waiting time for a request is expected to be very short, and the work to be done when the request is granted is also small, the overhead of context switching becomes too large in proportion.

A "spinlock" avoids the context-switching overhead and delay; the spinlock approach causes the thread to wait by "spinning" in a loop, checking repeatedly until the request is accepted. This has advantages and disadvantages. The advantage is speed, specifically avoiding the overhead of context switching, allowing quick wait/accept turnaround. The disadvantages include keeping the thread active while it is waiting, which consumes both a processor and power, the fact that it does not free up the thread to be used for other work, and, if the thread spins too long, the OS may still swap it out with context switching.

The best advice is to be aware of the needs of your application, be aware of the specific synchronization offerings available in your environment, and make the best design/performance decisions that you can. Discussion and coordination of the needs/desires between the high-level application developer's point of view, the OS group, and the hardware group on your project can help determine and specify the exact synchronization needs and solutions.

Condition Variables

A Condition Variable is an object that has two primary methods, wait() and signal() (or similar names according to the specific implementation, such as wait() and notify()). Code that calls the wait() method will block until some other thread calls the signal() method on that Condition Variable, causing the blocking wait() to return. Many implementations of Condition Variables specify their use with a "monitor" concept, and, if multiple threads are waiting, only one will be released per signal(), and the signal() method can only be called by the thread that owns the monitor. In my experience, Condition Variables are not commonly used by the application developer, but they are used for low-level synchronization primitives.

Test and set; compare and swap (CAS)

The typical mutex implementations are considered heavyweight, typically involving kernel-level context switching. With the widespread use of synchronization being used in more and more applications, heavyweight mutex implementations are being replaced or supplemented with much lighter-weight, but similar, capabilities. Application developers really don't care about the underlying implementation details; they are concerned with the capabilities and the performance results (consequences).

The "new generation" (a relative term) atomic synchronization offerings use a "new generation" of atomic processor instructions; test-and-set, or compare-and-swap. These instructions take three inputs, a value, a newValue, and an address. They atomically compare the passed-in value with a value in memory specified by the address, and only if they are the same will the instruction replace the memory value with the new value. These instructions also return a success/fail status, such as a Boolean value or the value that was read from the memory address — if that value is different from what it was, indicating that it was set by another thread and not by the current thread. The software-implemented synchronization code can then decide what to do in the case of failure, which will need to use the new value in memory and try the CAS again or return failure itself.

The CAS instructions typically are expensive compared to other processor instructions, yet they are extremely lightweight when compared to typical mutex implementations. The java.util.concurrency package that was introduced in Java 5 was considered a breakthrough in 2004 as it "effectively exposes atomic instructions such as CAS to the Java programmer" [3]. Look for implementations of synchronization offerings that take advantage of CAS operations in more languages and synchronization packages in the future.

Barrier

At the conceptual level, a barrier in parallel applications is used to synchronize work done in multiple threads, causing them all to wait at the barrier until all of the threads reach the barrier. There are both software barriers and hardware barriers.

Software barrier

The concept is like a semaphore, but in reverse: instead of allowing up to N threads to not block, a barrier will block all threads until a set number is reached, then all are released. An implementation will be initialized with a count and will decrement the count and block each thread, and, when the count is complete (reaches zero), all blocked threads are released. The Java class CountDownLatch gives this functionality and can be used for a variety of needed algorithms.

Memory barrier

A memory barrier, also known as a memory fence, a fence instruction, or a member, is an instruction on the CPU used to prohibit processor-queue reordering over any given fence instruction. Many processors have processor-queue reordering for performance optimization; any reordering is guaranteed to not affect the sequential logic of the operations. However, instruction reordering can affect multi-threaded applications, and having the CPU reorder your application's logic at the chip-level can break thread-safe code.

In high-level languages, which are our focus in this section, we do not have processor-level control and cannot (should not) have in-line assembly-language code embedded in our high-level code to do special things. (Programming in a mixture of C and assembly language is fun, but also tedious and not portable.) We need to rely on the synchronization primitives in the language and support libraries to give the proper low-level control, including correctly placing memory barriers in the generated code when required. This is one key example of the many things outside the control of the application developer that must be properly in place to be able to write thread-safe multi-threaded code.

Architectural design

The largest impact an application developer can have on writing thread-safe multi-threaded code is through the architecture of their application. Designing segments of the code that do not require or rely on multi-threading reduces the percentage of code that needs to be written with synchronization in mind. The enforcement of segments of your code that require thread-safe programming and those that don't can be done using comments and conventions; however, the more you can enforce that in the code syntax itself, the better − to reduce errors when the comments are not read and the conventions are not followed.

The techniques mentioned here are to give the reader a flavor of the types of architectural design elements that can be used to help create thread-safe code, but we cannot cover this topic in great detail here. It is suggested that the reader investigate this further for their language- and domain-specific requirements: there are a lot of excellent resources on the subject out there.

Architect to limit and contain synchronization code

Writing correct thread-safe multi-threaded code is not easy (hence the motivation for this text). Instead of having 100% of your code sprinkled with synchronization, it is sensible to limit the synchronization details to a few specialized areas of code. This compartmentalization of the synchronization details allows your thread-safe expert to write the key synchronization code while the rest of your development team can focus on your business logic and user-facing interface and interaction. (We see that, in most projects or teams, one person will emerge as the expert in thread-safe multi-threaded techniques, and many others on the team do not want to know the gory details.)

Use immutable objects

In object-oriented programming, an immutable object is an object that is assigned values to all of its data upon creation, but whose data cannot be changed. Once created, that object can be accessed by any thread since the data's value (state) will not change. To change the value of an immutable object, you must create a new immutable object with the new value and then use the new object in place of the old object. If you keep pointers to the old object, you would require a convention or mechanism to modify the pointer to now point to the new object. One approach is called an "object handle", which is a pointer to an object. When the immutable object is changed, you would also change the object handle to it; all subsequent accesses using the object handle would get the new immutable object. Care must be taken to not store any direct pointers to the object, which is known as a "data escape" (also known as "escaping references") and can thwart the best-intended thread-safe architectural design.

Data's scope limited to one thread

If the scope of a piece of data can be limited to be read or changed only in a single thread, then access to that data does not require synchronization. Data requests to access the data can be sent to the data-owning thread and snapshots or copies of the value can be returned. Similarly, requests to modify the data's value can be sent to the data-owning thread, which again will handle the request. Since only one thread is managing and directly touching the data, atomic operations or transactions are easy to implement. Care must be taken to avoid data escapes; all returned values should be done using either copies of the data or immutable objects.

Avoiding data escapes and understanding copies

To protect the integrity of a piece of data (an object) and to ensure that its scope is indeed limited to a single thread, we must ensure that that data (object) does not escape, which would allow direct access by other threads or other parts of the code. As mentioned above, a typical scheme to avoid a data escape is to only return a copy of the data which is intended as a read-only copy, but there are different types of copies to be aware of. (Care must be taken to understand data, and pointers to that data. A "data escape" occurs when there are multiple pointers to the same piece of data, and one or more of these pointers exists outside the control of the object that owns that data. Once that escape occurs, the controlling object cannot ensure the data is valid.) To change the data, the controlling object would implement a setter method to change the data, but care must be taken with the passed-in value; the value should be copied into the data so that an external pointer to that value does not now exist in the data. This would result in multiple pointers pointing to the data, one of which would be out of the control of the controlling object.

There are both shallow copies and deep copies of data, and you need to understand the differences and consequences.

The original object A is shown in Figure 12.1(a). A has three values: the number 5.7 and pointers to the objects B and C, which have values of

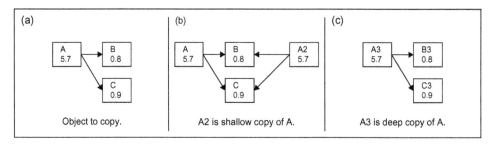

Figure 12.1:
Showing (a) the original object to copy, (b) a shallow copy, and (c) a deep copy.

0.8 and 0.9, respectively. The values in B and C are important to the object A and are considered part of the state of A.

A shallow copy of A, as seen in Figure 12.1(b), will create a copy A2 of the object A, but will also replicate any pointers to objects (B and C) so that both A and A2 are pointing to the same B and C objects. In this shallow copy, if the number 5.7 in A2 is changed, the number 5.7 in A is not changed. This is good. However, if the value in the first pointer from A2 is changed, which will change object B, then the object A will also see that change since both A and A2 are pointing to the same object B. The same is true for the pointers to object C. Thus, in a shallow copy, the data values defined directly in the object (A or A2) are not shared between the original and the copy, but any data (objects) referenced by pointers (i.e., B and C) are shared. Both A and A2 rely on the values in B and C, and a change in either B or C will change the resulting states of both objects A and A2. So even though A2 is returned to avoid a data escape, the data for objects B and C has indeed escaped.

A deep copy of A, as shown in Figure 12.1(c), will copy all objects needed to define A; thus the value 5.7 is copied to A3, the objects B and C are copied to objects B3 and C3, and the A3 pointers refer to these copies B3 and C3 instead of the same objects that A is pointing to. Thus, a change in the value of B3 will affect the state of A3 but will not affect the state of A; the same holds true for C3. Where a deep copy gets deeper is when any of the pointed-to data objects have pointers themselves (i.e., if B pointed to E, F, and G). Then B3 would need to point to copies of E, F, and G (i.e., E3, F3, and G3). In turn, if E, F, or G had pointers, we would continue diving deeper. But the goal is to have the copy, A3, completely disconnected from object A and any pointed-to objects from A.

There are, however, two notable exceptions to this. The first is if the pointer is to an immutable object, whose value, by definition, cannot change; then it is OK for two objects to point to the same object. The second exception is if the pointer is to an object that is needed for reference but that does not define the state of A — for example, a pointer to the main GUI window that allows A to refresh the window when its data is changed. Both A and A3 can point to the same GUI window object. The key take-away here is, to avoid data escapes, you should make deep copies of returned data, but you should be careful in nested objects to make the correct deep copy.

Segregate synchronization code into a synchronization package

To keep the majority of your code free from needing to be multicore-aware, you can move all of your application's synchronization code into a particular directory or package so that none of your application's other code directly calls any synchronization method. Some of the methods in your synchronization package may be just simple cover methods (that is, single-line "wrappers" for synchronization code) with a single lower-level synchronization method call. Other methods may be more complex with interaction between multiple synchronization primitives and supporting data structures. Using a synchronization package standardizes the use of the synchronization primitives and can also be used to limit the set of synchronization primitives your application should use (don't provide cover methods for things you don't want used).

Another key benefit from having all synchronization code in a synchronization package is that you can instrument the synchronization package with many cross-checks and diagnostic code, such as tracking all acquires and releases and looking for non-matching use; being able to easily know which threads are blocking on a lock at any given time; knowing which thread owns a lock at any given time; and so on. Diagnostic code can be disabled or turned to an error-only mode in the release version of the application, but it is a wonderful capability to have in your system to quickly identify and diagnose problems.

Thread-safe frameworks and libraries

Many development teams use thread-safe frameworks or libraries. Be they from outside the project or developed within the project (or developed by the project to be leveraged by other projects), a thread-safe framework or library provides proven thread-safe capabilities that can give your application access to the multiple cores available. As we have seen, the details and implementation of thread-safe and multicore-aware software is not easy, so there is a large benefit in using proven and hardened capabilities from a mature thread-safe framework or library.

A thread-safe library typically provides synchronization primitives and support to give the developer higher-level synchronization offerings than the programming language and fundamental environment provide. When using a thread-safe library, the application developer must still thoroughly understand synchronization issues and details, and care must be taken to use the library correctly.

A thread-safe framework goes further by providing a cohesive set of thread-safe capabilities, typically for a specific domain such as matrix functions, database operations, or graphics processing. When using a thread-safe framework, the goal of the framework is to provide thread-safe multicore-aware capabilities to the average developer without requiring the developer to know synchronization details beyond the need to use the framework to do things correctly. This allows the thread-safe experts in the project to package their knowledge in a framework for the project to use, while protecting the project from thread-safe mistakes made by well-meaning but inexperienced developers.

There is a lot of work being done on thread-safe libraries and frameworks; it is good to search around for mature capabilities to leverage on your project before developing your own.

Specific conditions requiring synchronization

There are a number of specific conditions or situations that require proper synchronization, and these have their own specific definitions. As in many things, defining something is easy, but finding these situations or conditions in existing code, or when writing new code, is not always clear.

Data races

A data race is when two or more threads access and change a shared data value at the same time. There are numerous instances of data races, and they can be difficult to spot; a simple example here will demonstrate the problem. Without proper synchronization two of three threads will read the same value, and they will subsequently both write their calculated new values. One write will occur after another, and since the initial value for both threads was the same, the contribution of the first thread to finish writing to the resulting value of the data will be lost.

In the example in Table 12.11, there are three threads, each adding 5 to the data value x, which was initialized to 100. The end result should be a value of 115 when all three threads are finished. However, as we can see in the details, both threads 1 and 2 get the data value of 100 in steps 1 and 2, so they both write values of 105 as seen in steps 3 and 4. At this point, things are wrong: one of the increments of 5 has been lost. The last thread gets the just-modified value of 105, adds 5, for a resulting value of 110 instead of 115.

With proper synchronization around the access and writing of the data value x, as seen in Table 12.12, the three threads will work with the data value in a sequential manner instead of in parallel. All three threads try to acquire the lock at the same time, but only one will be successful (and not necessarily in order of threads 1, 2, then 3). The remaining threads will be blocked in the "acquire lock" call until the lock is available and one of the waiting threads acquires it, and so on.

Table 12.11: Three Threads Adding 5 to x, Showing Data Race with Unsynchronized Code

Operation #	Data value x	Thread 1	Thread 2	Thread 3
0	100			
1	100	get x (100)		
2	100	$x = x + 5 = 105$	get x (100)	
3	100	put x (105)	$x = x + 5 = 105$	
4	105		put x (105)	get x (105)
5	105			$x = x + 5 = 110$
6	105			put x (110)
7	110			

Table 12.12: Three Threads Adding 5 to x, Fixing Data Race
Synchronized Code

Operation #	Data value x	Thread 1	Thread 2	Thread 3
0	100			
1	100	acquire lock	acquire lock	acquire lock
2	100	get x (100)	acquire lock	acquire lock
3	100	$x = x + 5 = 105$	acquire lock	acquire lock
4	100	put x (105)	acquire lock	acquire lock
5	105	release lock	acquire lock	acquire lock
6	105		get x (105)	acquire lock
7	105		$x = x + 5 = 110$	acquire lock
8	105		put x (110)	acquire lock
9	110		release lock	acquire lock
10	110			get x (110)
11	110			$x = x + 5 = 115$
12	110			put x (115)
13	115			release lock

Please note that the resulting time required for all three threads to work with the data value x has increased from seven operations to 13, including the acquiring and releasing of the lock. The idea of operations is notional; the time for the different operations can be quite large and the synchronization operations are typically quite expensive. There are a few things to realize from this simple example.

- When synchronizing to achieve thread-safe code, the resulting multi-threaded performance will be degraded from the non-synchronized version, but the answer will be correct.
- When multiple threads are all trying to acquire the same lock, all but one of them will block until they can in turn have the lock.
- The simple sequential code to give the same functionality is much quicker and simpler than the multi-threaded version, as seen in Table 12.13.

At this point the "obvious" take-away from this simple example could be to not make anything multi-threaded as the resulting code will run slower than the simple sequential version. That is the wrong message, yet a key realization in becoming a multicore thread-safe expert and implementing efficient multi-threaded systems.

Table 12.13: One Thread Adding 5 to *X* Three Times, Showing Sequential Code for the Same Functionality

Operation #	Data value x	Thread 1	Thread 2	Thread 3
0	100			
1	100	get x (100)		
2	100	x = x + 5 = 105		
3	100	x = x + 5 = 110		
4	100	x = x + 5 = 115		
5	100	put x (115)		
6	115			

Multiple threads working with common or shared data can cause a severe bottleneck; the real benefit in parallel threads running on multicore is when the amount of thread-local work on each of the threads is maximized between the need to work with shared data or resources. Any bottleneck between threads will cause the threads to block, wasting their ability to accomplish work in parallel. This is the cause of a classic dilemma implementing thread-safe multi-threaded code for multicore systems: the easy and obvious sections of code to farm out to a large number of threads also can be dependent on the same data, thus the parallel implementation could be rife with bottlenecks and not achieve the desired speedup; it might even result in a much slower implementation compared to the simple sequential implementation. The challenge, now that you understand this multicore dilemma, is to design your application to be able to have multiple threads working for you while minimizing the synchronization costs as well as minimizing the thread-blocking bottlenecks. Some of the tool-assisted techniques discussed in the Partitioning chapter include the performance cost of synchronization when estimating the impact of a particular partitioning strategy.

Put in a more quantitative way, you can classify the type of work a thread will do as either thread-local (W_{loc}) or needing a shared resource (W_{sr}). (Much more detailed quantitative models can be created, but this simple model will be useful here.) We want to maximize the ratio of W_{loc}/W_{sr} to keep the threads as busy as possible without suffering from contention of shared resources. When the ratio is low, the threads spend the majority of their time blocked waiting for shared resources.

Deadlocks

Deadlock is a condition where two threads are both waiting for the other thread to complete or to release a shared resource. The classic example is when a thread must obtain locks A and B to proceed. Thread 1 has lock A and thread 2 has lock B (or vice versa) and both threads will now wait forever to obtain the second required lock. A less obvious but still fatal example is two threads needing to each lock two tables in a database to be able to do their work. Thread 1 has locked the Salaries table while thread 2 has locked the TotalPay table. They both now need to lock the other table to be able to do their database updating. Classic deadlock!

The solution to deadlock is to avoid the requirement to acquire multiple resources in an unexpected sequence. For example, it could be policy that lock A is always acquired before lock B, which would avoid the example above. This assumes that if a thread has acquired lock B it won't discover a requirement for lock A, or that if that is the case, it will release lock B then acquire locks A and then B in that order. This may sound doable here — even simple — but in a large complex application the clearness of the synchronization details may become blurred and hidden. Other "fixes" are to try acquiring the needed locks and have time-out code in case of failure that will release all already-acquired resources and then will try again. As mentioned earlier, instrumenting your code to track all synchronization-specific things, like acquiring and releasing locks, will give you the diagnostic information needed to identify a deadlock and to determine how and where in your code to fix it.

Issues with deadlock are a good reason to keep your synchronization code isolated to part of your application with clear policies to implement your designed architecture. Even better is to implement a system that enforces your synchronization standards and architecture to help keep your team from making mistakes and causing synchronization issues.

Livelocks

Livelock is similar to deadlock but the threads involved are not all blocked waiting for each other but are continuing to consume processing power doing non-useful work while waiting for the locks. A naive implementation to avoid deadlock could be to check if the second lock is available before acquiring it.

```
1)    lockA.acquire();
2)    while( !lockB.available() ) {
3)        doSomethingWhileWaiting();
4)    }
5)    lockB.acquire();
```

This implementation is fraught with problems, and does not solve the issue of failure. A "better" attempt could be:

```
1)    while( !(lockA.available() && lockB.available()) ) {
2)        doSomethingWhileWaiting();
3)    }
4)    lockA.acquire();
5)    lockB.acquire();
```

Even though line 1 will block in the while loop until both locks A and B are available, which is good, there is still a problem. Unfortunately by the time lines 4 and 5 are executed, some other thread may have acquired one or both of the locks, so the locking problem we were trying to solve has not been solved. Generally, the more complex and convoluted your synchronization code becomes, the more hidden and unexpected problems are there; when your solution is simple and elegant, it is generally correct.

Non-atomic operations

As mentioned above in section 1.2 and Table 12.2, atomicity is the ability or expectation that an operation or a set of operations take place as a single unit without being interrupted, also known as an atomic transaction. As also detailed above, there are many operations in a high-level language that may appear to be atomic but are not. Other important and subtle factors affecting atomicity include the underlying hardware architecture (32-bit or 64-bit, or with small embedded systems even 8- and 16-bit) and assigning data values or pointer values: assigning values that have sizes larger than the underlying hardware architecture may take multiple CPU operations and may be interrupted before completing the assignment.

No matter what the cause for needing to ensure that a set of operations be completed in an atomic manner, in multi-threaded applications synchronization is required to ensure automaticity.

Data caching

Variables defined in your high-level language will actually exist somewhere in the hardware, from a register in a core to a memory cache on a core, a memory cache shared among cores, main system memory (plus many other device-specific possibilities and devices not yet designed) or even swapped out to disk in a virtual-memory system. It is sometimes useful for a developer to imagine a variable just living "in memory" in a single place and always being available to any code, but having a better understanding of the true situation (memory model) occurring and the performance trade-offs that happen is key for the thread-safe programmer. For example, if a core suspects it might read something from memory, it could pre-fetch the value to be ready to use it.

There are a lot of performance optimizations that occur both in the hardware and also in the compilers, but the stated correctness for such optimizations is that they will not affect the sequential logic of the operations. (This works in the old era of sequential computing, but we need a lot more in this new era.) These optimizations include operation reordering, as discussed in section 4.3.2 on barriers, but also include many techniques dealing with variables. A loop that is reading and writing a variable may be optimized to read the variable once into a local register, then while in the loop the value will be updated just in the local register, and the new value will only be pushed back into memory at the end of the loop. This works for sequential logic, but will break if you have two or more threads all reading and writing the data value at the same time. This is a type of data race, but it is made worse by the compiler optimizations and the changing location of the data.

There are options in some languages to force "deep" reads and writes of variables that will flush or reload caches and registers, such as the "volatile" and "transient" attributes, and there are ways to influence the scope of a variable such as the "static" attribute or the placement of the variable in the source file. (C/C++ has the register keyword, which asks the compiler to only store the variable in a register for performance.) The use of these language-specific keywords should be done carefully since the desired affect may not actually be realized. (I have seen language standards implemented differently in different compilers, some of which were outright buggy. Also, fairly obscure compiler features tend to be much more problematic due to lack of testing.)

Another way to enforce deep reads and writes of variables is to use synchronization in your code. Using a synchronization directive will cause variables used within the synchronization block to be deeply read and written (if changed). Depending on your language, the exact synchronization primitives available and the exact guaranteed side-effects will vary — be careful in choosing and using synchronization primitives, and be sure that they give you the exact behavior you need. Care must also be taken to clearly document your program's behavior and your use of synchronization primitives so the code can be understood and maintained in the future.

Conversion for endianness

If different cores or processors in your system (on-board, multi-board, distributed, networked, etc.) employ unlike byte-ordering architectures (big-endian versus little-endian) or floating-point representations, then any transfer of data among these components needs to be properly converted. This need touches on concepts from both atomicity and data caching, and care must be taken.

How to implement synchronization

In this section we have explored a variety of circumstances where synchronization is necessary to achieve thread-safe code. We hope you appreciate the wide range of situations that require synchronization and can see these situations within your own code. Once you identify a synchronization need you need a plan to properly manage it.

How to avoid a data race

To avoid a data race, all reads and writes involving that variable need to be synchronized with locks. This can take fairly simple code and make it much more complex, and, unless done carefully, you cannot guarantee that the synchronization that is implemented cannot be bypassed. (I tend to not trust conventions and documentation, but prefer compiler or IDE (integrated development environment) enforcement.) To avoid having the same synchronization code scattered throughout your application, it is good practice to use accesser methods (also known as getter and setter methods). Following are examples in C, C++, and Java to protect the double-typed value from data races and also from atomicity problems. (An assignment may take multiple operations, not a single operation, and can thus be interrupted.) A variable with the double data type is typically 8 bytes long, but can vary depending on your architecture. For atomicity, even a read of the value must be synchronized, since an assignment to the value may take two operations, and a read from the value when only one assignment operation is done will cause a very strange number in value, as demonstrated in Table 12.3.

C/C++ data race example syntax

```
#include <pthread.h>
pthread_mutex_t valueLock = PTHREAD_MUTEX_INITIALIZER ;.
double value = 0 ;
mutex_t valueLock ;
void init() {
   pthread_mutex_init( & valueLock, NULL ) ;
}
/** Get value using synchronization. **/
double getValue() {
   double temp ;
   pthread_mutex_lock( &valueLock ) ;
   temp = value ;
   pthread_mutex_unlock( &valueLock ) ;
```

```
      return temp ;

}

/** Set value using synchronization. **/

void setValue(double newValue ) {

   pthread_mutex_lock( &valueLock ) ;

   value = newValue ;

   pthread_mutex_unlock( &valueLock ) ;

}
```

C# data race example syntax

```
private double value = 0 ;

/** Get value using synchronization. **/

[MethodImpl(MethodImplOptions.Synchronized)]

public double getValue() {

   return value ;

}

/** Set value using synchronization. **/

[MethodImpl(MethodImplOptions.Synchronized)]

public void setValue(double newValue ) {

   value = newValue ;

}
```

Java data race example syntax

```
private double value = 0 ;

/** Get value using synchronization. **/

public synchronized double getValue() {

   return value ;

}

/** Set value using synchronization. **/
```

```
public synchronized void setValue(double newValue ) {

   value = newValue ;

}
```

As we can see from the examples, each language implements synchronization using different syntax and different primitives. For C and C++ we need to use an external synchronized library — pthreads is a common choice — to add synchronization capabilities to the language. Both C# and Java have synchronization constructs built into the language, and, as you can see, this allows synchronization to be implemented with much less and much cleaner code.

An alternate technique to avoid a data race for read/write access to a single variable is to use a data object that uses the CAS non-locking methodology. Java has a good offering of atomic types in its java.util.concurrent.atomic package.

Java data race example syntax using an atomic variable

```
private AtomicLong value = new AtomicLong() ;

/** Get value using Atomic Variable. **/

public double getValue() {

   return Double.longBitsToDouble( value.get() ) ;

}

/** Set value using Atomic Variable. **/

public synchronized void setValue(double newValue ) {

   value.set( Double.doubleToRawLongBits( newValue ) ) ;

}
```

There are a few interesting aspects with this atomic-variable example. First, if the data type was an `int` or `long`, then cover methods would not be required as all of the CAS protection is built in, and any code and thread that would access the value would be protected. To work with the data type `double` (as Java did not provide an AtomicDouble class) we need to map the 64 double bits into the 64 bits of a long, and back again.

To place this conversion code in only one place (I had to look up how to do this; you can't expect developers to know how to do this every time value is accessed) we include it in the getValue() and setValue() methods. The conversion along with the set/get will definitely span multiple operations so even though we are using a CAS building-block we still need to synchronize the method to ensure atomicity in the transaction.

How to avoid deadlock/livelock

Avoiding deadlock and livelock is not as simple as just synchronizing the access methods, since these conditions are the result of synchronizing poorly or not following your synchronization architecture. (Your application does have a detailed, documented, and known synchronization architecture, right?) The best advice is to:

- limit the number of locks your application requires (use CAS-based capabilities for synchronization whenever possible);
- limit the number of simultaneous locks any code requires (never more than one is the goal);
- limit the scope of each lock (don't use a single lock for multiple purposes);
- acquire locked resources in a careful predetermined and orderly manner (a simple scheme such as alphabetic can work well and be easy to follow); and
- use lock-leveling with a defined hierarchy of locks.

The bottom line with avoiding deadlock is to have a carefully planned and strictly adhered-to policy for dealing with locks in your applications.

How to guarantee atomicity

As we have seen in section 1.2, things that look atomic or that we assume are atomic from a high-level language point of view may not, in fact, be atomic. There are other cases where multiple fields of data must be changed in an atomic operation or transaction.

For single pieces of data that must be atomic, the CAS synchronizations are the preferred way to guarantee atomicity since they do not use locks and are generally much more efficient. For larger data-types and also for multiple data fields, synchronization through locks is the way to go for changeable data; but immutable objects can be a much better way to go if you can change pointers to point to a new immutable object using a CAS operation. However, in many cases locks must be used, so extreme care should be taken to avoid deadlock and livelock problems.

Language support for implementation

Intro

Some languages have no support for threading and synchronization, and others have built-in support, while yet others are somewhere in between. Of course, you can write a thread-safe properly running multi-threaded parallel application using any language, but some languages help you more than others. All of the synchronization primitives discussed above can be implemented in any language, on any system. However, unless the language and development environment you are using has specific support to help to enforce and manage these ideas and provides a robust offering of the needed synchronization primitives, the only way to be successful is through extreme discipline in your code.

This coding discipline is not easy to achieve and can be easily thwarted by a seemly innocent change in a single line of code. In industry, it is not sufficient to just write a thread-safe properly running multi-thread parallel application; the meaningful challenge is to write a thread-safe properly running multi-thread parallel application that can be easily maintained through the life of the application, which could be decades, and that easily (automatically) scales to run efficiently on future hardware.

Fortunately there is support for languages (add-on or built-in) that provides some help here, from a little help, which still requires strict discipline to correctly use, to more help, which helps to force the developers into correct usage. (I have not yet seen a language that prevents a "clever" developer from circumventing the desired architecture.) At this point in time, even the established languages that now strongly and natively support multi-threading must be used with

care and discipline to hope to create thread-safe multi-threaded code —
we are still in the era where a seemly innocent change in a single line of
code can still break a well-written thread-safe multi-threaded application.
(There is a constant exploration in new languages to "solve" this
problem, but we have not yet seen one emerge as *the* language of choice
for industrial-strength programming.) The only hope we have is an
intelligent and well-trained development team that thoroughly
understands these concepts and needs; thus the reason for us offering
this text.

We are also seeing an emergence of libraries and frameworks addressing
various synchronization needs, from a collection of primitives the
developer must carefully understand and use to self-contained
frameworks that hide all of the synchronization details from the
developer. Expect to see a lot of growth here in the future.

Language features and extensions

What follows is a brief summary of programming languages and
extensions and support for threading and synchronization. Please refer to
the chapter on Programming languages for a more in-depth discussion.

C

C has no native support for threading or synchronization.

C++

C++ has no native support for threading or synchronization. However,
the "new" C++11 (known as C++0x until it slipped into the 2010s)
adds a threading library and synchronization support:

> *"Unquestionably, the most important addition to C++11 from a
> programmer's perspective is concurrency. C++11 has a thread
> class that represents an execution thread, promises and futures,
> which are objects that are used for synchronization in a concurrent
> environment, the async() function template for launching concurrent
> tasks, and the thread_local storage type for declaring thread-unique
> data." [4]*

According to Bjarne Stroustrup, the creator of C++, in a January 2012
C++11 FAQ [5], the current state and availability of C++11 is:

"Availability of C++11 compilers: currently shipping compilers (e.g. GCC C++, Clang C++, IBM C++, and Microsoft C++) already implement many C++11 features. I expect more and more features to become available with each new release. I expect to see the first complete C++11 compiler sometime in 2012, but I do not care to guess when such a compiler ships or when every compiler will provide all of C++11."

Cilk++

An extension to C and C++, Cilk was developed at MIT from 1994 where it has matured and has found its way to be offered commercially as Cilk++ and Cilk Plus from Intel. It adds keywords to C/C++ to let the developer/programmer indicate areas of their code to be executed in parallel while leaving the scheduling details to the underlying Cilk system.

Objective C

Objective C has no native support for threading or synchronization.

C#

C# has a rich offering for parallel programming, using the Task Parallel Library (TPL) in their .NET Framework.

Java

Java has a rich offering for parallel programming, using their `java.util.concurrent` package.

Libraries

There are many available libraries that support multi-threaded programming; these are detailed in the Synchronization libraries chapter of this book.

Patterns

Design patterns have been popular in software development since the 1994 "Gang of four" text was published [6], which presented the concepts in a language-neutral format with code examples in C++ and SmallTalk. Design patterns are expert solutions to recurring problems in a certain domain; they expedite the development process by providing tested, proven paradigms. Design patterns also provide a standard vocabulary to communicate in the development community. There have been a number of publications detailing these design patterns in specific languages such as C# and Java.

With this new era of computing mandating multicore-aware programming, texts have been written detailing design patterns in this domain. (For example, *Patterns for Parallel Programming* by Mattson *et al.* and *Concurrent Programming in Java: Design Principles and Patterns* by Lea). These texts help the reader tackle the challenges of multicore programming by showing techniques and best practices condensed into thread-safe multicore-aware design patterns for easy consumption, replication, and extension. Learning and understanding the design patterns for thread-safe multicore software can help in the understanding and partitioning of your code as you convert it or design it for multicore.

The primary concept with design patterns is to use the same patterns, and similar implementations of these patterns, in all areas of your code so that any developer looking at any part of the code can quickly see common themes and can quickly understand the algorithms in the code. It is also possible, with well-organized code based on design patterns, to refactor many similar sections to use common and shared building blocks of code that can be used to implement the chosen patterns.

The Mattson text breaks the topic into the following four categories.

Finding concurrency design patterns

This category assists in an analysis of the problem and existing code, looking for high-level architectural concepts or practices that can be used as a model for implementation. Without this 100,000-foot view of your system, it is difficult to architect a good easy-to-build and easy-to-maintain solution.

- Task decomposition: how can a problem be decomposed into independent tasks that can execute concurrently? Which tasks are computationally intensive? Which tasks can be implemented in a scalable algorithm?

- Data decomposition: how can a problem's data be decomposed into units that can be operated on relatively independently? What is the granularity of the data? Can the decomposition scale?
- Group tasks: how can the tasks that make up a problem be grouped to simplify the job of managing dependencies? The groups are formed according to CAN, MAY, or MUST tasks be run concurrently. Pick a grouping that simplifies the dependencies.
- Order tasks: do tasks in a group require ordering in sequence, must they run concurrently, or are they truly independent?
- Data sharing: given a data and task decomposition for a problem, how is data shared among the tasks? Is this data access read-only or read/write access?
- Design evaluation: is decomposition/dependency analysis good enough to move on to the next design space, or should we revisit the decomposition and refine it?

Algorithm structure design patterns

These are the common design patterns used to implement parallel algorithms.

- Task parallelism: after decomposing a problem into a collection of tasks that can execute concurrently, how can this concurrency be exploited efficiently? Consider tasks, dependencies, and scheduling.
- Divide-and-conquer: suppose the problem is formulated using the sequential divide-and-conquer strategy. How can the potential concurrency be exploited? Can other logic be posed or restated as a divide-and-conquer algorithm, and can common divide-and-conquer strategies be used throughout the code?
- Geometric decomposition: can an algorithm be organized around a data structure decomposed into concurrently updatable "chunks"?
- Recursive data: with operations on a recursive data structure (list, tree, or graph) that appear to require sequential processing, how can operations on these data structures be performed in parallel?
- Pipeline: when performing a sequence of calculations on many sets of data, how can the potential concurrency be exploited?

- Event-based coordination: is it possible to organize groups of semi-independent tasks interacting in an irregular, but data-flow-ordered fashion?

Supporting structures design patterns

These are small structural pieces for your code to implement parallel algorithms.

- Single-program, multiple data (SPMD): all units of execution (UE) execute the same program in parallel, but each has its own set of data. Different UEs can follow different algorithmic paths.
- Master/worker: a set of tasks is managed by a master thread, which tracks dependencies and assigns tasks to available worker threads.
- Loop Parallelism: given computationally intensive loops, translate the loops into a parallel program while retaining needed dependencies.
- Fork/join: tasks are created dynamically (forked) and later terminated (joined), supporting a varying number of concurrent tasks.
- Shared data: this involves understanding, limiting, and managing needed data access that is shared between concurrent tasks.
- Shared queue: concurrent units of execution that safely access a common queue.
- Distributed array: an array partitioned between multiple units of execution in a manner that is both readable and efficient.

Implementation mechanisms

These are the specific mechanism(s) used to implement your chosen parallel strategy.

- Unit of execution management: the creation, destruction, and management of the processes and threads used in parallel computation.
- Synchronization: the enforcement of the constraints on the ordering of events occurring in different units of execution. This is primarily used to ensure that shared resources are accessed by a collection of units of execution in such a way that the program is correct regardless of how the units of execution are scheduled.

- Communication: the exchange of information between units of execution, ensuring that the results are correct regardless of how the units of execution are scheduled.

Side-effects of synchronization

As we have seen, synchronization is needed to write thread-safe multicore-aware software. In an ideal world, synchronization would work with no side-effects (it would be easy to just synchronize every method, data access, or line of code; compilers could do that), but alas, there are unwanted side-effects. That is the challenge of the experienced thread-safe multicore software developer: to use synchronization effectively to get all of the benefits while minimizing the side-effects.

Incorrect synchronization

Perhaps the largest problem with synchronization is the assumption that the synchronization logic is correct, or that if you call a method that is synchronized, then there is nothing else to consider or worry about. For example, suppose, for the purposes of automaticity as shown in Table 12.3, you have a method for setting the address that is synchronized as it sets all required fields that comprise an address. Now consider reading an address and the opportunities for side-effects or errors to occur:

- The read address is not a copy but a pointer to the data, and while reading the address the data is changed and garbled.
- The read address is not a deep copy, and while reading the address the pointed-to data is changed and garbled.
- There are also getters for the individual fields, so getCity() followed by getCountry() could result in "Seattle" "Australia" even though all of these methods are synchronized. The expectation would be to call both getters as an atomic operation without the country changing after you have retrieved the city.
- The read address is a deep copy, but the same deep copy object was cached (for efficiency — a clever programmer could erroneously think it efficient to cache the deep copy if the data had not changed) and then returned two different times. If this deep copy is changed

by one of the recipients, that address was just changed for the other recipient too. (It is possible to make an immutable proxy deep copy of an object to avoid data escapes.)

• Everything is synchronized "correctly" (correctly for expected behavior, but errors and exceptions happen and every possible error and exception possibility needs to also be taken into account and handled correctly), but a read in another thread crashes for some reason (perhaps from an out-of-memory error trying to make immutable proxy deep copies of every data value returned), which blocks all other reads as the MUTEX is locked and never is unlocked.

• Other side-effects that are specific to your implementation and system.

There is no magic formula to ensure you have synchronized correctly. It takes lots of experience, walking through real-world synchronization problems and solutions step-by-step (see the Debugging chapter), and seeing the plethora of side-effects to know what to look for and to quickly "see" the root-cause of a synchronization side-effect you are witnessing. Even knowing or expecting that a synchronization problem could be the cause of a program bug is a huge step in the right direction.

Program execution

The goal of using synchronization is to achieve the desired program logic, or execution, but at a much higher rate of wall-time execution than running the same logic sequentially. If this is not the case, why would you consider anything other than a simple sequential algorithm? Here we discuss the correctness of the program execution; later we discuss the performance.

There are new types of error that only occur in multi-threaded programs, and some of those only manifest themselves on multicore hardware. Consider deadlock and livelock, which were discussed in detail earlier. Both of these conditions occur due to contention for locked resources. If your program is sequential, there would be no need to lock any resource, since the single execution thread would know the exact state of everything at all times. It could not be blocked by other work since it is the only thread doing work. Similarly, there would be no possibility of a data race.

The addition of multiple threads — and thus the addition of needed synchronization to a program — adds a new dimension of complexity to the architecture and coding and, perhaps with even more of an impact, to the maintenance of that program. The dynamics of new developers coming on board the team, developers experienced in the program moving on to other assignments, and the program being migrated onto other hardware platforms can break the synchronization architecture and conventions if they are not clearly understood. Any of these changes in the dynamics of the program team and environment can wreak havoc on the correct execution of the program.

Priority inversion

Priority inversion can occur in parallel logic when a lower-priority task (thread) acquires a resource that is currently free, then while holding that resource a higher-priority task (thread) requests that resource but, due to synchronization, that higher-priority task is blocked by the lower-priority task. This is not a trivial or obscure problem; the Mars Lander problem in 1997 was an example of priority inversion.

Performance

As stated above, the goal of using synchronization is to achieve the desired program logic, or execution, but at a much higher rate of wall-time execution than running the same logic sequentially. However, as we have seen, besides adding complexity to your program, synchronization adds code, and that code needs time to execute. We have seen that the classic mutex implementations are considered heavyweight, typically involving kernel-level context switching. The overhead of synchronization can be hundreds or thousands of clock cycles per switch. If the performance gain is not greater than the cost of synchronization, then the resulting performance of the program, even though it is running on multiple cores, will be slower than running sequentially on a single core.

Another term for looking at this is the "granularity" or "coarseness" of the parallel logic. In fine-grained parallel logic, a task is spawned that does a small amount of work: using order-of-magnitude numbers, it could take 1000 cycles to spawn a task that does 100 cycles of real work. However, for large-grained parallel logic, for the same 1000 cycles to spawn a task that task could perform 1,000,000 cycles of real work. Let's define P_s as the number of cycles to spawn parallel tasks, W_r as the real work done per task, W_t as the total work to be accomplished, and N_c as the number of cores available to do the work. The resulting wall-time execution times T_w would be $P_s + W_t/N_c$. Let's consider the wall-time execution for fine-grained versus coarse-grained parallel logic for different numbers of cores.

We see in Table 12.14, for fine-grained logic, you would need at least four cores to get an improvement in wall-time execution, and two cores would cause a degradation in wall-time execution. Coarse-grained logic, however, sees a significant improvement in wall-time execution, yet as more cores are added, the improvement-per-core diminishes. Of course, given different values for these variables, you will get different results, and P_s is really a function of N_c so the diminishing effect is actually worse, but the trends are there in the reality of the equations.

The key take-away here it that coarse-grained parallel logic can more easily result in an improvement in wall-time execution and has the potential to show a much greater improvement.

Code complexity

It should be clear to the reader at this point that adding synchronization logic to a program adds complexity, perhaps significant complexity. The challenge is to add the synchronization to make the application thread safe with improved performance while not making the code so complex or convoluted that no one except the person who wrote the code can understand it. I have seen simple and clear algorithms that were unrecognizable when converted to run in parallel.

Table 12.14: Fine-Grained Versus Coarse-Grained Parallel Logic for Different Numbers of Cores

	1 core	2 cores	4 cores	8 cores	16 cores
Fine-grained 1600 cycles of work	$T_w = P_s + W_t/N_c =$ $0 + 1600 = 1600$	$T_w = P_s + W_t/N_c =$ $1000 + 800 =$ 1800	$T_w = P_s + W_t/N_c =$ $1000 + 400 =$ 1400	$T_w = P_s + W_t/N_c =$ $1000 + 200 =$ 1200	$T_w = P_s + W_t/N_c =$ $1000 + 100 =$ 1100
Coarse-grained 160,000 cycles of work	$T_w = P_s + W_t/N_c =$ $0 + 160,000 =$ $160,000$	$T_w = P_s + W_t/N_c =$ $1000 + 80,000 =$ $81,000$	$T_w = P_s + W_t/N_c =$ $1000 + 40,000 =$ $41,000$	$T_w = P_s + W_t/N_c =$ $1000 + 20,000 =$ $21,000$	$T_w = P_s + W_t/N_c =$ $1000 + 10,000 =$ $11,000$
Fine-grained speedup		$= 1600/1800 =$ $0.89 \times$	$1600/1400 = 1.14$ $\times$	$1600/1200 = 1.33$ $\times$	$1600/1100 = 1.45$ $\times$
Coarse-grained speedup		$160,000/81,000 =$ $1.98 \times$	$160,000/41,000 =$ $3.90 \times$	$160,000/21,000 =$ $7.62 \times$	$160,000/11,000 =$ $14.5 \times$

Here are some suggestions and best practices to help reduce the complexity (or confusion) in your code while addressing the requirement of thread safety and improved performance:

- Use the same synchronization patterns: establish specific ways to synchronize your code, including method and variable names, so that, once a developer knows that pattern, he or she can easily recognize it anywhere in the code.
- Document synchronization conventions and usage: clearly document your application's synchronization strategy and conventions in your code, and keep that documentation current.
- Put synchronization building-block code in one place: as an application matures, you often create code (classes) that handle non-trivial synchronization tasks. This code should all reside in a known place in your project for easy access and discovery, thus reducing the tendency for a developer to write that code again.
- Make synchronization building-block code general and reusable: don't create special synchronization code (classes) that only work for one special case, but design them to be usable for a set of cases, both what is currently needed and also what is expected to be needed.
- Communicate within the team: one developer may clearly understand the synchronization details, but other developers need to know the conventions, patterns, and building-blocks so they can avoid breaking the synchronization, implementing incorrect synchronization, and duplicating existing code, which all result in more complex code.
- Leverage synchronization libraries and standards: instead of writing a lot of custom synchronization classes, find a mature synchronization library to fit your needs. Writing this type of custom code is just the start; debugging it, making it bullet proof, and optimizing it takes a lot of time and experience.
- Use global synchronization settings: define a set of global (read only) synchronization variables, such as the number of threads to use, the coarseness of parallel tasks, the number of cores to use, and the depth for divide-and-conquer algorithms. Having all of your synchronization code refer to these variables will make tuning and

optimizing your synchronization implementation fast and easy. Manually fixing hard-coded numbers in each location in your code where synchronization occurs is tedious and error-prone.

- Use extensive testing and monitoring: design your code to make it easy to test often for thread safety and performance. This will help validate the correctness of your synchronization code, and it will also be a great benefit in fine-tuning your global synchronization settings. (This can lead to building run-time tuning for performance optimization, but implement the simple things first.)

With reduced code complexity in mind, modifying an application for thread safety and improved performance can produce a resulting application that is simpler and easier to understand than the original sequential implementation.

Software tools

One last word on software side-effects of synchronization is that the tools used to achieve synchronization (languages, compilers, libraries, vendor offerings, etc.) are all racing forward to be able to solve your synchronization problems. However, just like you, the developers building these tools are constantly learning this new synchronization domain and any software tool will have problems. The more complex the problem the tool is trying to solve, the more likely the tools will have unintended side-effects. Let me diverge into two real-world stories to illustrate this. The examples themselves aren't multicore-related, but rather illustrate the nature of the challenges you can expect to encounter.

Compiler errors over page faults

Years ago I experienced a problem with code on a Cray Y-MP. It gave incorrect answers in certain cases, yet when debug code was added, the effect moved around. After quite a chase and many long nights studying Cray assembly-language core dumps, the problem was discovered. An obscure addressing operation had an error in the page calculation when the multi-byte instruction itself crossed a page boundary (the calculation was based on the memory page that the instruction started on, yet the on-chip address-indexing occurred based on the memory page the address-reference in the instruction was on — or the other way around, yet for the impact of the story this detail is immaterial).

Unfortunately, this error was in a vendor product (I could not fix it), so I had to change the high-level logic in the code to avoid generating that obscure addressing operation. I also implemented testing to search all resulting compiled code to ensure that this obscure addressing operation was not generated in other parts of the code.

The key take-away here is that any tools you use may have problems, and the less-used features of these tools experience less use, less testing, and are therefore more likely to have hidden problems lurking in them.

Vendor software updates change things

Around the same time as the Cray compiler problem, the same code, when compiled on an Apollo computer, started experiencing various errors and run-time hanging. Again, spending many a long night tracking down the problem through our code and examining the Apollo assembly-language dumps, the problem was tracked to numerous FOR loops that were failing to terminate. This was caused by the integer being used as the index in the FOR loop being only a 2-byte integer instead of the expected 4-byte integer. Further obscuring the problem was the fact that only certain FOR loops were set-up with 2-byte indexes, based on the initial value of the loop. It turned out that a new version of the compiler defaulted to use 2-byte integer indexes to save a little on memory (in the early 1990 s memory was a scarce commodity), without our knowledge, and caused the problem. Luckily there was a compiler option to disable this "feature", but we didn't know to use it, and didn't know that this default was changed, until we had spent considerable time diagnosing the problem.

Again, the key take-away here is on tools, but this time on the tool trying to do something good for you while actually doing harm.

Hardware and OS effects on synchronization

While the software's domain is the software, hardware details can greatly affect the synchronization results. Time spent optimizing a program for a specific hardware platform may be counter-productive when taking that program to a different hardware platform. I have seen projects spend significant time optimizing to a very specific hardware target and achieving good results, only to have the hardware platform change a year later, and the current program was unacceptable on this new hardware and had to be significantly redone. Optimizing for a specific number of cores, specific cache sizes, or a specific version of an operating system may yield good short-term results, but may also cause severe long-term grief. You need to judge carefully optimizing for specific hardware.

Number of cores

As we saw in Table 12.14, a parallel program will operate differently on different numbers of cores. Adjusting the coarseness of parallel logic to optimize for a certain number of cores may cause problems if a different number of cores become available. You could query the system for the number of available cores and adjust your algorithms accordingly, but this leads to additional complexity and also increased risk in testing. It is also impossible to test today's programs with expected future hardware configurations, yet targeting your programs for future expected hardware may give you're a significant marketing advantage. (Of course there are emulators and virtual machines, but these are themselves complex software systems and are they to be trusted 100%?) You need to weigh the risk versus the benefit and the possible synchronization problems.

Memory, caches, etc.

As we have seen in the chapters on hardware, there are many configurations and options possible. Each of these parameters in the hardware configuration can have an impact on the synchronization side-effects. For example, a parallel FOR loop may perform exceedingly well with a certain cache size, as the cache hits are well-planned. However, the same logic with a different hardware cache size may cause massive cache misses, cache contention, and a resulting huge performance impact. Note that the software is the same, only the hardware has changed.

It is desirable to separate out the software details that relate to the specific hardware, but, in most cases, the specific hardware will greatly influence the software performance, so the tendency would be to adjust and optimize all of the software to that specific hardware. This should be approached cautiously.

Thread scheduling

The scheduling of your threads has a major impact on your synchronization work. In most cases, control of the scheduling details is not available to your application as the operating system (OS) typically is responsible for threads and thread scheduling. Thread scheduling is discussed in more detail in the OS chapter, but suffice it to realize that different thread scheduling can greatly impact your synchronized code and even impact the correctness of your code if your code is not bulletproof.

Garbage collection (and other system-level globally synchronized operations)

The last word on hardware and OS effects is that of garbage collection. Garbage collection occurs when objects are created and discarded: if enough of them have accumulated and have fragmented your physical memory, then the system needs to comb through memory and clean things up to recover memory to use for newly created objects. Different languages have different levels of program control and program responsibility for garbage collection, from totally program-controlled to totally automatic, with various offerings in between.

The typical garbage collector will stop all other threads when it combs through memory, since these other threads may be changing the memory that the garbage collector is working with. This periodic "freezing" of all program threads to do periodic garbage collection can be unexpected and can significantly impact the performance, especially time-critical and real-time sections, of your application.

As discussed above, one technique to protect your code and to make it bulletproof is to only return copies, preferably deep copies, of all values.

This practice can significantly improve the robustness of your code in one way, but can also cause a lot of grief (side-effects) by eating up memory quickly and thus cause frequent garbage collection — another very bad side-effect. It should be clear that these two desires are at odds with each other.

The key take-away is to be aware of these effects and to try to minimize them in your program. For example, to minimize garbage collection, keep a cache of discarded objects to be reused instead of continually re-creating new objects when needed. Care must be taken, however, to avoid caching too much memory. Using weak references for pointers is a useful technique for this since weak references can be disregarded if the memory is needed, but are kept in the meantime. (Examples are the WeakReference class in Java and .NET, weak_ptr class template in C++, PyWeakref objects in Python.)

The knowledge that the underlying hardware and OS may be doing periodic operations, like garbage collection, that can impact your program is important for helping your diagnosis of problems.

Problems when trying to implement synchronization

Here is a sampling of typical problems encountered with synchronization. It is by no means a complete or exhaustive list, but is intended to give a flavor of the errors, mistakes, and consequences seen, and to inspire you to exercise due diligence in writing your synchronization code and in testing your synchronization code.

Inconsistent synchronization: not synchronizing all access methods

Many times all of the set methods are synchronized, but the following problems can occur:

- get methods are not synchronized since the get looks atomic (we've seen how even simple assignment states may not be atomic),
- a new set method was added and was not synchronized,
- a new get method was added and was not synchronized,
- and many other subtle ways to miss proper synchronization.

Data escapes

Even with 100% correct getters and setters, if the getter returns a pointer to an object that requires synchronization for its state, then your code has become broken in terms of synchronization. As detailed earlier, data escapes can cause unexpected down-stream synchronization problems.

Using a mutable shared object with two different access steps (i.e., init() and parse())

An example is an Address object that has an init() method to prepare it to parse a specific kind of address by setting up instance variables for the expected parsing, then a parse method that takes input in the expected form and sets its various instance variables. This is broken up as two methods since the initialization may be expensive and, once it is initialized for a specific address type, it can be used to parse multiple addresses without re-initialization. The problem here is the two-step process, init() and parse(), and the problem of another thread changing the object's state unexpectedly.

Options are to not share such a mutable object between threads, to make the init() method only callable once, or to combine init() and parse() into one atomic operation and suffer the expense each time.

Cached "Scratch-Pad" data

This is an interesting problem where an object requires a large amount of data to calculate and return a value, for example calculating the intersection of two 3D surfaces using a method `surface.intersect (surface)`. Rather than allocate the needed memory each time the intersect method is called, the clever programmer creates an instance variable that contains the needed scratch-pad memory. This is considered good, since we are saving on garbage collection, avoiding out-of-memory errors if the needed scratch-pad memory cannot be allocated, and so on.

The problem is when two threads both invoke the intersect method on the same surface; they are now both using the same scratch-pad memory, but for different results. The logic that depends on the scratch-pad memory becomes quite confused, fails in unexpected ways, and is quite difficult to debug, since who would think to consider a synchronization problem in the midst of the intersection equations?

The obvious answer is to synchronize the intersect method since there is common data involved, thus necessitating an atomic intersect method. However, this algorithm is expensive and takes considerable time, so the result could be a lot of blocking in this now-synchronized method.

Multiple lock objects created

This is a common problem in trying to enforce the singleton pattern; the instance variable that is the synchronization lock is changed or set multiple times. Imagine code in which all access to the object checks a synchronization variable. Due to desired "lazy evaluation", actual creation of the synchronization variable is deferred until actually needed. The synchronization variable is initially null and the first access checks to see whether the synchronization variable is null, and, if it is, the synchronization variable is created and set.

The problem is that the check-and-set on the synchronization variable is not itself synchronized, and it is possible for two threads to try to access this object, resulting in the creation of two different synchronization variables. They are both set, but only the last one remains. Since these two threads were accessing the object at the same time, your synchronization of that object has failed.

Trying to lock on a null-pointer

The synchronization locks must be created at some point, and before one is created and set, the lock variable will be null. Any access to this variable before it is properly set will cause a null-pointer error. Note that trying to avoid this situation can easily cause the previous problem.

Double-check locking errors

"Clever" programmers, usually new to synchronization, will implement a double-checked lock to try to avoid the previous two problems, resulting in code something like:

```
class ThingWithAddress {

    private Address address = null ;

    public Address getAddress() {

        if(address == null) { // to avoid null-pointer error

            synchronized(this) { // try synchronization

                if(address == null) // really make sure it is still
null!

                    address = new Address();

            }

        }

        return Address;

    }

}
```

This innovative code can break for a number of reasons, including operation reordering by the compiler, the processor, or the optimizer. Once this is seen to break, the "fix" is to add more code to shore it up, but that only makes the problem worse.

The advice here is to (1) be very careful with "authoritative" solutions found on the internet, and (2) find a good source to learn the proper singleton-pattern-implementation for your language and environment since the good ideas you read in this book may at some point be superseded by even better ideas.

Simple statements not atomic (i.e., increments, 64-bit assignments)

This has been detailed before, but is needed in this quick-reference list of things to look out for. Many of the seemingly simple and seemingly obviously atomic statements are not. Beware!

Check/act logic not synchronized

There may be many places in your code where you check whether an instance variable is null, and if it is not, then the code will access a method on that variable. You are not doing a get or set, not changing the data, so why synchronize? Because if that instance variable that you are checking for null before access can be modified in any other thread, then after you see it is not null and you are about to access the method through it, it may become null and that just-checked-for non-null access will throw a null-pointer violation. You can stare at those two lines of code:

```
if( data != null )
    data.doSomething();
```

and not understand how can the call to doSomething, through a just-checked non-null data pointer, fail with a null-pointer exception. (I know I have done this, and I have seen this in vendor code too. I now know to quickly consider synchronization.)

Synchronization object used for many unrelated things

A common problem is creating a single lock for an object, and then synchronizing every method in that object on the same lock. Suppose there are 20 accesser methods in that object. Do all of the other 19 accesser methods need to be blocked if any one accesser method is in use? In this case, the error would unnecessarily restrict access to methods when there is no reason to.

Summary – synchronization problems

Unfortunately there is no silver-bullet approach for proper synchronization, no single solution that will solve all the synchronization problems. We hope this short list discussing some of the typical problems encountered with synchronization will be an aide in helping you learn proper synchronization, and seeing where in your code you may have problems.

The last piece of advice, alluded to in places above but perhaps not stated directly, is to avoid synchronization where you do not need it. If implementing a multi-threaded algorithm gives you no measurable benefit, keep the algorithm sequential, thus eliminating the complexity and the opportunities for synchronization errors. Use immutable objects wherever possible, since a primary need for synchronization is to protect the atomicity of an object's state changing. And beware of system libraries, frameworks, and GUI code that may introduce multiple threads without you knowing it.

References

[1] Herb Sutter, The free lunch is over: a fundamental turn toward concurrency in software, Dr. Dobb's Journal March (2005).

[2] Java Language Specification, 3rd edition.

[3] Available from: http://www.javamex.com/tutorials/synchronization_concurrency_6c.shtml.

[4] Available from: http://www.softwarequalityconnection.com/2011/06/the-biggest-changes-in-c11-and-why-you-should-care/.

[5] Available from: http://www2.research.att.com/~bs/C++0xFAQ.html.

[6] Gamma, Helm, Johnson and Vlissides. *Design Patterns: Elements of Reusable Object-Oriented Software.* Addison-Wesley, ISBN-10: 0201633612.

CHAPTER 13

Hardware Accelerators

Bryon Moyer[a], Yosinori Watanabe[b]
[a]*Technology Writer and Editor, EE Journal,*
[b]*Senior Architect, Cadence Design Systems, Berkeley, CA, USA*

Chapter Outline

Introduction

In the endless pursuit of more and more computing speed, there are a number of tools that can be brought to bear depending on the speed

Real World Multicore Embedded Systems.
DOI: http://dx.doi.org/10.1016/B978-0-12-416018-7.00013-4

required and the trade-offs. The very use of multicore represents a major step in trying to get more performance while keeping power dissipation reasonable.

Software is a great way to implement algorithms because it's flexible. You can write and rewrite easily, and you can update a system even after it's deployed. So, many developers adopt a software-whenever-possible approach. But some compute-intensive algorithms can bog down a system to the point where it makes sense to move to a hardware implementation.

While things like multicore can provide performance increases that are limited by Amdahl's law, making it fortunate if you can get $4 \times$ performance in a four-core system, hardware accelerators can boost the speed of their implementations by much more $- 10 \times$ or $100 \times$ or more, depending on the application. In addition, hardware implementations can improve power efficiency by the same order of magnitude.

These kinds of performance increases can make it worth the relative inflexibility of using hardware. Some ICs may benefit from a dedicated hardware circuit for a computationally intensive function like encryption. Other platforms may make available an FPGA that can be used by the end user to configure custom hardware functions. What that means to you as a developer is that, in some cases, you will be making use of an existing accelerator; in other cases, you may be specifying your own accelerator.

Either of these cases means that a new block will participate in the execution of a program. This adds some complexity to the system, but, managed properly, that complexity will be nominal. Managed poorly, an accelerator can introduce new bugs and smear the line between hardware and software, making debug and, ultimately, product release more difficult.

The purpose of this chapter, therefore, is to describe hardware accelerators at a system level, focusing on architectural alternatives and how accelerators can be invoked by software programs.

Note that this chapter will not discuss the process of creating hardware accelerators. That is a hardware design topic that is covered in numerous other works, and it is outside the scope of this book. We will treat accelerators as black boxes and assume that some hardware engineer did a good job implementing either manually or using an automated tool. The aspects we will cover will be architectural options, how to specify requirements to the hardware engineer who will create the accelerator, and how to integrate the resulting accelerator into both the system and the software.

Architectural considerations

The first thing we will look at is the accelerator in relation to the rest of the system, from both a hardware and a software standpoint. Naturally, the amount of work required to use or implement an accelerator will depend on the overall system context.

Blocking vs. non-blocking

The first major factor to consider is whether or not the accelerator is "blocking". An accelerator will be called by some software program, and there are two options for what happens after that call. On the one hand, the software program might simply wait for the results of the accelerator before proceeding. Such behavior is called "blocking" because progress in the software routine is blocked by the accelerator (Figure 13.1).

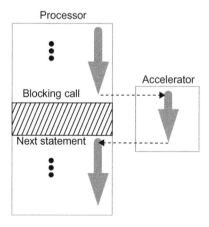

Figure 13.1
A blocking accelerator call.

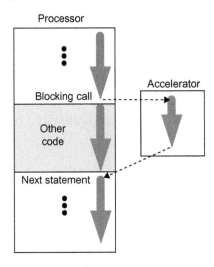

Figure 13.2
A blocking accelerator call where some other code runs in parallel with the accelerator.

It might seem like some performance is lost by having the processor just sit around and wait for the accelerator, but, in reality, many systems will see the OS swap in a new task during that wait time (Figure 13.2) so that the processor isn't completely idle. Even if the processor does spend some idle time, the idea here is that the speed-up afforded by the hardware is significant enough to be worth the slight wait time.

The other alternative is to have the software routine continue after calling the hardware (Figure 13.3). This is referred to as "non-blocking". Of course, if there comes a point where the software needs the hardware results and they're not ready yet, you may end up with what is effectively a delayed blocked call, so any gains in performance depend on the actual algorithm. It also means that there must be some way for the accelerator to alert the program that it is ready with results. It's therefore a more complex setup — a key reason why it might be easy in some cases to live with a blocking implementation.

However, for the purposes of this book, only the non-blocking approach involves true concurrency, where both the calling program and accelerator operate in parallel. Blocking implementations are simply a hand-off back and forth, and don't involve concurrency. They are

Processor

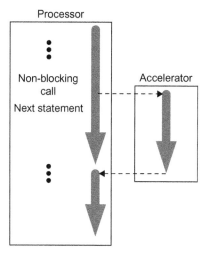

Figure 13.3
A non-blocking accelerator call. There may or may not be a wait if there comes a time in the code where it needs the accelerator result and it's not ready yet.

therefore of less interest to this discussion. While some of the coverage of things like interface specification may apply to both blocking and non-blocking accelerators, the focus of the discussion will be on the non-blocking case.

Shared or dedicated

The next major architectural factor to consider is whether an accelerator is dedicated to a specific processor or is available for use by multiple processors (Figure 13.4). The obvious implication of multiple processor access is that the accelerator now becomes a shared resource – it's essentially a hardware critical region. The good news is that, typically, an OS can't swap tasks for an accelerator mid-stream – once the accelerator starts, it will normally finish before picking up a new task. This simplifies synchronization of atomic tasks.

The process of granting access, however, remains, and whether that's done on a simple first-come basis or something more elaborate with, for example, a VIP queue, is an application-dependent architectural decision.

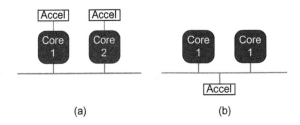

Figure 13.4
(a) Dedicated accelerators; (b) shared accelerator.

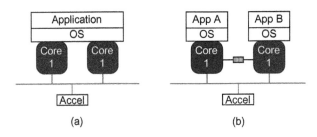

Figure 13.5
(a) SMP setup; (b) AMP setup with IPC.

SMP vs. AMP

Another major architectural consideration relates to whether the overall system is configured as SMP or AMP (Figure 13.5). With an SMP setup, every part of the program — including, in theory, the accelerator — can have direct access to every other globally declared part. By contrast, in an AMP configuration, different cores execute different processes, so visibility must be accomplished by communicating between processes — an approach that introduces latency.

Copying data — or not

Many accelerated processes involve some computation over a large amount of data — for instance, manipulating graphics. This may mean that a large block of data has to be transferred. Such transfers are typically avoided as much as possible using "zero-copy" approaches, where only a pointer to the data is moved around. Occasionally an application may have some other portion of the program that needs access to the data while the accelerator accesses the same data, running the risk of contention; if the performance hit from that contention is bad enough, it may then make sense to create a separate copy of the data. However, once a separate copy of the data exists, if more than one process is writing to the data, then steps must be taken to ensure that the data copies are coherent.

When copying the data, many systems will have a dedicated DMA resource for doing the copying; the processor assigns the job to the DMA, freeing the processor for other work. Smaller systems, however, may not have a DMA capability, making the cost of copying higher due to the impact on the processor.

Note that solving the contention issue by making a copy only applies if the issue is contention within that memory block. If the issue is bus contention in general, then creating a copy might actually make things worse since the bus will likely be used much more heavily during the copy process.

Signaling completion

With a non-blocking accelerator, the rest of the system carries on with its work until the accelerator signals that it is finished. There are two ways of having the accelerator alert the system that it has completed its task: having the accelerator fire an interrupt when complete, or having the application poll a register to see if the results are ready.

If the application polls the accelerator directly, then application resources or focus must be applied to that polling. That takes away from whatever else the application is doing. Less frequent polling means increased latency since the accelerator may finish and sit around longer waiting to be polled.

Using an interrupt service routine, or ISR, allows you to decouple the application from management of the accelerator. The driver can handle the accelerator details and then, when the accelerator is finished, it can fire an interrupt and the ISR can handle the results. This makes the application and accelerator somewhat mutually asynchronous, and provides some slack between the two so the application doesn't have to do its work plus monitor the accelerator.

The interface: registers, drivers, APIs, and ISRs

The interface is, conceptually, the same as a function call template. It specifies all of the information that the accelerator needs in order to accomplish its task. For a more complex accelerator, it may more resemble a software object, with elements serving as initialization (constructor), member variables, and methods.

The level at which a software program deals with an accelerator depends on the system. You will ordinarily want to work at the highest possible level of abstraction. This provides the best portability and system consistency. It also moves you far away from the hardware details — a good thing unless the utmost performance is needed, in which case you may want to give up some abstraction.

Hardware interface

At the lowest level, there is a hardware interface, with registers that are used to hold the values. While a very specific accelerator can have a simple interface, more often, given the amount of work that goes into creating an accelerator function — especially if it's going to be sold as commercial IP — numerous options are provided, all of which are captured in interface registers. In addition, a status register is needed for return conditions and errors. On such IP, there may be as many as 100 registers in the interface (Figure 13.6).

Along with the hardware is a protocol for using the accelerator. That will typically involve some combination of writing data to registers, copying data (or instructing a DMA to copy data), and starting the actual execution. If a DMA is used to copy the data, that function could be done by higher-level code before kicking off the accelerator function, or it could be built into the accelerator itself, offloading even that work from the processor.

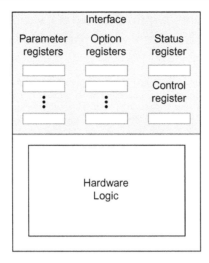

Figure 13.6
The interface for a hardware accelerator can have 100 or more registers.

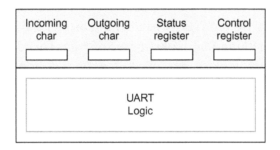

Figure 13.7
Simple UART interface.

We're going to use a simple fictitious UART as an example in a few cases, so let's set that up here. The simplest UART would have a register for an incoming character, one for an outgoing character, a status register, and a control register (Figure 13.7). This is the essence of the hardware interface.

If the hardware interface is actually accessed directly by the software, then it becomes a hardware-level API, and the software programmer must understand the details of what the hardware expects and in what order. The hardware API for the UART would simply consist of the memory-mapped addresses of the four registers, along with any constants required to define various status and control conditions. For example, given some base address UART_BASE, you could define:

```
uartSend    = (*((volatile char*) (UART_BASE + 0x10)))    //for send data
uartReceive = (*((volatile char*) (UART_BASE + 0x14)))    //for receive data
uartStatus  = (*((volatile char*) (UART_BASE + 0x18)))    //for reading status info
uartControl = (*((volatile char*) (UART_BASE + 0x1C)))    //for writing control info
```

The volatile designation is important because these registers will be written by the accelerator, outside the scope of the program. Without that, the compiler might see that at least some of these are never written within the program and might optimize them away.

Drivers

It's generally undesirable for an application program to deal directly with the hardware interface. A stripped-down or bare-metal system may have to, but otherwise a driver is needed to abstract the hardware interface into something that looks more software-friendly. Higher-level abstraction layers may exist, but when creating a custom accelerator, it's most likely that you will have a driver.

The driver is a piece of software that offloads the application program from the details of interfacing with the accelerator. There are two reasons for doing this:

- it simplifies and encapsulates the accelerator interface in a manner that makes the system more robust;
- it makes the accelerator available to other programs.

Software API

The means by which the software program interacts with the driver forms the software API. The accelerator driver is likely to be one of several drivers in the system, and system architects are placing a higher level of value on API consistency within the system. If you are designing your own custom accelerator and driver, then it is easy, in the planning stages, to design the API to conform to the system standard.

On the other hand, if you're re-using an older accelerator or purchasing one that comes with a driver, you may need to wrap or modify the driver to conform to the system standard. This ensures that, if a different driver or accelerator is swapped in at some point, then only the wrapper needs to be modified and then the entire system will work. Otherwise, every driver call in every software program would have to be edited to fit the new driver.

An example of API consistency can be seen with Analog Devices's Blackfin processor. The driver files all have names that follow a specific format, which includes the processor variant, the operating environment, debug version status, and workaround status. Memory, handles, return codes, initialization, and termination are all done the same way for all drivers. The driver commands themselves are the same for all drivers:

- adi_dev_Open   for opening the driver
- adi_dev_Close   for closing the driver
- adi_dev_Read   for reading from a buffer
- adi_dev_Write   for writing to a buffer
- adi_dev_SequentialIO   for reading or writing sequential data
- adi_dev_Control   for accessing various accelerator parameters

Note that this consistency is enforced across a family of processors. A system, on the other hand, may have one or more processors and various other devices, and if they come from different sources, they may all follow different standards. This means that the system architect will need to select one to use for the system (probably the main processor) and then wrap any other drivers so that they conform.

Note that this issue of driver consistency isn't particular to non-blocking drivers or multicore; it applies to any such systems. In fact, in some cases, the device operation can be made blocking or non-blocking according to a parameter in the interface.

ISRs

If you are using an interrupt to indicate completion or communication from the accelerator, then you need an interrupt service routine that executes when the interrupt fires. This is in contrast to polling, and a look at both approaches here will make the difference in approach clearer.

If you use an interrupt approach, then you are disconnecting the calling program from the accelerator function, especially in the non-blocking case. You start the accelerator going and then let it do its thing, coming back to it only when — and if — you need, since the ISR takes care of any necessary work when the accelerator needs attention, without being "noticed" by the calling program.

Let's use our fictitious UART to illustrate things a bit more concretely. A polled read operation would look simply like the following:

```
uartPolledRead () {
    while (<uartStatus shows empty>)
        continue ()              //loop while there's no data
    return (uartReceive)         //return the read data
}
```

Note that this only returns a single character, so the calling program will need to keep reading until all the desired characters are read or there are no more to read. (I used pseudo-code for the empty condition above to avoid getting bogged down in the details of how the empty condition is indicated.)

On the other hand, if a character arrives before the application program needs it, then, without an ISR, the application has to be there to capture the character before it is overwritten by another one. This means the application constantly has to be watching for characters. So, in fact, the code above would be intermingled with the actual application work — not very efficient. By letting the ISR, an independent routine, handle placing newly arrived characters in a buffer, the application can focus on its processing, knowing that the needed characters are being captured.

In order to use an ISR, we can define a buffer rxBuffer (Figure 13.8) to hold up to 256 characters as a circular buffer, meaning that, when the

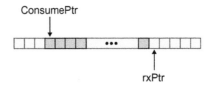

Figure 13.8
Character buffer separates incoming character capture from consumption by the application.

end is reached, it restarts at the beginning. The ISR will place characters in the buffer; concurrently with that, the application program will be consuming those characters. But the reception and consumption will not be synchronized, so we need two pointers — one for where the next received character should be buffered, one where the next buffered character should be consumed. We'll call those rxPtr and ConsumePtr.

This decoupling between reception and consumption means that, for as long as there is room in the buffer, the ISR can capture the characters coming in without the application program being right there to use them up before any get lost. Of course, if the application program is on hiatus for more than 256 characters worth of reception time, then you will lose characters because the ISR will overwrite characters in the buffer that haven't been consumed yet. The architecture of the driver arrangement is shown in Figure 13.9.

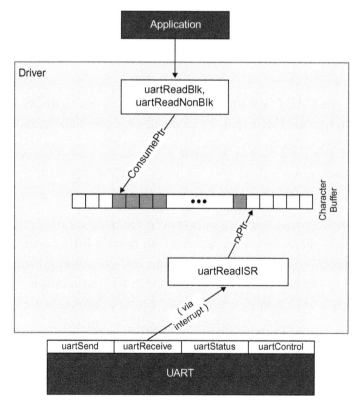

Figure 13.9
UART driver architecture.

The read function that the application program calls simply tries to consume characters from the buffer. If both pointers are pointing to the same location, then the buffer is empty, and the read function waits until that condition isn't true. Once there's data in the buffer, then it consumes the next character. I've used pseudo-code for the incrementing just to avoid the detail of the circular behavior.

```
char uartReadBlk() {
    wait (rxPtr != ConsumePtr);        //block until the pointers are different
    char temp = rxBuffer[ConsumePtr];
    <increment ConsumePtr>
    return temp;
}
```

If the application semantics allow the program to continue even without any received characters in the buffer, then a non-blocking version would look as follows:

```
char uartReadNonBlk() {
    if (rxPtr = ConsumePtr) return;    //return if buffer empty
    char temp = rxBuffer[ConsumePtr];
    <increment ConsumePtr>
    return temp;
}
```

Meanwhile, the UART will fire an interrupt each time a character is received, and it will trigger the ISR to put the received character into the buffer.

```
uartReadISR() {
    RXbuffer[rxPtr] = uartRecieve
    <increment rxPtr>
{
```

As mentioned, this was an example where the application would block until expected characters were received, so there's little concurrency except perhaps for concurrent writing into and consuming from the buffer. But to the application program, the read instruction is blocking; program flow won't continue until one or more characters are read.

Let's modify this slightly to illustrate a different application where the UART read can be non-blocking. In other words, even if there is no new character, the application won't care and will carry on, checking again later. We'll use a simple nonsense application program that does nothing but loop and write out how many characters have been read. The same

UART will be used, but instead of consuming the characters, only their count will be incremented.

The heart of the application program would look as follows:

```
...
uartCountChars();                    //start the counting of characters
while (<not some exit flag>)
    printf("Received %i characters\n", uartCharCount())
...
```

The uartCountChars function would simply clear the count:

```
int count;

void uartCountChars() {
    count=0;
}
```

The ISR, meanwhile, will simply count characters as they come in.

```
void uartCharCountISR()   {
    count+=1;
}
```

And finally, the uartCharCount routine simply returns the current count:

```
int uartCharCount() {
    return count;
}
```

In this example, once the character counting has started, the application program no longer interacts at all with the UART in any way. It only reads the count. The reading of characters by the UART and the printing by the application program are happening completely concurrently. The only impact the UART has on the printing is that, mysteriously, the count will occasionally increment. That incrementing is completely asynchronous with respect to the printing loop.

Initialization

On power-up, the system must be made aware of the accelerator. We'll briefly review what's required at a high level, but this process is required for any hardware resource that can be programmatically accessed, and so isn't specific to multicore systems.

There are two pieces to the start-up code. The first is a routine that resides in the BIOS ROM. This tells the OS anything it needs to know about the hardware: memory mapping, which resources are used, any power domain information, and other such low-level details that allow the OS to access and control the hardware.

The second piece of the start-up is boot code that the OS will use to load necessary ISRs, give those ISRs IDs, set any static access permissions, and initialize accelerator register values.

Once both sets of code are run, the accelerator is ready for use by the software − either application or driver.

Operating system considerations

In the simplest case, an accelerator simply acts as a hardware incarnation of a software function, the hardware interface is built from the software function prototype, and everything proceeds as if the function were being done in software. But in complex systems that require careful balancing of performance and power, the operating system is taking more control of resources to ensure efficient operation.

As an example, Android has a feature that keeps watch over various system resources, shutting down those that have no applications requiring them. It also has a work/wake API to bring various resources up or down. In the longer term, resources fall under the purview of the operating system and they are considered normally off. So before any application can make use of a resource, it has to request the resource from the operating system, which will wake the resource if it's powered down, shutting it back down again when the resource is released by all users.

Accelerators typically do not execute system calls, so in such a system the calling function would need to get access to the resources before passing control to the accelerator.

Coherency

One of the unique challenges of multicore is the fact that a single piece or block of data may be written to by more than one process at exactly the same time. This most commonly affects caches, and coherency strategies and protocols have been developed to handle that. This is a complex topic in its own right, and it is covered in more detail in the Memory and coherency chapter.

But if an accelerator is going to make changes to memory, then coherency concerns remain. What's more challenging is that the accelerator may not be linked into the coherency infrastructure used by the caches, meaning that you might need to manage it manually. The best solution to these kinds of issues is simply to avoid the need to maintain coherency if at all possible when designing the high-level architectures and protocols. But that may not always be feasible.

There are two cases to consider. The first is where the accelerator is changing a shared memory location. In this case, the accelerator will need to inform the rest of the system that a change has been made so that any cache lines that contain the changed location can be invalidated.

The best situation here is one where the accelerator can hook into the overall coherency infrastructure. Then it can directly invalidate cache lines. If this isn't possible, then the only real option is for a core to do the invalidation through an ISR − possibly even by having the core do the writing so that the coherency aftermath is automatically handled.

The obvious downside to this, of course, is that you lose some performance as control leaves the hardware accelerator and passes to the core and back. If the core does the writing and the ISR simply handles coherency, then the accelerator can continue its work without waiting for the core to complete its coherency task.

Figure 13.10 illustrates how this might be done in more detail. There are two possible flows here. The first has the accelerator writing directly to memory, with two alternatives for coherency. The benefit here is that the accelerator doesn't have to wait for the core to continue.

1a. Write.
2a. Invalidate cache lines through coherency infrastructure if possible,
     OR
2b. Fire an ISR so that a core can invalidate the cache lines.

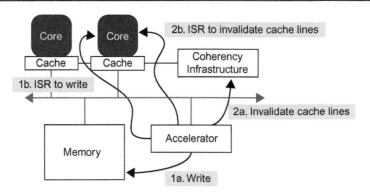

Figure 13.10
Options for handling cache coherency with shared memory.

The second flow involves less explicit coherency work because a core does the actual writing. However, if that step blocks the accelerator, then it could slow performance.

1b.   Fire ISR to write to memory (coherency handled automatically).

Much more complicated is the situation where the accelerator takes a copy of the data and then makes changes that must flow back to the original copy – and potentially any other copies that might exist. It could be argued that one would never want to be put into this situation, but the following describes what would be necessary – and potentially helps motivate against this kind of design.

When working on a copy of data, the cache coherency infrastructure is no longer helpful. This isn't an issue of caches being out of sync with their original memory lines; this is an issue of different places in main memory being kept in sync – the caches then follow from that.

This has to be managed manually. One solution would be to build a table wherein all instances of a particular chunk of memory are recorded. An ISR that fires on a write to one of the copies can then trigger a core to go through the table and update that equivalent memory location in all the other copies – assuming the original location is stored in a register and that translation to the equivalent location in the other copies can be readily calculated (which must be well managed).

With each of these writes, of course, the caches must then be made coherent, so a second wave of updates passes through the caches. Because a core updates the copies, resulting cache updates will be handled automatically. However, because the accelerator did the original write, then cache updates due to that write must still be explicitly handled, either directly through the coherency infrastructure or by having the core invalidate any cache lines.

Figure 13.11 illustrates these steps in the following flow. The idea is that some core runs a coherency manager that handles the updating tasks.

1. Accelerator writes to its copy.
2. Accelerator fires an ISR for coherency management.
3. Coherency manager scans table and determines which locations must be updated.
4. Other copies are updated.
5a. Caches of updated copies implicitly handled.
5b. Must update caches due to original write by hooking into coherency infrastructure, OR
5c. ... by having the coherency manager handle that.

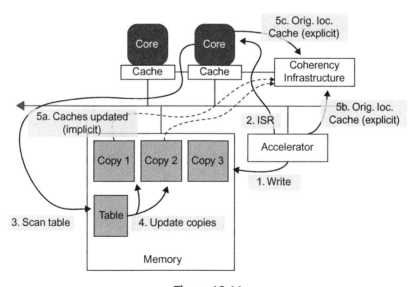

Figure 13.11
Keeping multiple copies coherent is more complex.

Where this becomes more difficult is when considering performance: the accelerator will presumably be operating much faster than the cores — that's why it exists. So there may be a significant speed mismatch between the accelerator doing its job and the rest of the system trying to maintain coherency. If the accelerator were doing frequent writes, then the cores would be frantically trying to keep up with the changes. Clearly, this would not be a good idea — were that to happen, then a different strategy would need to be used to avoid the need for keeping different datasets coherent.

Note that this more complex example also holds without an accelerator. The same problem would exist if two cores had individual copies of data that had to be kept coherent. Those cores would also need to manage the coherency explicitly. The only difference would be the fact that the cores would be operating on a similar timescale. However, the possibility of extreme overhead inefficiencies still exists, further arguing against this kind of approach if at all possible.

Making the architectural decisions

How these architectural decisions are made depends dramatically on how an accelerator is being designed into the system. There are three ways in which to do this: make use of an accelerator that's already available in your system (for example, an existing encryption engine); select IP (which may come from a third party or, in larger companies, from an in-house stash); or have a custom accelerator created (either in logic on an SoC or in an FPGA).

In the first two cases, all of the accelerator architectural decisions have already been made. If the accelerator already exists in the system, then a driver will also typically be available, and the only decision necessary is whether or not to use it.

When selecting IP, it falls to the system architect to review the options and pick one that integrates best into the system. If the APIs that come with the drivers are inconsistent with the rest of the system, then they may need to be modified or wrapped. Whether that's done by the IP provider or by someone on the design project is a matter for negotiation.

By using a driver, the details of the hardware interface and interrupt details are hidden from the programmer. Not only does this abstraction simplify programming and reduce the chances of low-level bugs, but it also makes it easier to switch IP vendors if necessary with minimal impact on the higher-level software.

Figure 13.12 shows a generalized flow diagram that summarizes the decisions and actions that have to happen in order to incorporate an accelerator — either existing or newly custom-designed — into an embedded system.

Video example

We can use a video function as an example to illustrate how this works. It's very common for a video to be resized — it may stream in using a given format, but need to be changed frame by frame to enlarge or shrink it. In an actual application, this function will typically be done along with the actual video decoding algorithm, so separating it out is somewhat artificial. But it's an application that's easy to understand, and so can serve its purpose here. Experts in video processing may strenuously object to the approach taken; real-world refinements to the oversimplified example will be briefly described after we get through the main points.

A real-world example where this might be required is in a tablet or phone for use when downloading videos, which are unlikely to be sized for the specific display on which the videos will be watched.

When a still image or video frame is resized — particularly when enlarging — the quality of the image can be degraded significantly if care isn't taken regarding interpolation of points and colors. Numerous algorithms exist for this, some better than others depending on the kind of image and resizing. Regardless of the algorithm specifics, they all tend to be computationally intensive, and might benefit from acceleration in a platform with limited computing power.

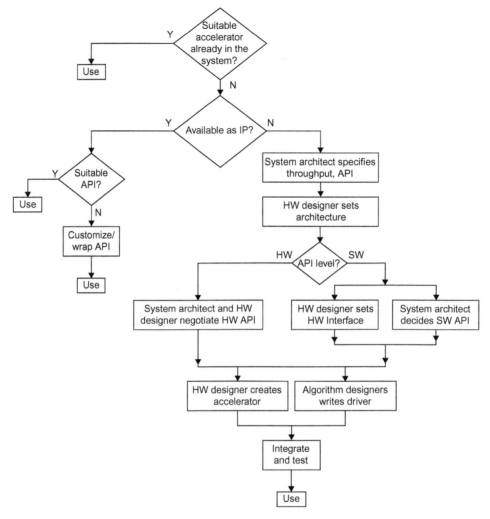

Figure 13.12
Decision flow for using an accelerator in a system.

The description that follows will use a more simplistic approach than would actually be typical, just for illustrative purposes. For example, full-frame buffers will be used, where, in reality, smaller chunks of data — for instance, a few lines from the frame — would be buffered. After we describe the simplified approach, we'll look at more realistic modifications to this approach.

We can structure the application as a pipeline for processing full frames (Figure 13.13). Note that some of the issues with pipelines are described

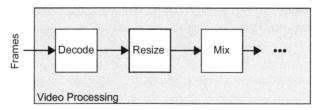

Figure 13.13
Simplified video processing pipeline.

Decode	Decode	Decode	Decode	Decode	Decode	Decode

	Resize	Resize	Resize	Resize	Resize	Resize

	Mix	Mix	Mix	Mix	Mix

Figure 13.14
Once the pipeline is "full", all stages are busy concurrently.

more thoroughly in the Bare-metal chapter, so we won't worry about them here. For simplicity, three main operations are assumed: first the frame is decoded, then it is resized. Following that, we'll add another operation: let's assume that the resizing is being done for a picture-in-picture application, so that, following the resizing, the image is now mixed into another image.

We want as much concurrency as possible, so we don't want each frame to go all the way through the process before the next frame starts: we want to start decoding the second frame while the first frame is being resized, and start decoding the third frame while the first frame has the mixing operation performed (Figure 13.14).

Our focus will be on the resizing portion of that pipeline. The resizing function won't simply be taking a video frame buffer and modifying it; it will create a new frame buffer and place the contents of the modified frame in it. While it sounds straightforward to modify a single frame, a video stream consists of an ongoing series of frames. So having a fixed interface assuming fixed locations for the starting and ending frames will be inefficient. The starting frame will be the result of the video decoding operation, and the resized frame will then be used by the next step in the video process. Implemented as a pipeline, each stage hands off its result to the next (Figure 13.15).

The interface

This means that the system itself will need to provide a collection of frame buffers and allocate those to the stages of the pipeline. So instead of the accelerator having fixed starting and resized frame locations, those locations become part of the interface, to be specified by the calling routing. This lets the functions proceed in parallel without stepping on top of each other.

The interface for such an accelerator would need at least the following, as shown in Figure 13.16:

- Location for the starting frame.
- Location for resized frame — for a more general-purpose accelerator, this could be made optional.
- Which of various algorithms to use.

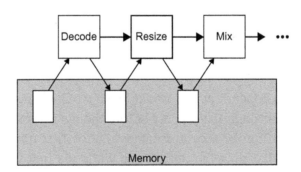

Figure 13.15
The outgoing buffer from one stage becomes the incoming buffer in the next stage.

Figure 13.16
Video frame resizing accelerator interface.

- Optionally, whether or not to copy data. Since the original frame is only being read, it can likely be accessed without copying.
- Status register.
- Control register.

In more general application, other information will be needed as well, such as the encoding standard, header information, and sampling ratio. If the coefficients on filters are programmable, there would be registers for storing them. We won't bother with those here in order to keep the example simple.

Much of the interface information will remain constant for some long string of frames, and so it only needs to be set up occasionally. We will assume a simple "setupResizer()" driver function or an initialization routine that handles that as needed; our focus will be on frame-to-frame operation below.

The application

We'll start by looking at the high-level application itself, followed by the driver, which includes the ISR. A naive way of approaching the application is simply to loop through frames, decoding, resizing, and mixing them (Figure 13.17).

This can be implemented by the following psuedocode:

```
while (1) {
    <if frame ready> then {
        <decode frame>
        <resize frame>
        <mix>
    }
)
```

But a closer look shows that this won't give us the pipelined behavior we want: the second frame won't get processed until the first frame has gone completely through the loop; we want the second frame to be decoded while the first frame is being resized.

There are two possible ways to handle this. The first, simplest, and most traditional is simply to have the functional call set up the accelerator registers and return immediately in a standard non-blocking call. This is fast enough that you can get through the loop quickly even though the

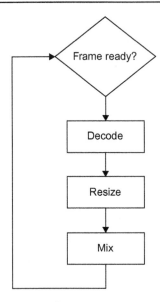

Figure 13.17
Simplistic frame-processing loop.

function calls for the three different operations won't, strictly speaking, occur concurrently. You do need, however, a way to see if there is a frame ready for any of the three operations independently since, for any given pass through the loop, anywhere from 0 to 3 of the operations may be ready for a new frame.

```
while (1) {
    <if frame ready> then <decode frame>
    <if decoded frame ready> then <resize frame>
    <if resized frame ready> then <mix>
}
```

Another way to approach this is to use three independent loops, all running at the same time, each looking for a frame that's ready to process (Figure 13.18). This means creating three threads, one for each stage of the pipeline, and having them join when all of the frames are complete.

In addition, we'll need to start and stop the driver code (which we'll discuss shortly). So the critical portion of a very short,

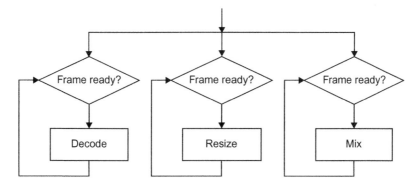

Figure 13.18
Using three threads to process concurrently.

oversimplified application (using pseudo-code to avoid thread syntax complication) is:

```
long startPtr, decodeBuffer, resizeBuffer, nextopBuffer;
...
startResizer(<parameters>);                    //initialize and start resizer running
...
create Decode thread {                         //start decoding loop
    while <there are frames to process>  {
        startPtr=<get the pointer to the next frame >
        decodeBuffer =<get the next empty buffer location >
        decode(startPtr, decodeBuffer)         //do the decoding
    }
}
create Resize thread {                         //start resizing loop
    while <there are frames to process >  {
        resizeBuffer =<get the next empty buffer location >
        resize(decodedFrame(),resizeBuffer)    //resize the decoded frame
    }
}
create Mix thread {                            //start next operation loop
    while <there are frames to process >  {
        mixBuffer=resizedFrame
        mix(resizedFrame(),mixBuffer)          //do the mixing
...
            }
}
<join threads>
...
stopResizer();
```

In this code, each of the threads gets a pointer to the location where the finished result will go; the actual creation of that finished result is part of the decode(), resize(), or mix() function.

Note that we need four driver functions, startResizer(), stopResizer(), resize() and resizedFrame(), which will be described below. Similar `decode()`, `decodedFrame()`, and `mix()` functions are assumed for the decoder and mixing drivers; we won't discuss them in detail.

The driver

Figure 13.19 provides an overall view of the driver structure, and the following discussion will describe how it works.

In order to provide the asynchronous interplay between the application and the driver that we saw with the UART, we also need buffers here. The first buffer will be for incoming new frames that need to be processed. There will be two pointers, one for the location of the unprocessed frame, and one for the location of the frame after resizing,

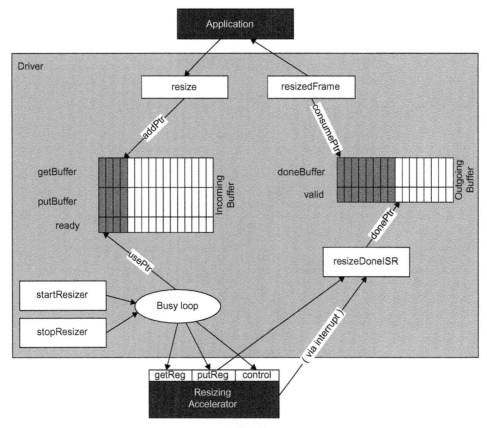

Figure 13.19
Video resizer driver architecture.

called `getBuffer` and `putBuffer`, respectively. Because several pieces of data have to be in place before the resizer can start working, we'll add one more flag called "ready", which will be the last one set. This will indicate to the resizer that it can start on the frame.

We could add a flag to indicate when the resizer has finished processing a frame, but it's cleaner to have a separate outgoing buffer to store finished frames. That way the incoming buffer can be freed up as soon as the resizer has started processing rather than after the application has consumed the frame. This outgoing buffer only needs a pointer to the finished frame, called `doneBuffer`, and a flag indicating that the frame is valid.

Given these two sets of buffers, the API to the application is very simple: a `resize()` call places the incoming frame data into the incoming buffer, and a `resizedFrame()` call retrieves the next completed frame from the outgoing buffer.

```
long getBuffer[15], putBuffer[15]  //frame pointer buffers for before and after
bool ready[15];                     //to ensure both pointers in place before start
doneBuffer[15];                     //for completed frames
bool valid[15];                     //indicates valid resized frame ready
int addPtr;                         //points to next open buffer location
int consumePtr;                     //points to next full buffer location

//puts frame descriptors on a buffer for the accelerator to work on
void resize(get, put)  {
    wait(!ready[addPtr])        //wait until location frees up if wrapped around
    getBuffer[addPtr] = get;    //store decoded frame location
    putBuffer[addPtr] = put;    //store result location
    ready[addPtr] = TRUE;       //mark as ready to use
    <increment addPtr>;         //circular
}

//used by application to consume the next resized frame location
long resizedFrame()  {
    wait(valid[consumePtr]);    //make sure there is a ready frame
    long framePtr = doneBuffer[consumePtr];
    valid[consumePtr]= FALSE;
    <increment consumePtr>      //circular
    return framePtr;
}
```

Within the driver, a loop will poll for frames to process, and, upon finding them, will send them to the accelerator so that it can start processing. This loop must be started before any frames can be processed, so a `startResizer()` function handles that as well as initialization of the buffers. When the whole thing is done, a `stopResizer()`

function stops the loop. Note that the loop is part of the `startResizer()` function. Because we intend it to loop throughout, we can't just create a simple loop within the function or the function will never return. So we need to create a new thread within which the loop can run endlessly without blocking the rest of the application. That thread will never merge back into the main program; as soon as the `done` flag is set, the loop stops and the thread is destroyed.

```
int addPtr, usePtr;          //ptrs for frame adding, in progress
int donePtr, consumePtr;     //ptrs for logging and consuming frames
bool done;                   //kill busy loop

//assumes frames from stream have the same parameters
//sets up parameters and flushes buffers; starts loop to feed frames to accelerator
void startResizer (<parameters>) {
    <write parameters to registers >
    addPtr=0;
    usePtr=0;
    donePtr=0;
    done=FALSE;
    <set all ready flags to FALSE >;
    <set all valid flags to FALSE >;

    create new thread {
        while (!done) {                    //start busy loop
            wait(<resizer free>);          //wait until accel is available
            wait(ready[usePtr]);           //wait until there's data ready
            <write buffer locations to registers > //write parameters
            <write control register to Go >        //start accelerator
            ready[usePtr]=FALSE            //free up buffer
            <increment usePtr >            //circular
        }
        <destroy thread>
    )
}

//stops the busy loop
void stopResizer () {
    done=TRUE;
}
```

Finally, we assume here that the accelerator itself fires the interrupt indicating that it's done. When that happens, the following ISR takes the buffer location from the accelerator and puts it into the `doneBuffer`.

```
//puts the completed frame location in a buffer for consumption
void resizeDoneISR () {
    wait(!valid[donePtr]);              //don't overwrite if wrapped
    doneBuffer[donePtr] = <put buffer addr. Reg.>  //get finished buffer address
    valid[donePtr] = TRUE     //mark as ready for consumption
    <increment donePtr >      //circular
}
```

Real-world refinements

Having gone through a simplistic example, there are a number of real-world considerations that affect how this particular function would normally be handled. Some of the refinements could apply to other functions; some are more specific to video resizing.

Chaining accelerators

The first thing to note is that the application makes each accelerator call, albeit indirectly through the buffers in the driver. But if the decoding, resizing, and mixing are all accelerated, then, in a sense, you're popping back and forth between hardware acceleration and software for calling the accelerators. While that may sound academic, hardware operates orders of magnitude faster than software, so having software intervene between three consecutive hardware accelerators is inordinately inefficient (Figure 13.20).

What's more typical is to group the accelerators so they can be chained directly together. You would typically have an "agent" that would act as the interface between the software world and the accelerator group, isolating the details of the accelerator from software. This arrangement is shown in Figure 13.21. In this way, once the first accelerator is started, you stay in hardware until they are all complete.

Smaller data granularity

More efficient use of hardware can also be achieved by using a smaller unit of data than a full frame. With a full frame, the entire frame has to

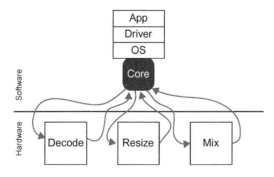

Figure 13.20
It's inefficient to return to software between consecutive hardware accelerator calls.

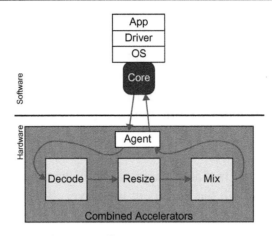

Figure 13.21
Chained accelerators, with an "agent" as interface to software.

be decoded before resizing can start. By using a smaller chunk — a few lines of video data from the frame, for example — then, once that chunk is complete, resizing can start on it while decoding continues on the next chunk. Because each stage of the pipeline can get started sooner, the overall frame latency is reduced (Figure 13.22).

DMA

As mentioned much earlier in the chapter, DMA is often engaged to move data around without burdening the processor. In our simple example, data was pretty much managed through software, with the exception that the accelerator took unresized data from one block and placed the resized data in another block. We really didn't deal with the details of how data ended up in the blocks originally.

More typically, the accelerator would have access to the DMA so that the DMA can handle the data transfer, bypassing the processors. Figure 13.23 illustrates the difference.

Burying the functions in a DMA copy operation

Specific to this function, in fact specialized DMA engines can do the decoding and resizing more or less as a side-effect while copying data. This, of course, means that there's no accelerator dedicated to resizing only, but rather that the hardware that performs that function is a part of the overall DMA hardware (Figure 13.24).

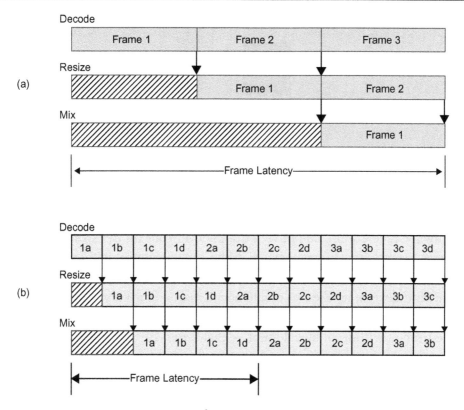

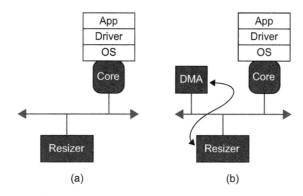

Figure 13.22
(a) Processing a full frame before handing to the next stage increases latency as
compared to smaller chunks (b), in this case, 1/4 frame.

Figure 13.23
(a) The CPU has to manage data copies; (b) a DMA, controlled by the accelerator,
handles data copies.

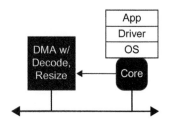

Figure 13.24
Decoding and resizing done "on the fly" while copying the data.

Summary

Hardware accelerators can have a dramatic impact on the speed of critical operations. But, especially in the case of non-blocking accelerators, they are yet another layer of concurrency in the system. Hardware design, as used to create accelerators, is a very different discipline from software design. In particular, hardware involves massive concurrency within the accelerator – that's part of why it's so much faster. So the considerations and paradigms used by software engineers, who encounter little to moderate levels of concurrency, are different from those of hardware designers.

For that reason, careful planning in the specification of the accelerators is required to efficiently partition the functionality for minimal interaction with other variables. When creating the supporting software, care must be taken to allow the best use of the hardware and processors, making sure that everything is busy and nothing is waiting around for something else to finish. Careful design of the driver, including any interrupt service routines, will decouple the operation of the hardware from that of the application software, maintaining a layer of mutual asynchronicity. That gives both sides the flexibility they need to do things in their own time, without one having to wait for the other.

CHAPTER 14

Multicore Synchronization Hardware

Jim Holt

*Systems Architect for Freescale's Networking and Multimedia group,
Austin, TX, USA*

Chapter Outline

Chapter overview

Synchronization is an essential capability for programming
multicore systems. This requirement spans operating systems,
virtualization software, device drivers, system libraries, run-times, and

Real World Multicore Embedded Systems.
DOI: http://dx.doi.org/10.1016/B978-0-12-416018-7.00014-6

481

application software. However, despite the existence of software-only solutions such as Peterson's algorithm [1], it has been shown to be impractical to achieve these capabilities in multicore systems with software alone. Thus, it is essential for multicore systems to provide hardware support in order to ensure correctness and efficiency for synchronization in concurrent software. This chapter provides a detailed look at the hardware/software interface for synchronization and presents underlying hardware support commonly found in multicore systems.

Programmers find themselves in need of two basic forms of synchronization: *data synchronization*, and *control synchronization*. Data synchronization prevents race conditions and corruption by coordinating concurrent access to shared data. A typical mechanism for data synchronization is some form of *lock*. A process must obtain a lock before accessing shared data, and then release the lock when finished. Control synchronization coordinates parallel algorithm execution to ensure correctness. A typical mechanism for control synchronization is a *barrier*. A process or thread enters the barrier, waits for the other processes or threads, and eventually all processes or threads leave the barrier together. Despite these apparent differences, for most current multicore systems, the lowest level of software implementation for both types of synchronization will generally depend upon a combination of locking and shared data to provide the higher-level synchronization semantics.

Several decades of research and development have been devoted to providing hardware synchronization mechanisms. Today, hardware support for synchronization is ubiquitous, spanning from massively parallel supercomputers to deeply embedded multicore systems. Most modern computer instruction sets include operations for synchronization, and all processors adhering to these instruction sets provide underlying hardware to ensure correct and atomic operation. (It is not always necessary to have strictly atomic operations in a multicore system; sometimes it is sufficient to guarantee that an operation appears atomic from the point of view of software. Many of the mechanisms discussed in this chapter are *speculatively atomic*, e.g., allowing software to detect when atomicity has been violated. In these cases it is up to software to recover from atomicity violation.) Many different idioms for atomic synchronization operations have been invented, but in recent years three primitive operation classes have emerged as the mainstream: *test-and-set*, *compare-and-swap*, and *load-reserved/store-conditional*.

A number of additional synchronization operations can be defined using these few instruction set primitives, and compilers also provide a variety of intrinsic operations built upon them. The section "Instruction set support for synchronization" examines each of the three most common primitives, provides examples of how to create additional synchronization operations using them, and illustrates the use of compiler intrinsics.

Sometimes it is desirable to avoid synchronization. This may seem counterintuitive, and, while synchronization is an effective and time-proven approach, using locks and shared memory can be quite inefficient when there is significant contention for shared data. This is unavoidable as the number of processes or threads increases and as the number of available cores increases. Therefore, many multicore systems provide additional mechanisms to avoid synchronization altogether. These *lock-free* techniques can yield higher performance and better power efficiency in certain situations. The "Hardware support for lock-free programming" section will examine a number of these approaches for both data and control synchronization and discuss their applicability.

Instruction set support for synchronization

Low-level hardware synchronization mechanisms are typically available to programmers via function calls in system libraries. These functions encapsulate architecture-specific instructions that rely on underlying hardware mechanisms to provide atomicity. The fundamental capability required for synchronization is the ability to read/modify a shared memory location atomically. Processor architectures may provide this ability as true blocking operations, or, alternatively, by splitting the read and modify transactions while providing speculative atomicity. Whichever approach is taken, the goal is to guarantee that only one thread or process is active in a critical section of code at a time — thereby avoiding race conditions and deadlocks.

Most modern processors provide instructions that implement one or more forms of a few common idioms. These idioms include blocking atomic approaches such as *test-and-set* or *compare-and-swap*, and speculatively atomic operation pairs such as *load-reserved/store-conditional*. Regardless of which semantics an architecture provides, the underlying principle is the ability to atomically read/update a memory location. In the strictest sense, atomicity means that nothing will intervene between reading the value from a specific memory location and updating the value stored at the same memory location. In practice, there are a few ways to achieve this in hardware, but we will defer that discussion until the "Hardware support for synchronization" section.

When implementing synchronization routines from instruction set primitives, it is important to understand that the primitives themselves are not necessarily atomic in the strictest sense. In many cases the primitive operations only provide software the means to request atomicity for a causally related pair of operations (one for the memory read operation and one for the memory update operation) and a mechanism to detect when atomicity has been violated. Therefore, when programming routines using speculatively atomic operation pairs, a programmer must be cognizant of how to detect and recover from atomicity violations.

Given the complexities of managing speculative atomicity, the benefit of wrapping instruction-level primitives in library routines is twofold. First, it is clearly difficult to build and test a synchronization library. Once this has been done successfully, the result should be reused as much as possible! Second, encapsulating processor-specific instructions within a high-level library interface provides for better application portability. Portability is further aided by the fact that many variations on high-level synchronization routines can be built from a few low-level primitives. Thus, even when a processor does not directly provide certain operations in its instruction set, it is still possible to implement equivalent functionality in library routines.

Test-and-set

The test-and-set operation modifies the contents of a memory location and returns the old value from that location in a single atomic operation. This operation is called "test-and-set" because it allows the programmer to atomically swap values between a register and memory (*set*) and then to evaluate the data retrieved from memory to make a programming decision (*test*). This will become clearer with an example of using test-and-set to create mutexes.

It is easy to imagine how this can be used to obtain a lock. For example, we can define a memory location representing a lock to have the value 1 if locked and the value 0 if unlocked. The SPARC architecture provides these semantics via the `ldstub` instruction [2], while the Intel architecture provides them via the `bts` instruction [3].

There are certainly subtle differences between instructions that provide test-and-set semantics. For example, the `bts` instruction operates on a byte, while the `ldstub` instruction operates on an unsigned byte. Hence, it's imperative to use correct datatypes when wrapping assembly language within another language such as C, and furthermore it is highly desirable to hide these differences from higher-level programming constructs beneath an abstraction layer.

Setting low-level differences aside, the generalized semantics of test-and-set are illustrated by the psuedo-code in Figure 14.1. This operation is called "test-and-set", because it allows the programmer to atomically swap values between a register and memory (*set*), and then to evaluate

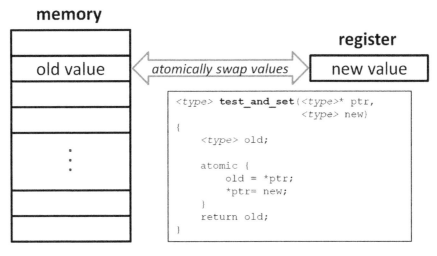

Figure 14.1
Pseudo-code for test-and-set operation.

the data retrieved from memory to make a programming decision (*test*). This will become clearer with an example of using test-and-set to create mutexes.

The implementation of a software `test_and_set()` routine itself is straightforward in assembly language; however, as discussed above, the specifics will vary by architecture. This is illustrated in Figure 14.2 using the SPARC `ldstub` instruction [4]. Note the extra steps taken to ensure that the 32-bit register is zero-filled since `ldstub` operates on unsigned byte data.

Given an atomic test-and-set primitive, it is straightforward to implement high-level language routines for locking and unlocking a mutex. This is depicted by the C code in Figure 14.3. A mutex is first initialized to zero. Then, the mutex can be locked by calling `lock_mutex()`. The implementation of `lock_mutex()` will remain in the `while` loop as long as another process has the mutex locked (e.g., as long as the return value of

```
test_and_set:
        ldstub          [%o0],  %o0         ! atomic load + set
        sll             %o0,24, %o0         ! zero fill
        retl                                ! result register
        srl             %o0,24, %o0         ! return
```

Figure 14.2
Assembly code for test_and_set() operation using SPARC `ldstub` instruction.

```
typedef struct { int val; } mutex_t;

void init_mutex(mutex_t* mutex) { mutex->val = 0; }

void lock_mutex(mutex_t* mutex) {

    while(test_and_set(&mutex->val,1) == 1); /* wait */

}

void unlock_mutex(mutex_t* mutex) { mutex->val = 0; }
```

Figure 14.3
Simple C code for mutex with test_and_set() operation.

calling `test_and_set()` is one). It will return successfully once the lock is obtained (e.g., when the return value of calling `test_and_set()` is zero, indicating no other process has the lock).

You can see that this simple mutex interface provides a suitable abstraction layer that can hide the differences between a `bts` or `ldstub` implementation from the programmer. However, a production-quality implementation would need to take other considerations into account, such as how to minimize wasted processor cycles in the `while` loop (since the process is effectively blocked while waiting to obtain the lock) or disabling interrupts in situations where there is the possibility of deadlocks (for example, in operating system spin-locks).

Revisiting the code fragment implementing the test-and-set operation (Figure 14.2), it is important to note that, in a multicore system, it will sometimes be necessary to insert a memory barrier or an instruction to lock the bus when executing an "atomic" instruction. This depends on the semantics of the particular machine and will be discussed in more detail in "Hardware support for synchronization" below. Regardless of the finer machine-specific details, the code that manages these semantics should be contained within the library routines implementing the locking primitives, thus ensuring correct operation for all users of the library.

Compare-and-swap

The compare-and-swap operation atomically compares the contents of a memory location with a user-specified value contained in a processor register and − *if they are equivalent* − modifies the contents of that memory location to a new user-specified value contained in another register (Figure 14.4). This operation is available in the SPARC architecture via the `cas` instruction, and in the Intel x86 architecture via the `cmpxchg` instruction. When using compare-and-swap, it is up to the programmer to test whether the swap actually occurred, and to wait or retry as necessary.

As with the test-and-set primitive, it is straightforward to provide a high-level programming interface to the low-level compare-and-swap operation. Encapsulation of compare-and-swap within a C library routine

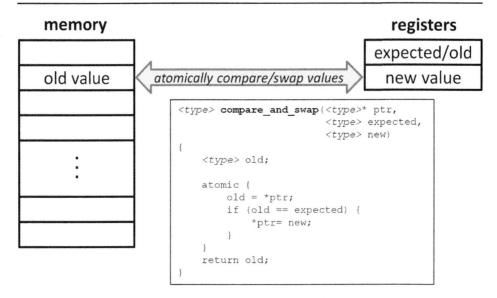

Figure 14.4
Pseudocode for compare-and-swap operation.

```
unsigned char compare_and_swap(int *ptr, int old, int new)
{
    unsigned char ret;

    /* Note that sete sets a 'byte' not the word */

    __asm__ __volatile__ (
        "    lock\n"
        "    cmpxchgl %2,%1\n"
        "    sete %0\n"
        : "=q" (ret), "=m" (*ptr)
        : "r" (new), "m" (*ptr), "a" (old)
        : "memory");

    return ret;
}
```

Figure 14.5
Inline assembly code for compare_and_swap() operation using Intel cmpxchgl instruction.

is illustrated for the Intel cmpxchg instruction in Figure 14.5. Note that the cmpxchg instruction requires the expected value to already be in the processor's accumulator register (EAX). If the value at the destination location in memory is equal to the value in the EAX register then the

```
void lock_mutex(mutex_t* mutex) {

    while(compare_and_swap(&mutex->val, 0, 1) == 1); /* spin */

}
```

Figure 14.6
Modified C code for mutex lock with compare_and_swap() operation.

destination location is updated with the new value. Otherwise, the EAX register is loaded with the original value of the destination location. The *lock* prefix shown in Figure 14.5 is used to explicitly lock a shared bus, as will be discussed later in more detail.

Given the existence of a simple mutex library such as the one used for the code in Figure 14.3, only a small modification is required in order to use compare-and-swap instead of test-and-set (Figure 14.6). This example illustrates the portability benefits of a software abstraction layer used to hide the details of the underlying hardware semantics.

Again, as with test-and-set, use of compare-and-swap will likely involve other machine-specific considerations, especially for multicore systems, as will be detailed in "Hardware support for synchronization" below.

Load-reserved/store-conditional

For multicore systems, there will be performance implications when using atomic operations like test-and-set or compare-and-swap. The reasons will become clear from the discussion of underlying hardware support later. Recognizing this scalability concern, many modern architectures provide a speculative means to read/update memory via separate load-reserved and store-conditional operations. These operations depend on the system coherence mechanism to determine when an operation that occurs between the load-reserved and the store-conditional breaks the logical atomicity of the pair. A system which implements this paradigm will provide a means for software to detect an atomicity violation so that the operation pair can be retried. The two operations are depicted in Figure 14.7 and Figure 14.8. Architectures that provide these operations include Power Architecture® (lwarx/stwcx instructions [5]),

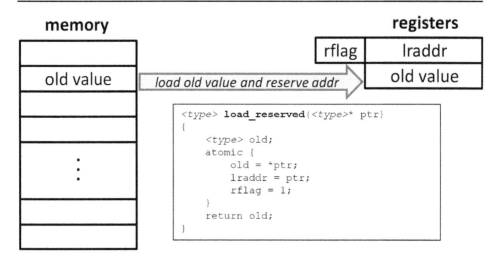

Figure 14.7
Pseudo-code for load-reserved operation.

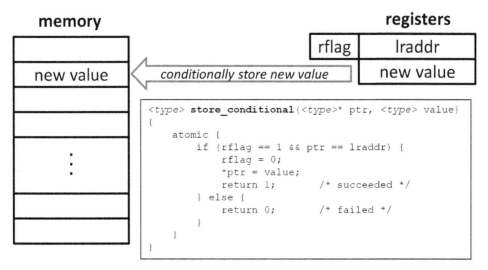

Figure 14.8
Psuedo-code for the store-conditional operation.

ARM (`ldrex/strex` instructions), MIPS (`ll/sc` instructions [6]), and Alpha (`ldl_l/stl_c` instructions [7]).

When a load-reserved operation is executed, the hardware will save the reserved address and will set a flag to indicate its reservation. When a subsequent store-conditional operation is executed, the flag and address

```
void lock_mutex(mutex_t* mutex) {

    while(1) {

        while(load_reserved(&mutex->val) == 1) ; /* wait */

        if (store_conditional(&mutex->val,1) == 1)

            return;  /* locked */

    }

}
```

Figure 14.9
Modified C code for mutex lock with load_reserved() and store_conditional()
operations.

will be compared and, if there is a match with the most recent load-reserved, then the store can proceed. However, whenever the system coherency mechanism sees a store transaction to the reserved address, the reserved flag will be cleared. If this were to occur after the load-reserved operation but before the store-conditional operation then the store-conditional would fail.

Implementing a mutex lock with load-reserved and store-conditional is illustrated in Figure 14.9. The power of separating the load and update operations into a pair of speculative operations is evidenced by the ability to insert additional code between the operations. This ability allows one to create additional atomic primitives with ease (see below), and provides opportunities to better manage overall system performance and efficiency.

In the following sections we will see how to build additional atomic primitives using the load-reserved/store-conditional operations. This is followed by a brief explanation of how compilers provide many of these operations as intrinsics and discussion of the pros and cons of using intrinsics for synchronization.

Creating new primitives

It is straightforward to create additional "atomic" primitives using load-reserved and store-conditional. The examples in this section will illustrate this using the Power Architecture® instructions lwarx (*load word and reserve*) and stwcx (*store word conditional*) [5]. The functional equivalent of the test-and-set operation defined earlier is illustrated in Figure 14.10. Of course this code could be quite inefficient if there are many processes or threads contending for the mutex lock. In this case the library implementer would probably add additional code to interact with the operating system's scheduler to cause the calling process to block for a short period of time before retrying rather than greedily consuming processor cycles merely busy-waiting. Note that the example code in Figure 14.10 does not take into account the memory semantics of Power Architecture. This would require a memory barrier following the stwcx instruction to guarantee correct operation. This will be elaborated upon in "Using memory barriers and synchronizations" below.

It is also possible to implement compare-and-swap semantics using load-reserved/store-conditional operations. This is depicted in Figure 14.11. This example implements exactly the original IBM System/370 style of compare-and-swap.

Other potentially useful atomic operations are *fetch-and-add* and *fetch-and-and*, as well as all their arithmetic and logical cousins such as

```
loop:      lwarx    r4, 0, r3      # load and reserve

           stwcx.   r5, 0, r3      # store new value if still reserved

           bc       4, 2, loop     # loop if lost reservation
```

Figure 14.10
Assembly code for test_and_set() operation using load-reserved/store-conditional.

```
loop:      lwarx    r6, 0, r3      # load and reserve

           cmpw     r4, r6         # first 2 operands equal?

           bc       4, 2, exit     # exit if not

           stwcx.   r5, 0, r3      # store new value if still reserved

           bc       4, 2, loop     # loop if lost reservation
exit:      or       r4, r6, r6     # return value from memory
```

Figure 14.11
Assembly code for compare_and_swap() using load-reserved/store-conditional.

```
fetch-and-add

loop:        lwarx   r5, 0, r3     # load and reserve

             add     r0, r4, r5    # ADD r4 to r5, place in r0

             stwcx.  r0, 0, r3     # store r0 if still reserved

             bc      4, 2, loop    # loop if lost reservation


fetch-and-and

loop:        lwarx   r5, 0, r3     # load and reserve

             and     r0, r4, r5    # AND r4 with r5, place in r0

             stwcx.  r0, 0, r3     # store r0 if still reserved

             bc      4, 2, loop    # loop if lost reservation
```

Figure 14.12
Fetch-and-add and fetch-and-and.

subtract, *multiply*, *or*, *xor*, etc. Two of these are illustrated in
Figure 14.12. It is left to the reader to expand these into other variations.

Compiler intrinsics

Compilers such as gcc [8] also offer programmers access to basic
synchronization primitives via intrinsics. Intrinsic functions, sometimes
referred to as built-ins, are functions which the compiler has special
knowledge of and whose implementation can thus be optimized
intelligently by the compiler. Intrinsics can therefore be thought of as
similar to inline functions with a great likelihood of optimal
performance.

Because compilers such as gcc support multiple target processors, using
intrinsics can lead to portable code. However, one should be aware that
synchronization primitives provided by compiler intrinsics may not have
the same performance properties as synchronization primitives in system
libraries since an intrinsic may not cooperate with the kernel scheduler in
the same way that a run-time library would. For example, if the
`lock_mutex()` routine were to be implemented using a compare-and-swap
routine from a system library, the operating system has the ability to

```
gcc intrinsic

bool __sync_bool_compare_and_swap (type *ptr, type oldval, type newval)


void lock_mutex(mutex_t* mutex) {

    while(__sync_bool_compare_and_swap(&mutex->val, 0, 1) == false); /* spin */

}
```

Figure 14.13
Mutex lock routine using gcc intrinsic.

block the calling process on an unsuccessful compare-and-swap call. The process would eventually be rescheduled and hopefully could obtain the lock at that time. An instrinsic-based version would have to busy-wait in the lock_mutex() routine shown in Figure 14.13, potentially wasting considerable cpu cycles.

Hardware support for synchronization

An instruction set defines the hardware/software interface for synchronization operations. Underlying hardware must efficiently and correctly provide the semantics guaranteed by this interface. A number of techniques have been developed, and these techniques have varying trade-offs between ease of implementation (e.g., complexity) and performance. In general, as multicore systems continue to provide increasingly more cores on a single die, hardware support for synchronization will take various steps to mitigate both complexity and performance issues. This section explores these various mechanisms and trade-offs.

Bus locking

Single-processor systems can guarantee that atomic operations such as compare-and-swap will in fact be performed atomically (since there are obviously no other processors attempting to utilize the interconnect and memory at any given time). The easiest way to ensure atomicity of such operations in a shared-bus multicore system is to lock the bus for the duration of the transaction. Some architectures provide automatic bus locking in the hardware for certain atomic instructions (for example the Intel `xchg` instruction). Some architectures also provide an explicit means for the programmer to lock the bus from software. In Intel architectures, this is done by specifying the *lock* prefix to atomic operations such as `cmpxchg` (see "Compare-and-swap", and Figure 14.5).

In multicore systems with explicit software locking, the programmer must beware: an atomic instruction such as `cmpxchg` must be combined with the lock prefix to guarantee the operation will be performed atomically! The micro-architectural details can be subtle, and semantics will vary depending on the processor implementation. For example, older Intel processors always assert the bus lock signal during execution of an instruction with the lock prefix, even when the data to be operated on is in the processor cache. However, in newer Intel processors, when the data to be operated on is in the cache and the cache is configured as write-back, then the processor implementation may choose not to assert the lock signal on the bus. In this situation, the processor may instead modify the memory location internally to its cache and then allow the cache coherency mechanism to provide underlying atomicity. Clearly these subtle but important differences can lead to quite different performance characteristics in synchronization code.

Load-reserved and store-conditional

In a multicore system that implements bus locking, the bus must remain locked for the duration of an atomic read/modify operation. This can be a performance bottleneck even for simple buses with few cores, and it can be quite a significant issue in split-transaction buses with many cores. This has motivated modern architectures to provide split operations such as load-reserved and store-conditional. The mechanisms for managing and monitoring the reservation of a load-reserved/store-conditional operation pair can vary considerably in hardware. Regardless of how it is implemented in the hardware, the essential requirement is to guarantee that a reservation for a store-conditional must still be valid for the store to succeed. For this to work correctly, the hardware must maintain a reservation context and track accesses to the target memory address.

The MIPS R4000 architecture includes the LLAddr register and the user-transparent LLBit for this purpose [6]. When an ll instruction is executed, the target address will be retained in the LLAddr register and the LLBit will be set (refer to Figure 14.7 and Figure 14.8). A subsequent sc instruction execution will cause the hardware to check that the store address matches the contents of the LLAddr and that the LLBit is still set. The target register will be loaded with the value 1 in the case of a successful store or 0 otherwise. In general, the LLBit will be cleared when any local or system coherency event occurs that indicates a potential change to the data at the target address. This sequence of events is complicated by the fact that a processor may not have the target address cached at the time of the store-conditional operation (for example, the cache line in question may have been replaced after the ll instruction executed but before the sc instruction executes). Therefore, coherency protocols must be carefully designed and verified such that the reserved address and linked bits are properly maintained. In some coherency implementations (for example directory-based protocols), this responsibility may be assigned to an external hardware agent. In this case, the processor will be required to communicate additional information to the external agent to ensure that tracking of the LLBit can be delegated outside the processor. Interactions between the processor core and an external agent are transparent to software and are handled automatically by the hardware.

Hardware support for lock-free programming

The act of waiting for a lock can be either *blocking* or *busy-waiting*. Blocking is not typically provided by hardware but, instead, must be coordinated in software between the locking routine and the operating system scheduler. Thus blocking can lead to extra overhead due to context switching or non-deterministic execution time due to scheduling variations. On the other hand, busy-waiting can waste a lot of processor cycles polling for lock acquisition, affecting both performance and energy consumption.

In some cases hardware can provide alternate mechanisms that allow the programmer to avoid explicit synchronization altogether. This is typically accomplished by exploiting either *decoupling* or *speculation*. Decoupling techniques remove the need to acquire locks by moving or buffering data in hardware. This approach requires no explicit synchronization between processes that wish to access shared data.

Examples of decoupling techniques include *hardware queues*, *direct memory access* (DMA), and *decorated storage operations*. Each of these provides a way for a producer of work or data to proceed independently of its consumer, thus avoiding explicit locking.

Speculation techniques assume a low probability of interference between processes sharing data. Under this assumption, execution can aggressively proceed. Speculation is complemented with the ability to defer committing any changes to architectural state until it is certain that there was in fact no interference between processes attempting to access shared data. Hardware support for speculation has been a research topic until very recently. The primary technique that has emerged as a candidate for inclusion in commercial multicore chips is *hardware transactional memory* (HTM) [9].

Messaging is a special case of decoupling that, when used, can pervade the programming model visible to the application programmer. A message between processes can achieve both communication and synchronization in a single action. While a message-based programming model can be implemented purely in software using locks and shared memory, some multicore systems are beginning to provide hardware-level support for messaging which can be used to efficiently implement lock-free programming.

The following sections will take a deeper look at various lock-free mechanisms provided by hardware, and the associated software considerations.

Lock-free synchronization with hardware queues

One decoupling approach to lock-free synchronization is the use of hardware queues. An example of this is the Freescale P4080 device, which has a highly configurable hardware queue manager [10]. A common application for this type of hardware queue in multicore systems is load balancing. In this scenario, one core can be assigned the task of receiving work items and then distributing them evenly across multiple worker cores for processing. Systems with these capabilities will often include autonomous I/O devices that can enqueue and dequeue items to the queue manager as well. Figure 14.14 depicts work arriving from an autonomous I/O device such as an Ethernet controller. The controller places items in the queue assigned to the load balancer core (Core 0 in the figure).

Core 0 will evenly spread the work across remaining cores by enqueuing work items into queues assigned to each worker core (Figure 14.15); the worker cores will iteratively dequeue and process work items. In such systems an additional queue will typically be assigned for outgoing I/O

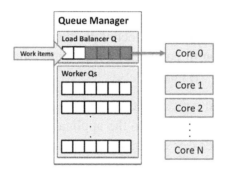

Figure 14.14
Work items enqueued for load balancer task.

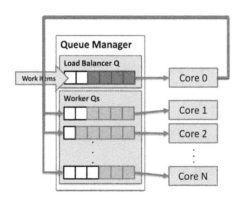

Figure 14.15
Work items enqueued for worker tasks.

as well. In the case of network packet forwarding software, for example, a worker core may finish a work item by enqueuing it to a hardware queue assigned to an outgoing Ethernet port, which would autonomously process egress of the packet. With this approach, locking can be completely avoided because each work item will be worked on by only one core at a time, eliminating context switches that might threaten thread safety.

Decorated storage operations

There are many situations in multicore programming when it is desirable to maintain shared data that serves as a counter or statistic of interest to the application. The nominal course of action to achieve this would be to use a lock variable to gain access to the counter value stored in shared memory and to update it within a critical section. This approach will certainly work, but in many situations it can have significant performance implications. Negative performance impacts are especially undesirable when the act of maintaining a counter is secondary to a requirement for achieving maximum processing throughput. An example of this situation is network packet-processing applications, wherein the main focus of the application is to route packets as fast as possible, and of secondary concern is a requirement to maintain statistics concerning the packets being routed.

Some multicore systems provide the means to maintain such counters using lock-free approaches. This is done via decorated storage operations, as in the Power Architecture l*dx (*decorated load family*), st*dx (*decorated store family*), and dsn (*decorated storage notify*) instructions. These instructions are uniformly defined from the processor point of view, but the semantics vary depending on the target device. For example, all decorated storage operations require two register arguments: register rA specifies the decoration semantics, register rB specifies the address. In addition, decorated store operations require a third argument: register rS, which contains additional data to be applied with the decoration. The decoration operation semantics will be defined according to the device targeted by the decorated storage operation. Examples include a decoration that increments the contents of a memory address

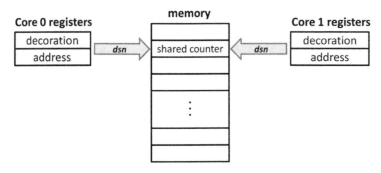

Figure 14.16
Decorated storage operations.

and a decoration that adds a fixed value (perhaps contained in rS) to the contents of a memory address.

Figure 14.16 depicts the scenario of incrementing a shared counter using a dsn instruction. Each processor wishing to increment the counter will load the counter's address into rB, and an *increment location* command into rA (where *increment location* semantics and format are defined by the memory controller in use in the particular multicore system). When the processor executes the dsn instruction, a transaction will be sent to the memory controller (or perhaps a cache controller that will eventually flush the data to main memory), which will then atomically increment the counter without locking. This occurs atomically because the target device serializes all transactions that it receives. Note that for the increment scenario, it is not important in which order the increment updates occur, since, after the two operations complete, the counter will be correctly incremented by two.

Messaging

Figure 14.17 illustrates how to achieve mutual exclusion synchronization using message passing. In this example, Core 0 owns a lock variable. Core 1 requests a lock via a message to Core 0 and is granted access to the lock via a message back from Core 0. During the time that Core 1 owns the lock, Core 2 also sends a request for the lock to Core 0. However, Core 0 does not grant Core 2 access to the lock until after it receives a message from Core 1 releasing the lock.

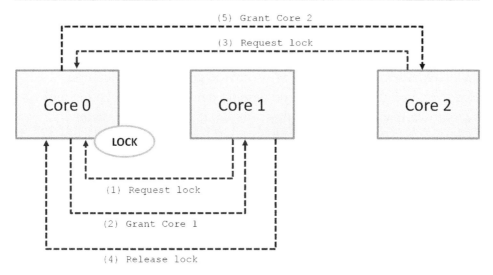

Figure 14.17
Implementing mutual exclusion synchronization with messaging.

Messaging can be used in similar fashion to implement various other synchronization models such as producer/consumer (e.g., software pipelines), barriers, and advanced synchronization techniques such as load balancing.

In general, multicore systems can provide messaging support via hardware queues, DMA engines, or by inter-processor interrupt facilities. The special case of multicore systems without coherent memory support must provide messaging using a combination of hardware features. Each of these is described in more detail in the following sections.

Inter-device messaging with hardware queues

Many multicore chips also include various devices that offload processor cores from processing certain well-defined specific protocols. Operating system drivers for these protocols are often implemented purely as code to be run on the processor core, or may utilize the offload processor but remain inefficient due to high synchronization overheads required to share the resource amongst cores. For systems that offer specialized accelerators, using these drivers unmodified can mean that performance will be left on the table while the accelerator resources are underutilized.

One example is the IPsec protocol, which is used to encrypt Internet traffic for secure transactions between a web browser and a server. IPsec is an essential and frequently used protocol in the Internet. Thus, mainstream operating systems will provide generic IPsec drivers. However, because of the high processing demands of encryption and decryption, many multicore chips provide dedicated acceleration hardware for IPsec. As described above, a customized IPsec driver is required in order to take advantage of this hardware feature.

Even with customized drivers, performance issues can arise when multiple processors are attempting to share a dedicated hardware acceleration engine. In general, the interaction with an accelerator is to exchange messages between the processor and the acceleration engine. These messages often take the form of job descriptors. For IPsec processing, these job descriptors will likely contain information such as a pointer to the buffer of data to be processed, the particular security algorithm to be applied, and a key. The hardware accelerator reads a message from software, processes the data as requested, and then passes a message back to software. The flow of messages between cores and accelerators is typically managed through some form of first-in/first-out buffers (FIFO) queue. If this queue is implemented purely in shared memory and managed by software, then processors must use locks to enqueue and dequeue messages. This synchronization overhead results in lower throughput and poor utilization.

Devices such as the Freescale P2020 address this issue by providing a number of hardware queues for descriptors [11]. Queues are assignable to cores or to hardware accelerators, and descriptors can flow lock-free through these queues (Figure 14.18). Experiments with off-the-shelf IPsec drivers modified to use hardware queues show that avoiding locks altogether can provide approximately 25%–30% throughput improvement.

Messaging with DMA

It is often desirable to offload the processor core from the overhead of message delivery. This can be done by utilizing dedicated hardware which is first programmed by the processor core and then processes the

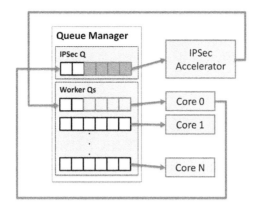

Figure 14.18
Messaging using hardware queues.

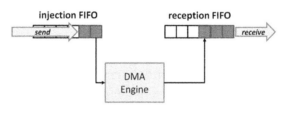

Figure 14.19
Use of DMA for messaging.

message autonomously. A common approach is to leverage DMA devices.

Examples of this approach are the IBM Blue Gene/P and Blue Gene/Q systems [12,13]. These systems include a DMA engine which can be dedicated to messaging. The provisioning of this DMA engine includes a collection of FIFOs for message injection and message reception. Figure 14.19 provides a high-level view of this approach. To perform a send operation, a message descriptor is placed in the injection FIFO. When the DMA engine processes the message descriptor, it will eventually result in an entry being appended to the appropriate reception FIFO. This entry will ultimately be read by software to perform the message receive.

The performance advantage of the DMA approach can be significant for large message sizes or for large numbers of messages. This allows the processor load for moving message descriptors to be quite small relative to the load on the DMA engine for moving entire messages. In addition, the FIFOs provide efficient decoupling of send and receive operations, allowing processors to avoid busy-waiting during send and receive operations.

The Blue Gene DMA hardware is complemented by the Deep Computing Messaging Interface (DCMF). DCMF provides a software interface to the messaging hardware, including support for *active messages*. Active messages differ from messaging models which require explicit receive actions to be issued by the receiver. Instead, active messages carry additional information in their header which identifies a software routine to be invoked on arrival of the header packet at its destination.

Inter-processor interrupt facilities

Some architectures include instruction-set support for messaging between processor cores via direct interrupt delivery. This approach to messaging completely bypasses the memory subsystem, thus requiring no locks. One example of this approach is provided by the Power Architecture `msgsnd` instruction. Processors initiate a message by executing the `msgsnd` instruction with a message type and message payload specified by the contents of register `rB`. Register `rB` is either a 32-bit or 64-bit register, so the payload is limited to a few bits of information indicating things such as a logical process ID or the specific processor ID for which the doorbell is intended. Despite limited payload size, a mechanism like this makes it is easy and efficient to implement a messaging strategy supporting arbitrarily large messages. To achieve this, a message is first copied into shared memory by the sender and then a doorbell is sent to the intended target indicating that the message buffer is ready, and the receiver can subsequently read the message from shared memory. This provides a lock-free message send/receive transaction as long as the sender and receiver can agree not to access the shared memory buffer out of turn.

Executing the `msgsnd` instruction causes the hardware to broadcast the message to all processors within a coherence domain in a reliable manner. The message is transported over the system interconnect using a

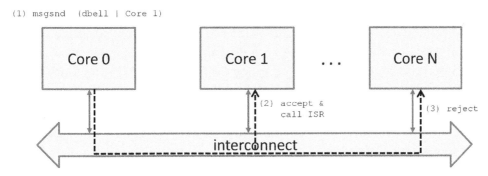

Figure 14.20
Message send via doorbell interrupts.

unique interconnect transaction type that can only be generated by the `msgsnd` instruction. This technique can work with any type of interconnect as long as the required attribute bits for the transaction type can be supported in transactions between master devices.

When a message is received by a processor, it is filtered in hardware based on the message type and payload. If the processor determines that it is a target for the message, an asynchronous interrupt service routine will be invoked to process the message (Figure 14.20).

Inter-core interrupt facilities are especially useful for implementing intra-operating system synchronization, as well as synchronization between a hypervisor and a guest operating system. To complete the hardware facilities for such interactions, the `msgsnd` instruction is complemented by interrupt enable/disable bits as well as a `msgclr` instruction that will clear any pending interrupt bits. These features allow the operating system or hypervisor software to fully manage the messaging facility to meet their demands.

Messaging in non-coherent systems

Anticipating an eventual move to manycore architectures, researchers at Intel have produced two generations of manycore chips under their "terascale" research program. The second-generation chip is called the Single Chip Cloud (SCC) [14,15]. The SCC provides no cache-coherent shared address space, so coherency must be completely software managed.

The SCC chip consists of 48 P5 cores arranged in pairs on 24 tiles connected by a 2D mesh interconnect. Each core has access to private memory, and each tile provides 16 KB of shared (but not coherent) memory known as the message-passing buffer (MPB). MPB memory is cacheable in the processor's private L1 cache, but bypasses the L2 cache. When data from the MPB is cached in the L1, the corresponding cache line is tagged as MPB data, enabling the memory subsystem to distinguish messages from other application data.

The normal Intel atomic instructions are not available in the SCC due to its lack of hardware coherency. Therefore, to provide a minimal locking capability, each processor has a `test_and_set` register which can be updated atomically by the hardware. In addition, the processor includes a new instruction (`cllinvmb`) to aid message passing. This single-cycle instruction is used to invalidate all L1 cache lines that are tagged as MPB data, causing subsequent memory accesses to go to backing store.

The SCC architecture defines conventions for interacting with the `test_and_set` registers, including specific data values used to indicate whether a lock has been set or cleared. Any SCC core can read any `test_and_set` register in the system, and this must be carefully managed in software. A core attempts to acquire a lock by reading a `test_and_set` register. The returned data value indicates whether the lock was obtained. Any core can reset a lock associated with any `test_and_set` register by writing the appropriate data value to the register. This capability provides a simple (but not highly scalable) capability to synchronize access to the data associated with MPB for the purposes of messaging. All other synchronization in the system is built upon this capability using the message approach depicted in Figure 14.17.

Given the basic synchronization mechanisms described above, messaging is then implemented in the SCC as follows. When a message is sent, the data is moved from core private memory through the L1 cache of the sending core. It is then moved to the MPB, and eventually to the L1 cache of the receiving core. The receiving process then reads the message from its L1 cache. The software responsible for negotiating the message exchange process stores various flags and other data in cache lines shared between the sending core and the receiving core (not to be confused with the MPB flags which are automatically handled by hardware). Access to the data in the shared cache line is guarded using locks managed in the `test_and_set` registers.

The SCC chip is a research vehicle, never intended for mass production. The lessons learned from this purely message-passing programming model should eventually materialize in production processors. Most SCC-based research has focused on how to make this programming model easily comprehensible by programmers while exhibiting good performance. It's important to remember when envisioning such future systems that performance will likely be evaluated using a throughput metric rather than a latency metric. We can expect such systems to run at lower frequencies for power efficiency reasons, and the combination of many processor cores on die and lower operating frequencies will affect how we think about programming and which algorithms we ultimately choose!

Hardware transactional memory

Another form of lock-free programming is hardware transactional memory (HTM). This approach is based on concepts from transactional databases. Much like concurrent programs interacting through shared memory, transactional databases allow concurrent accesses to data stored in their data files. Concurrent database transactions first snapshot the system state, then speculatively read and write data while logging the updates to non-permanent backing store, and finally commit the changes to permanent storage as long as no conflicting read or write accesses occurred during the transaction. If conflicts are detected, the permanent state is not changed and the commit fails (e.g., the transaction is *rolled back*).

With HTM, a transaction is defined as a block of instructions containing a non-empty set of memory accesses. There can be no memory-level parallelism within a transaction, but separate transactions can proceed concurrently. Furthermore, a transaction is guaranteed to execute as an atomic unit if no memory access conflicts are detected (Figure 14.21). If a conflict is detected, then all instructions in the group will fail to execute. Conflicts occur when accesses for one transaction causally

Memory

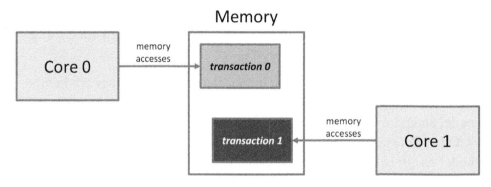

Figure 14.21
Concurrent memory transactions with no conflicts.

Memory

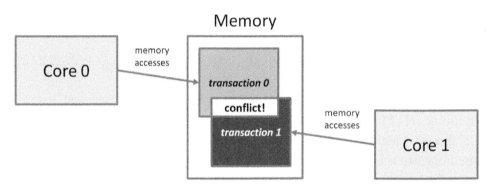

Figure 14.22
Concurrent memory transactions with conflicts.

depend on accesses from another concurrent transaction. In a typical multicore system, the coherency mechanism will take care of broadcasting all changes related to a commit.

When a conflict is detected amongst concurrent transactions then the attempt to commit an associated transaction will fail and system state must reflect the state it was in before speculative execution of the transaction began (Figure 14.22). This requires hardware to provide the equivalent of the transactional database checkpointing and rollback features. Thus, with HTM the hardware is responsible for determining the set of memory accesses associated with a transaction and for providing conflict detection. This can be done, for example, by modifying the cache and the coherency protocol. In addition, the complexities of managing exceptions and interrupts are likely to be mitigated by simply aborting transactions in flight when these events occur.

HTM is likely to become widely available over the next few years. One of the first commercial implementations to be announced is the IBM Blue Gene/Q system [16]. This system will employ a multi-versioning cache to support the transactions. Data in the cache will be tagged with additional bits allowing multiple versions of a memory location to be tracked using a scoreboarding approach. The scoreboard will be used to detect any load/store conflicts for a transaction. When a conflict is detected the system software will be notified and it will manage rollback and re-execution of the transaction.

The advantage of HTM is that programmers do not need to employ explicit synchronization. This removes significant control complexity and overhead from software implementation. However, the programmer must carefully determine transaction boundaries. When transactions frequently conflict there will be performance degradation, thus transaction granularity and degree of concurrency are important considerations. Similarly, large transactions will require the hardware to temporarily store more speculative state during transaction execution and subsequently broadcast more updates on a successful commit. Programmers must also be cognizant that it is not feasible to support arbitrarily large or arbitrarily many transactions purely in hardware due to the limited amount of resources available to store speculative state for open transactions.

Memory subsystem considerations

Most of the synchronization mechanisms presented in this chapter rely on shared memory and associated coherency of data. For example, ensuring atomicity of a load-reserved/store-conditional instruction sequence depends on the ability of the hardware to detect whether or not an intervening store has occurred to the reserved address (thus violating the reservation). The ideal guarantee that programmers would like from the hardware is *sequential consistency*, as defined below [17]:

> *A system is sequentially consistent if the result of any execution is the same as if the operations of all the processors were executed in some sequential order, and the operations of each individual processor appear in the order specified by the program*
>
> **Leslie Lamport**

However, for a variety of performance reasons, multicore processor architectures provide a range of memory consistency semantics. For example, caches can sometimes prevent the effect of a store from being seen by other processors. These multicore complications motivate architects to provide weak or relaxed memory consistency models. Unfortunately, this comes at a price for programmers, who must be aware of these relaxations in order for loads and stores to ensure that their code is functionally correct.

Memory ordering rules

Strongly consistent memory ordering rules require that all processes sharing data in a multicore system will observe all updates to the shared data in the same order (typically the exact order in which they occur). In contrast, weakly consistent ordering rules allow the order as performed by processors to differ from the order in which updates occur at main memory. Thus different processors may observe different update orders. In many cases, this poses no significant problem since the majority of memory accesses are likely to be non-conflicting.

The biggest effect of memory access reordering is manifested in operating system routines, libraries for synchronization, and device drivers. Fortunately, these are relatively well-contained and highly reused areas of code, and the performance benefits justify the additional programming and verification complexity. For completeness, architectures with relaxed memory order must also provide memory fence instructions to allow programmers to enforce serialization of memory accesses when it is essential for program correctness. Weakly ordered architectures include SPARC (stbar and membar instructions), newer Intel architectures (sfence, lfence, and mfence instructions), and Power Architecture (sync, mbar, lwsync instructions).

The sfence instruction guarantees that all store instructions executed prior to its execution will complete before the processor will execute any additional instructions. Similarly, the lfence instruction ensures that all load instructions executed prior to its execution will complete before proceeding. The mfence instruction combines the effects of sfence and lfence, ensuring that both loads and store executed before the mfence will complete before proceeding.

Power Architecture provides a rich set of options for the programmer to manage the effects of weakly consistent ordering rules. The sync instruction (also known as sync 0 or msync) ensures that all memory accesses executed before the sync instruction (Group$_1$ in Figure 14.23) will be completed before any accesses following the sync instruction (Group$_2$ in Figure 14.23), and, in fact, ensures that no instructions following the sync start their execution until the sync completes. The mbar instruction (also known as mbar 0) is similar to sync, but only affects load and store instructions. The mbar 1 variation achieves better performance by relaxing completion rules to cover only (i) stores that are to locations in write-through cache, (ii) loads and stores that are both cache-inhibited and guarded (as in data for device drivers), and (iii) stores to normal cacheable coherent memory. The lwsync instruction (also known as sync 1)

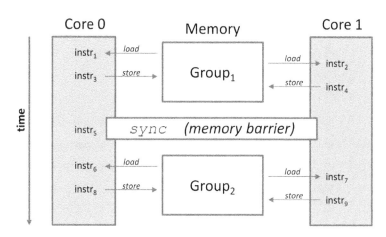

Figure 14.23
Effects of sync instruction.

performs a lightweight operation that will complete normal cacheable memory accesses but will not order a store followed by a load.

For certain applications, it is more convenient to designate behavior for a range of data rather than issuing synchronizing instructions within the code. This can generally be provided via configuration options at the system level or by designating attributes for physical pages in memory (in the memory management unit or elsewhere). For example, Power Architecture provides for a bit in a processor configuration register to specify at the system level that all cache-inhibited and guarded loads and stores shall be performed in order (HID0[CIGLSO] bit). And certain Intel architectures support memory type range registers (MTRR) that can be used to define memory ordering rules for specific regions of memory, while other Intel architectures provide page attribute tables (PAT) that can be used to strengthen memory ordering rules for specific memory pages.

Using memory barriers and synchronizations

Proper management of memory consistency only adds a few lines of code to important routines, but knowing when and how to use memory barriers is a very important consideration. In general, whenever an atomic operation will depend on coherency of the data then the programmer will need to insert some form of barrier into their code.

```
resource-lock

lock:      mfspr   r6, LR          # save Link Register

           addi    r4, r0, 1       # obtain lock:

loop:      bl      test_and_set

           bc      4, 2, loop      # retry until lock obtained

# Delay instructions below until instructions above complete

           sync    0               # or isync, mbar, lwsync, …

           mtspr   LR, r6          # restore link register

           blr                     # return
```

Figure 14.24
Assembly code for resource lock.

In the example Power Architecture code shown in Figure 14.24, a shared memory location is used as a lock to control access to a shared resource (such as an I/O device). The lock is available when its value is 0, and locked when its value is 1. This example code uses the `test_and_set()` routine from Figure 14.10 to obtain the lock. The lock routine contains a `sync` instruction since reads to the shared resource cannot be executed until the lock has been acquired. Because of the `sync` instruction, any speculative reads to the shared resource that have been issued by the processor will be discarded (such reads may have occurred, for example, due to a misprediction by the branch predictor).

A complementary resource-unlocking procedure must be created. This procedure will store a 0 to the lock location. In this case, the programmer will want to ensure that all stores to the shared resource complete before the lock is released (some of the stores may be buffered by the system for deferred write, for example). In the example shown in Figure 14.25 the resource-unlock procedure begins with a `sync` barrier to ensure this property.

In addition to functional correctness, the programmer must also be aware of the performance implications of managing weakly consistent memory. For example, to help guarantee forward progress in routines such as the resource lock/unlock described above, it is important to minimize looping on load-reserved/store-conditional transaction pairs. Code should be carefully examined to ensure that actions which would cause loss of the reservation are minimized; such actions cause more time to be spent in the lock acquisition loop than is strictly necessary. In other cases,

```
resource-unlock

unlock:     sync    0                # or isync, mbar, lwsync, …

# Delay instructions below until instructions above complete

            addi    r1, r0, 0        # prepare value

            stw     r1, r0, 0        # store 0 to lock location

            blr                      # return
```

Figure 14.25
Assembly code for resource unlock.

there may be a benefit to performing a regular load instruction to test a value before performing the load-reserved transaction (e.g., lwarx). Depending on the semantics of the target resource and the type of accesses made to it, the choice of which barrier instruction to use could range from the most heavyweight form (e.g., sync) to the most lightweight form (e.g., lwsync). The programmer must trade off correctness with performance when making this choice. These considerations are highly processor-dependent, so programmers should carefully examine their code with respect to the semantics of their system.

Conclusions

Synchronization is an essential element of multicore programming, and it comes in many forms. Underlying hardware must provide support for synchronization, and does so in the form of a few primitive atomic memory read/modify operations upon which many synchronization idioms can be built. For performance and complexity reasons, atomic synchronization primitives do not scale well as chips move to more and more cores. Therefore, instruction sets also provide various speculation mechanisms which can improve performance. These include paired memory read/modify instructions, which provide speculative atomicity with violation detection, and weakly ordered memory semantics, which allow loads and stores to be performed out of order. In addition, based on the recognition that synchronization can remain a bottleneck even with the most efficient underlying hardware support, several lock-free mechanisms have emerged that can further improve software performance by avoiding synchronization altogether.

References

[1] G.L. Peterson, Myths about the mutual exclusion problem, Inform. Process. Lett. 12(3) (1981) 115−116.
[2] D.L. Weaver, T. Germond, The SPARC Architecture Manual. 1994; Available from: <www.sparc.org/standards/SPARCV9.pdf>.
[3] Intel Corporation. Intel® 64 and IA-32 Architectures Software Developer Manuals 2011 [cited]; Available from: <http://www.intel.com/content/www/us/en/processors/architectures-software-developer-manuals.html>.

[4] A. Sood, Guide to Porting from Solaris to Linux on x86. 2005; Available from: <https://www.ibm.com/developerworks/linux/library/l-solar/>.

[5] Freescale. EREF 2.0: A Programmer's Reference Manual for Freescale Power Architecture Processors. 2011; Available from: <http://cache.freescale.com/files/32bit/doc/ref_manual/EREF_RM.pdf>.

[6] D. Chun, S. Latif, MIPS R4000 Synchronization Primitives. 1993; Available from: <http://www.mips.com/products/product-materials/processor/archives/>.

[7] A.A. Committee, Alpha Architecture Reference Manual, third ed., Elsevier, 1998, 952.

[8] The Free Software Foundation. GCC 4.6.2 Manual. 2011; Available from: <http://gcc.gnu.org/onlinedocs/gcc-4.6.2/gcc/>.

[9] S. Liu, J.-L. Gaudiot, Synchronization mechanisms on modern multi-core architectures, in: Twelfth Asia-Pacific Computer Systems Architecture Conference (ACSAC 2007). 2007.

[10] Freescale Semiconductor, I. P4080 product summary page. 2008 [cited; Available from: <www.freescale.com/files/netcomm/doc/fact_sheet/QorIQ_P4080.pdf>.

[11] S. Malik, R. Malhotra, Performance enhancement for ipsec processing on multi-core systems. In: 2011 International Conference on Communications, 2011.

[12] D. Chen, et al. The IBM Blue Gene/Q interconnection network and message unit. In Supercomputing 2011 (SC11), 2011.

[13] S. Kumar, et al. The deep computing messaging framework: generalized scalable message passing on blue gene/p supercomputer. In: twenty second Annual International COnference on Supercomputing, 2008.

[14] Intel Corporation. SCC External Architecture Specification. 2010; Available from: <http://communities.intel.com/docs/DOC-5852>.

[15] R.F. van der Wijngaart, T.G. Mattson, W. Haas, Light-weight communications on Intel's single-chip cloud computer processor, ACM SIGOPS Operating Systems Review 45(1) (2011) 73–83.

[16] IBM. IBM system blue Gene/Q data sheet. 2011; Available from: <http://www-03.ibm.com/systems/deepcomputing/solutions/bluegene/>.

[17] L. Lamport, How to make a multiprocessor computer that correctly executes multiprocess programs, IEEE Trans. Comput. 28(9) (1979) 690–691.

CHAPTER 15

Bare-Metal Systems

Sanjay Lal
Cofounder of Kyma Systems, Danville, CA, USA

Chapter Outline

Introduction

Most computer systems keep the programmer at arm's length from the details of the microprocessor hardware and other peripherals. The operating system (OS) provides a layer of abstraction that provides services to the programmer while managing the detailed interaction with the low-level components.

Real World Multicore Embedded Systems.
DOI: http://dx.doi.org/10.1016/B978-0-12-416018-7.00015-8

In the early days of computing, each computer was a giant box that inhabited the bulk of a room. The concept of each person having their own computer would have been laughable then: everyone had to share time on the computer. So one of the primary benefits of an OS was managing all of the jobs, keeping one person's program — and, in particular, its mistakes — from affecting anyone else.

On a personal computer, separating users is much less of an issue, but keeping programs separate is still important.

In embedded systems, however, not only do you eliminate the multi-user problem, but, in many cases, you also eliminate the multi-program problem since the system is designed to do only one thing — that is, to run only one program.

In such a system you have a trade-off. The OS still provides services and abstracts away the hardware, but it also comes with costs. First is the simple issue of performance: just as, in theory, a program can go faster if written in assembly language than in a high-level language, so, in theory, your system could perform faster if it didn't have any OS overhead.

More significant is the fact that an OS controls access to all of the hardware; the program itself can't take control. So, by design, a program running over an OS is subject to the whims of that OS. Some parameters are available for limiting the degrees of freedom that an OS can exercise, and RTOSes intentionally tame some of that behavior, but the fact remains that there is a level of unpredictability when you use an OS.

There are certain systems and programs — primarily in the packet-processing world — that, given the trade-off between OS services and performance, err on the side of performance primarily because they need to run as fast as the technology allows. What goes along with that is the fact that the sacrificed OS services aren't that critical for these programs: the services can be provided some other way.

Such systems eliminate the OS altogether, and have the program access the processor directly. This scenario has been colorfully described as having the program run "on bare metal" since there's nothing to cushion the ride (Figure 15.1). This chapter will describe bare-metal setups in more detail, describing how they work and how they can be built.

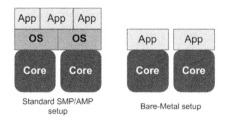

Figure 15.1
In a standard configuration, applications run on top of an operating system. In a bare-metal setup, applications run directly on the core.

We will also take a look at some intermediate solutions being developed that try to split the difference between the efficiencies of bare metal and the services of an OS. Note that graphics processors are often used without an OS, but such technology works differently from the bare-metal operation we will describe here.

What is a bare-metal setup?

Programs that run on bare metal typically follow a very simple format: they do one thing, more or less forever. This is epitomized by the fact that such programs are enclosed within an endless loop, implemented in C as `while(1)`.

```
While(1) {
    look for input
    if input then {
        graph input
        process it
        send to output
    }
}
```

This loop typically looks for work to do, does the work, and then comes back and looks for more work to do. When it finds work, it executes that work without interruption from anything else. This model of execution is referred to as run-to-completion (RTC).

Part of this ability to run uninterrupted comes from the fact that, without an OS, there will be no interrupts that cause the OS to change the thread of execution. Interrupts are still possible, but only if the program itself explicitly manages the processor's low-level interrupt infrastructure. In addition, such programs cannot make any system calls for things like memory allocation or performing device I/O: the program must somehow manage such services on its own. This also means that there are no debugger hooks into the software stack; you can't just plug in gdb to see where your program is going wrong.

Of course, in a multicore context, the overall system will consist of a group of cores or processors, each of which has a dedicated RTC program. Unlike a standard SMP setup, those programs are dedicated to their processors, and cannot be rescheduled or moved to other processors (Figure 15.2). Likewise, no other program can be made to share a processor that has an RTC program executing.

One practical result of this has to do with how a larger program might be decomposed into pieces that run in a bare-metal environment. With a standard OS environment, you can create multiple threads of execution that run concurrently. All of those threads belong to the same overriding process. Without an OS, however, there can be no threads, and there are no services available for creating and killing threads. In this model, each RTC component acts as a complete program; it is its own process, and it has no awareness of any of the other RTC processes it might be working with (Figure 15.2).

This also means that, if a single program is decomposed into multiple RTC programs, there is no common global namespace, so simple references between different RTC programs may need to be implemented using some form of inter-process communication (IPC). We'll discuss ways of managing this later in this chapter.

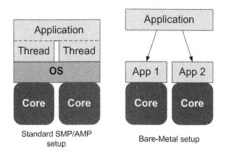

Figure 15.2

SMP setups spawn threads from a single application; in a bare-metal setup, an application is split into separate applications.

Who should use bare metal?

Bare-metal setups work best for algorithms that have straightforward chunks of work that needs doing. If the amount of work increases, you want to be able to scale the computing resources to match the higher demand. Because the RTC program can be designed to be self-contained by minimizing data structures that are shared with other cores, then, if you have twice the work, you can double the number of cores and handle the extra load with almost no loss of performance (Figure 15.3). The only extra overhead that may be required is a means of allocating work when there is more than one processor available — we'll discuss that more later.

In addition, a program to be targeted for bare metal would be one that would suffer from some of the non-deterministic behavior that an OS might introduce. The three primary sources of overhead are timing interrupts, page faults, and I/O.

Standard OSes like Linux are responsible for scheduling processes and threads. With a single-core system, the OS may pull one process or thread and swap in another. With multicore systems, not only might the OS suspend one thread in deference to another, but it may actually move that thread from one core to another thread.

Even if you use affinity and other controls to limit the OS's interference with execution, you are still left with timer interrupts where the OS

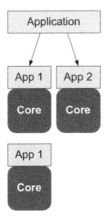

Figure 15.3
With well-defined RTC programs, simple scaling can increase capacity.

periodically pre-empts the running process to perform housekeeping duties. For a single-purpose program that won't ever cede to another program and will never be moved to another core, these timing interrupts are simply a nuisance and interfere with the run-to-completion nature of the program: the program can't run to completion because the interrupts break the flow.

More OS delays arise from page faults when accessing memory. Page faults can occur regardless of whether you have an OS running. The difference is that, without an OS, you can manage your TLB or page table directly so that you avoid page faults. You can also manage the cache directly, issuing data prefetches and directing the result to a specific cache, or locking parts of the cache down. In a bare metal system, you know that the OS won't mess up your careful work.

I/O interrupts can also waste time as the OS tries to distribute them over different cores. Strictly speaking, you can solve this without going full bare metal by using affinity to assign input and output interrupt servicing to specific cores, but it ends up being a natural part of the work and benefit of eliminating the OS entirely. By resorting to polling, a bare-metal application can avoid the disruptions and "jitter" caused by interrupts.

While, in theory, there may be many different application areas that meet these characteristics, there is, as mentioned, one primary application area that makes extensive use of RTC programs: packet processing.

The main motivator for networking is speed: you can never have too much, so removing every last ounce of wasted processor time yields benefits. But, especially in the early days, packet processing wasn't particularly complex; it was just that there were lots of packets to process.

So the model of "get the packet, do something to the packet, send it on. . . and repeat" works extremely well in the networking world. For this reason, the explanations and examples in this chapter will focus on packet processing for illustrating specifics.

Architectural arrangements

There are two fundamental ways that multiple bare-metal cores can be arranged to scale with workload, and, in fact, practical systems combine both. The two arrangements arise from the two kinds of parallelism identified in the chapter on concurrency earlier in this book: data and functional parallelism.

Data parallelism: SIMD

One of the most straightforward ways of scaling is simply by breaking up the amount of data that needs to be processed. This only works well if each unit of data can be processed independently of the other units.

This type of processing originated with vector processing, and is, in fact, common with graphics processing. At that low level, a vector – or an array of scalar numbers – is fed to an array of processors, ideally with the number of processors matching the length of the vector. At this level, these aren't full-on processors; they're simply arithmetic units scaled to handle vectors; each applies the same instruction. For this reason, this arrangement will often be referred to as "single instruction, multiple data", or SIMD.

A true multicore environment isn't just about a SIMD engine, however: it's an arrangement of multiple independent cores that happen to be running the same RTC program. For example, if you are processing packets and want to double the rate of packet processing, at a very simple level, you would simply add another core that was running the same packet-processing RTC program.

You still must account for the source and destination of packets, however. Assume that a single core processes packets as shown in Figure 15.4.

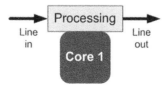

Figure 15.4
Basic packet-processing engine.

If the doubling of packets came from adding another input and output connection, then you might think that you could literally simply double the processing and the input infrastructure, as shown in Figure 15.5.

Assuming the processing is doing some kind of routing, however, even though the packets are coming in on independent lines, they may need to go on either of the output lines. So you would need some kind of routing logic following the packet processing to determine which of the output lines the packet would need to be sent to (Figure 15.6). Exactly where that router would run would depend on the processing headroom in the cores — perhaps there would be enough

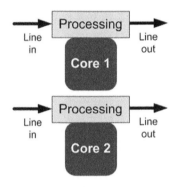

Figure 15.5
Simple doubling of processing capacity due to addition of input and output lines.

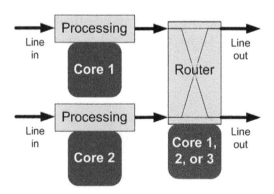

Figure 15.6
A practical doubling of capacity will likely require the addition of a routing stage.

spare cycles to accommodate routing; if not, then you might need another core for that.

On the other hand, if the doubling of the packet load came without changing the number of connections, then, after doubling the basic processing unit, you need to add some input logic to decide which packet goes to which core. If all packets were the same, then you could simply alternate which core would get a packet. If different packets required different amounts of processing, however, as is often the case, then you might want to dispatch packets to whichever core is least busy or has the least number of packets in its queue.

Meanwhile, at the output, you'd need to multiplex the two sources of outgoing packets back into a single stream (Figure 15.7). Again, processing capacity would be needed for both the scheduler and the mux, whether borrowed from the existing two cores or by adding another (presumably one would suffice for both scheduler and mux). Some packet-processing platforms may provide dedicated hardware versions of such things as schedulers for added performance and to reserve the cores for more sophisticated computing.

These examples illustrate that even though data parallelism can provide good scaling, it does come with some extra circuitry and processing. Packet throughput may double, but latency, or the time it takes from packet-in to packet-out, will increase slightly (Figure 15.8).

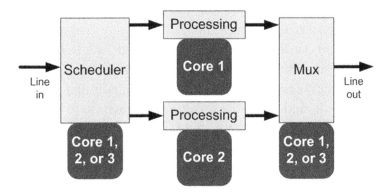

Figure 15.7
Increased capacity for a given line in and line out requires the addition of an input scheduler and an output mux.

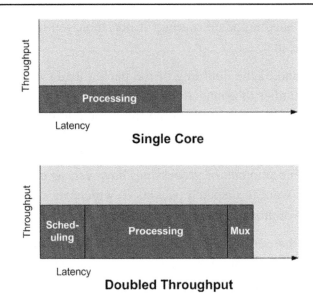

Figure 15.8
Doubling throughput may mean an increase in latency. Delays shown are not to scale.

Functional parallelism: pipelines

The other way to share the load of a big job is to invoke the old division of labor concept: implement an assembly line. If three steps are needed on a packet — say,

1. unwrap the packet to see where it's going,
2. look up the next hop destination, and
3. rewrap the packet for the next hop,

then you can split out those three tasks and assign one to each of three cores. (Bear in mind that this is an extreme simplification for illustrative purposes.) The first core will house an RTC program that simply takes a packet, unwraps it, and sends the resulting information to the next core. The second core takes the result of the first core, uses it to look up the next hop, and sends its results to the last core. The last core takes the next-hop information, wraps the packet with it, and sends it out the door. Figure 15.9 summarizes this setup.

This kind of arrangement is known in hardware circles as a pipeline for relatively intuitive reasons. In software circles it is known more

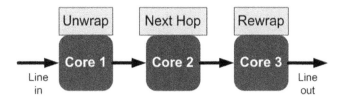

Figure 15.9
Three-stage pipeline for basic packet routing.

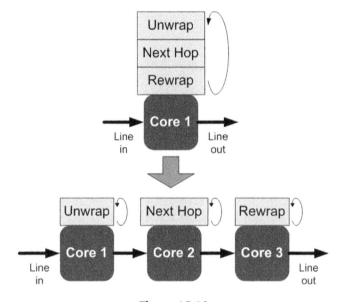

Figure 15.10
Loop distribution.

typically as a distributed loop, since the contents of one large loop has been split between multiple cores, each with its own loop (Figure 15.10).

The challenge with pipelines is that throughput is limited to what the slowest stage of the pipeline can manage. If, say, the middle stage takes 100 clock cycles to complete its work and the other two stages can complete theirs in 50, then those other stages will be idle for 50% of the time (Figure 15.11); the first one will have to wait to deliver its results until the middle one is ready; the last one will sit around waiting for the middle one to finish.

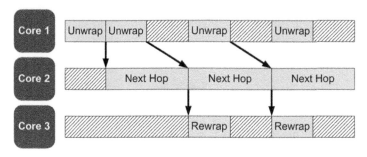

Figure 15.11
An unbalanced pipeline, with idle time shown by hashed portions.

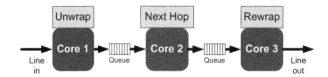

Figure 15.12
Buffers can help handle occasional imbalance that averages out to balanced.

If this is a function of different types of packets, with some taking more or less time, then you can help smooth matters by using queues between the stages to buffer the differences (Figure 15.12). But that still only works if, on average, each stage takes the same amount of time, and that any requirement for extra processing in some cases is balanced by less processing other times. The queues then provide some elasticity for that variation. But if the lack of balance persists, then, in this example, the queue for the middle stage will eventually overflow, and the queue for the last stage will never have more than one entry in it.

Balancing the pipeline perfectly is rarely achieved — it can be very hard to find the perfect place to split the process, even when each packet takes the same amount of time to process (which is not the case in practice). It's therefore much more typical to use a hybrid approach between the data-parallel and the function-parallel approach. In this example, where the middle stage conveniently takes twice as long as the others, you solve the problem simply by adding a second parallel middle stage, with the associated dispatch and multiplexing logic (Figure 15.13 and Figure 15.14). (This conveniently ignores where the scheduling and multiplexing take place to illustrate the main points.)

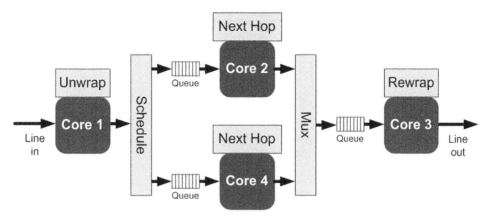

Figure 15.13
Doubling the middle stage to balance the pipeline.

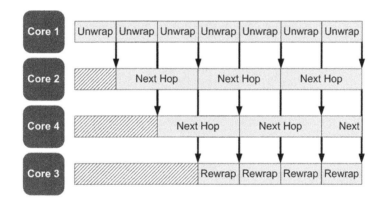

Figure 15.14
Doubling the middle stage eliminates idle times. Scheduling and multiplexing delays are ignored for simplicity.

The other way of dealing with a packet that's clogging up the pipeline is essentially to put a time limit on how long the packet gets to stay in that stage, presumably breaking up the work into milestones whose completion can be tracked. If the time limit is hit before the work is entirely completed, then the packet is ejected to let a different packet be handled. The incomplete packet is then fed back into the queue to be completed later (Figure 15.15).

It can be tempting to try to create a complex, long pipeline with finer granularity so that it's easier to balance. But this has to be weighed

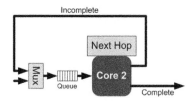

Figure 15.15
Overruns in one stage can be avoided by limiting processing time and then recycling incomplete packets into the queue.

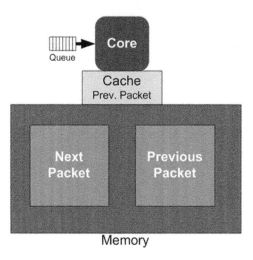

Figure 15.16
Switching from one packet to another will typically involve a cache miss as the new packet is in a different area in memory.

against the cost of handing a packet off from one stage to another. At the very least, handing off from a given stage involves taking a pointer and placing it in the queue of the next stage and then bringing in the next packet in its own queue. Very often, working with a new packet will involve a cache miss (Figure 15.16). So there's a benefit to doing as much as possible once you've got a packet set up to work on. We'll come back to this in the memory section below.

The host processor

These packet processing systems often include a larger "host" processor that's running a full-up OS (Figure 15.17); it's used for less-frequent activity that doesn't need the same speed; being able to handle such

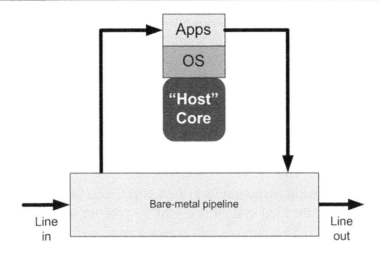

Figure 15.17
A host processor is often used in conjunction with a bare-metal pipeline.

situations outside the bare-metal pipeline means that the bare-metal programs can be simplified, with fewer oddball cases to deal with. The fewer decisions that have to be made in the RTC programs, the faster and more predictable the overall speed will be.

There's a bit of overlapping terminology with respect to this configuration. On the one hand, common packets are processed through the bare-metal section, commonly referred to as the "fast path" because it's optimized to speed these frequent packets along. Irregular packets are sent on the "slow path" through the host processor. Many of those irregular packets are control packets, not data packets, setting up the terminology "control plane" for the host processor and "data plane" for the bare-metal section. But, in fact, some irregular data packets may be handled by the host processor, so the control/data and slow/fast definitions don't overlap exactly, strictly speaking, but are typically used interchangeably in practice.

This host processor can help out with the bookkeeping required at initialization, but, from that point on, it isn't critical to the bare-metal section, and so we won't dwell on it in this chapter.

I/O

Additional architectural elements are required for data ingress and egress. Both of these are standard; I/O is different only in that it needs to place information on incoming data into the queue of the first stage rather than interrupting the OS, as would be typical in a standard system. Likewise, on the output side, the last stage would trigger the egress logic to send the result rather than the OS doing that. As a practical matter, this means that the first and last pipeline stages act as the input and output drivers.

Memory

One of the underlying assumptions of bare-metal design is that memory in a higher-end embedded system is cheap, so you use a lot of it as needed. You generally want to avoid copying data over and over, however – a huge time waster. The data (typically a packet) is placed in memory, and then, instead of actually passing the full packet from stage to stage, you pass only a packet descriptor, which includes some basic information about the packet along with a pointer to it (Figure 15.18). The descriptor typically contains some of the packet header information as well as other relevant "meta-data" like queuing information and quality-of-service (QoS) parameters.

This means that the memory has to be accessible to all of the stages, although, as we noted above, it also usually means that packet handoff from one core to another will result in cache misses in the new core,

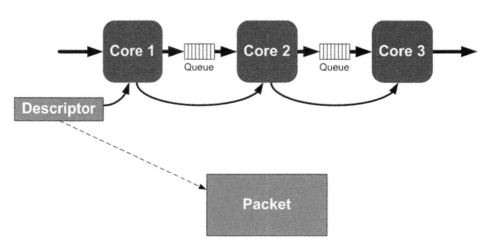

Figure 15.18
One copy of the packet is maintained, while a descriptor is passed from core to core.

with a trip out to L2 (or L3) cache. With an architecture like Tilera's, which has a "home cache" cluster concept, these cache misses can be avoided if the cores in the pipeline are put in the same cluster.

In general, memory can be managed to provide both private and shared data for each of the cores. When possible, one typically tries to align that data to cache line boundaries so that a cache miss invalidates as little unrelated other data as possible.

Managing memory and packet buffers is a critical management task when you don't have an OS to handle memory allocation. An OS manages dynamic memory allocation by maintaining a heap that is outside the scope of the program, and then taking blocks of arbitrary length as requested by a program through something like `malloc()` and making them available via pointers. But in a bare-metal system, you can't rely on `malloc()` because it's a system service that's no longer available, so you have to handle the problem differently.

The flexibility that `malloc()` provides comes at a cost of high overhead, so bare metal programs instead pre-allocate a block of memory and subdivide it into smaller units of a fixed size. This becomes a circular buffer of buffers. When memory is needed, a buffer manager, often implemented in hardware, takes the next available buffer (or buffers, if a larger chunk is requested) and gives it to the requester. The chunks are recycled when no longer in use. An example of buffer management for packets is shown in Figure 15.19. The actual buffer management function is often provided by dedicated hardware.

In order to avoid the need for any defragmentation or "garbage collection", multiple blocks can be allocated without being physically next to each other. This means that the data structure needs to allow for chaining of discontinuous blocks of memory; this is essentially a "next block" pointer at the end of each block. On some architectures, like the Cavium devices that we'll illustrate in an example later in the chapter, this is handled through hardware. On other architectures, it may be managed by a library provided by the processor vendor. If all else fails and you have to build your own memory allocation scheme, you may have to build this explicitly yourself. Regardless, you are free to construct things any way you wish, since it can all be done with

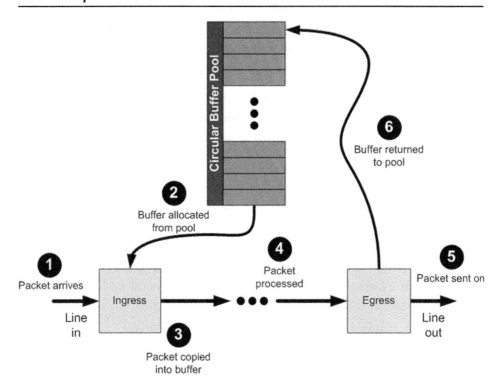

Figure 15.19
Fixed-size buffers are allocated from a pool and recycled when done.

software, so, for example, if your application requires only blocks of one size, then there is no need for such a chaining provision.

Inter-process communication

Finally, you may need to provide for cores to be able to communicate with each other. The simplicity and cleanliness of a pipeline will degrade severely if there's too much interaction with other cores – the whole RTC concept is somewhat predicated on "I don't need anyone else; I just do my job". But there may be cases where IPC is required; this uses the same infrastructure as was discussed in the prior chapter on communication APIs. If you need to share data structures, you can do that out of a region of memory that you set aside for sharing. You can use this region as well for messages, putting the messages in the shared region and queuing pointers to those messages.

We'll discuss the management of private and shared memory later in this chapter.

Software architecture

The most obvious notable characteristic of a bare-metal system with respect to software architecture is that, for the most part, there isn't any. Any required services that might normally be provided by software or firmware infrastructure have to be managed some other way.

When it comes to bringing in services for, say, IPC, you would normally be using a dynamic library (like a DLL on your PC); that won't work here because there's no OS to manage accessing such dynamic services. Everything has to be statically linked at build time. There are a number of attempts under way to provide some level of low-overhead system service without installing a full OS. One approach to this is to use a native (or Type 1) hypervisor. While these are typically used to host guest operating systems, in this case they can be used as is and relied on simply for basic services.

The fundamental benefits of using a hypervisor will be for "life-cycle management" – that is, start-up and tear-down; and "pinning", or deciding which tasks will run on which cores. In the ideal case, the hypervisor will step in at initialization, get everything up and running, and then disappear from view so that the cores can run unencumbered.

Companies with processors intended to run in bare-metal configurations often provide a "thin layer" of basic services in order to offload some of that work from the application (Figure 15.20). Such layers are, of course, proprietary, and code using them will be dedicated to that processor environment.

Two companies have made an attempt to move Linux into this space. Cavium has announced their MontaVista Bare Metal Engine (BME) as a part of their carrier-grade Linux edition; and Tilera has their so-called Zero-Overhead Linux (ZOL). Both of these try to maintain some of the standard services of Linux while drastically stripping down the overhead – in particular, making them "tickless" to minimize OS interrupts.

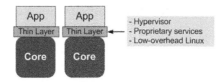

Figure 15.20
Limited system services may be provided by a number of "thin layers" that focus on critical needs while eliminating other services and interrupts.

Building the executable image(s)

As mentioned earlier, the functional result of your programming efforts will be different run-to-completion programs executing on different code. However, there are numerous ways of making this happen, the most common of which doesn't actually result in different programs.

One brute-force way of approaching the problem is to manually partition your original program into multiple programs and then compile and link them separately, arranging for each to be loaded on its appropriate core during system initialization (Figure 15.21). There are no specific automation tools for this; you must manage the process directly.

A less obvious, but simpler and more common, practice is to leave the program as a whole, with several RTC loops in it, one for each target core. Multiple images are then loaded, one for each core. The main() routine can first establish which of the cores is the "master" or "init" core; this can be done by statically assigning an init core, or by assigning that role dynamically to whichever core gets to that line of code in the main() routine first, or any other way you might think of; there is no standard way of doing this. The main() routine can also determine how many cores are running the code based on a low-level firmware call, or that number can be set statically.

That single init core will handle start-up duties; this is done using conditional code that is only run by that code. The other cores will execute any code in the main() routine that isn't conditioned on being run by the init core, and they then will typically hit a barrier and wait for the init core to complete its duties. At that point, each core will know, based on explicit conditional statements in the main() routine, which of the

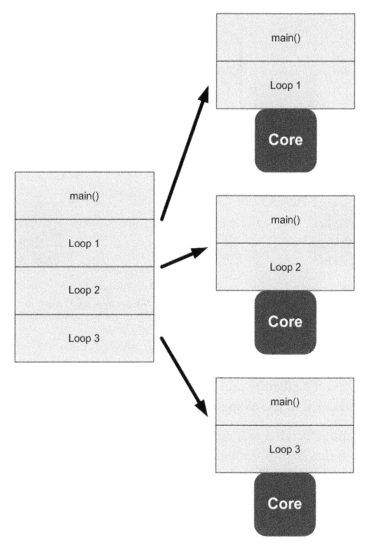

Figure 15.21
A single program manually split into multiple independent programs.

various loops to jump to (Figure 15.22). Once they start their loops, then, in general, they will stay there until execution stops.

While it may appear that memory is wasted for each core storing that portion of the image that's never run, it's common to load the image into shared read-only memory, meaning that, in fact, there is no wasted memory space.

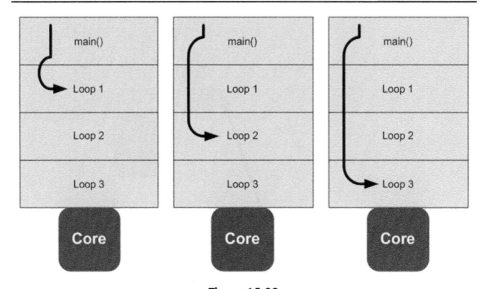

Figure 15.22

A single image, with one core performing initialization and each core jumping to its own loop.

It bears noting that none of this process is standard; it is all done explicitly, and it's all software, so you can do whatever you want. You can see some of it in the example code we show later in this chapter.

Most modern embedded systems have cores with some sort of memory management that allows virtualized addressing. So, while each of the programs in Figure 15.22 looks the same, the loader can, based on boot-up directives, load each one and set up the MMU or TLB tables transparently so that each process uses its own space in memory.

A critical consideration is the allocation of memory for data according to whether specific variables are to be shared or private. In general, most of the working memory for any given core would involve private variables that are not to be accessed by other cores. However, for reasons mentioned above, there may be variables or other entities that you want shared by more than one core. Unlike the private data, which is virtualized by the MMU and mapped to different physical memory regions, the shared portions for each core must be mapped to the same physical memory.

This is typically handled first by identifying which data is to be shared, usually through pragmas or other directives to the build tools that you

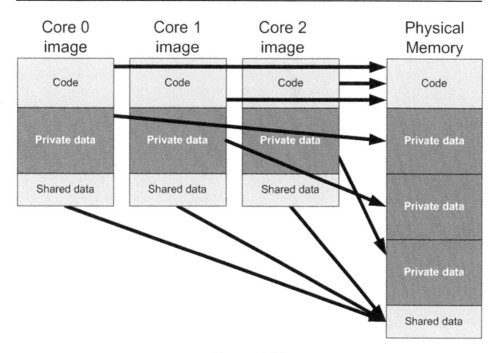

Figure 15.23
Mapping of virtual memory regions to physical memory.

include in your code. The loading firmware then allocates memory and mapping according to the different regions of the program image. This is illustrated in Figure 15.23.

In this example, all of the code segments are mapped to the same shared physical (read-only) memory. This is followed by the private data segments, which are each mapped to their own separate regions in physical memory. Following the private memory comes the shared memory, and each of those blocks is mapped to the same physical space.

It is important to identify all data that needs to be shared. Because the entire program operates in a single namespace, the compiler will not know which variables will end up in different cores when actually running, and it will assume a single execution as shown in Figure 15.24.

Here a piece of data is produced in Loop 1 and then consumed in Loop 2. The compiler has no way of knowing that, once Loop 1 starts, it will run until forcibly ended, and it will happily compile the program. If that variable is identified as shared, then it will be placed in shared memory,

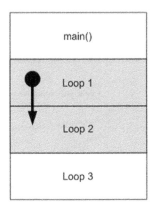

Figure 15.24
A single program with a variable produced in one loop and consumed in another.

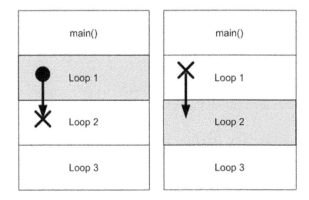

Figure 15.25
Data produced is not properly consumed if not placed in shared memory.

and both running programs will resolve to the same memory location to write and read that data.

If, however, you neglect to identify that data as shared, the compiler will not complain, but the handoff of data from producer to consumer will not proceed properly, as illustrated in Figure 15.25.

Here we see that, on the left, Loop 1 is executing and produces the data, but it stores the produced value in private memory for consumption by its copy of Loop 2. But Loop 2 never executes on that core, so the data is never used. Meanwhile, on the other core, Loop 2 consumes the data from a private location, but Loop 1 isn't running on that core, so the correct data hasn't been produced. These kinds of errors can be frustrating to pin down.

Note that if separate processes had been manually created as shown in Figure 15.21, then, depending on where the variables were declared, you might get a compilation error for Loop 2 since it's using a variable that doesn't exist in its namespace because Loop 1 has been removed. But if the variable was declared globally, then you would still compile successfully, and no tool would notice that one program has data produced but not consumed (which a smart compiler might even optimize out) or that another program is accessing data that is never explicitly written.

Example: IPv4 forwarding

We will illustrate all of these concepts using a specific example, probably the most widely used bare-metal algorithm: implementation of packet forwarding using the IP protocol, version 4. IPv4 will at some point be crowded out by IPv6, which accommodates many more IP addresses, but it's more complex, and the goal here isn't to provide an understanding of packet forwarding as much as it is to provide a concrete bare-metal example. So simple is good.

Packet forwarding is typically done in backbone routers, where, historically, a light touch has been used on packets in order to process as many as possible as fast as possible. Edge routers and terminals may implement more sophisticated processes, managing flows and sessions and peering more deeply into packets for security and management purposes. Again, we are trying to avoid these complications in order to keep the illustration straightforward.

We will first describe the problem in general, and will then provide two implementations, both on a Cavium OCTEON processor. In the first case, we will rely primarily on software; in the second case, we will make more use of the many hardware accelerators that this device provides for speeding up common critical tasks.

Because packet forwarding is being used to illustrate a more general technique, we assume the reader may not be intimately familiar with how packet forwarding works, and so we will describe it in some detail. That description will be a review (or unnecessary) for experienced packet-processing practitioners.

Packet forwarding

Packet forwarding is conceptually quite simple. A backbone router might be located anywhere between the origin and destination of a packet. The packet will take many hops between intermediate nodes to arrive at its destination, and the router we are implementing is simply one of those nodes. The job of that router is to figure out where to send the packet next. There are typically many choices, some more efficient than others.

Figure 15.26 illustrates a network, with many possible paths from source to destination. One such path is highlighted (thicker gray arrow), and one of the backbone routers is highlighted with a circle.

The route highlighted isn't the most efficient − the dotted line indicates a shortcut that's technically possible. But the router goes based on tables that are filled in ad hoc according to background communication that occurs between nodes, where routes are updated over time. So if that direct connection had been down for a while, then the table would reflect the longer route through the additional node. Once the shorter connection

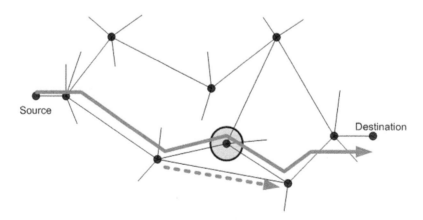

Figure 15.26
An example network and packet path, highlighting one backbone router and an alternate path.

came back up, the routing table may eventually be updated, but the timing isn't deterministic.

This updating of tables is managed by a host processor in the router, and is specifically not part of the pipeline through which the majority of packets are sent. The communication of routing table updates happens through management packets, but they are infrequent, and when they arrive, they are processed through the host processor (the slow path), not the pipeline (the fast path). That keeps the pipeline efficient and focused on the characteristics of the most frequent packets so that they can be handled quickly (Figure 15.27).

Forwarding is accomplished by looking at the destination address of a packet, which is a part of the packet header. It's extremely unlikely that the router will know about the exact destination, but it will know something about the "neighborhood" to which the packet is being sent, even if very generally. This is based on the Classless Inter-Domain Routing (CIDR) scheme, which specifies a network, or subnet, and a final destination. The first part, the network, can be thought of as the neighborhood in which the destination lies. The size of that neighborhood depends on who owns the network — it could be an enterprise or an internet service provider or some other entity.

A big neighborhood will contain many final destinations; a small one will contain fewer. The IP address has four eight-bit fields (for IPv4). Some number of those bits — the first ones — specify the network. This

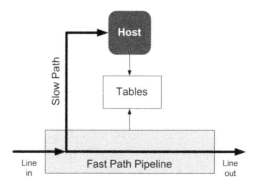

Figure 15.27
Table updates happen through management packets that take the slow path.

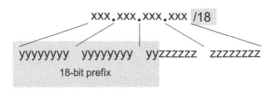

Figure 15.28
IP address with an 18-bit prefix.

is called the prefix (Figure 15.28). The rest of the bits provide the actual identity of the destination. The goal of the router is to get the packet to its network, and from there it will be sent to the appropriate final node. With CIDR, however, prefixes can be long or short, and the length of the prefix is included in the address.

The format for the addresses is "IP address/Prefix length". For example, 192.183.3.0/18 means that the first 18 bits of the 32-bit IP address specify the network; the final 14 bits locate the specific user within that network (Figure 15.28). Shorter prefixes specify bigger networks; longer prefixes narrow it down more specifically − to use an analogy, a short prefix might specify "New York City", a longer one "Manhattan", and a yet longer one "Greenwich Village". Note, however, that the networks are not specifically geographical.

The router must take this destination address and determine where to send it next − the "next hop". The tables that the router maintains will provide information on which packets should go where. But those tables aren't perfect − a packet may arrive for which no information exists in the table, in which case it gets routed to a default next node (and that next node will hopefully have more information about the destination). So the router is trying to do its best to nudge the packet in the right direction.

The overall process is as illustrated in Figure 15.29. First the header information is read from the incoming packet; this is referred to as decapsulation. That header will have numerous pieces of information including the packet type and, for "standard" packets, the destination. It is here that management packets, for example, can be identified and sent to the host instead of the fast path. It is also where the destination address for packets entering the fast path is determined.

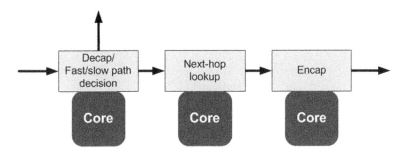

Figure 15.29
Basic IPv4 forwarding pipeline.

The next step is to determine the next hop — this is the most time-consuming step, and there are several algorithms and approaches to this, one of which we will look at in more detail. Finally, once the forwarding engine has determined where the packet should go next, a new header is put on the packet (encapsulation) and it is sent on to the next router.

Next-hop lookup: longest prefix match

The general lookup approach is referred to as "longest prefix match" (LPM). The idea is that the longer the prefix is, the more specific the address is. Using our New York analogy, if you are looking for the next hop for a packet destined for Manhattan and you know the next hop for New York City and for Manhattan, then you might as well go for the Manhattan next hop since it gets you closer to the final destination.

In terms of IP address, then, if the routing table includes 192.85.20.0/16 and 192.85.15.0/24, then the first one covers the space 192.85.xxx.xxx (so we'll abbreviate this entry as 192.85/16) and the second one the space 192.85.15.xxx (shortened to 192.85.15/24). The address space covered by the second one, with the longer prefix, is a subset of the address space covered by the first one, with the shorter prefix.

A simplified conceptual table is illustrated in Figure 15.30. Here we show the two entries (amongst many others). The first one indicates that the next hop should be to router A; the second entry shows a next-hop router B.

If a packet comes in with destination 192.85.11.25, the first 16 bits match that of the first entry, but the first 24 bits do not match that of the

IP Address Prefix	Next Hop
⋮	
192.85/16	A
⋮	
192.85.15/24	B
⋮	

Figure 15.30
A conceptual routing table showing two entries whose next hops are A and B, respectively.

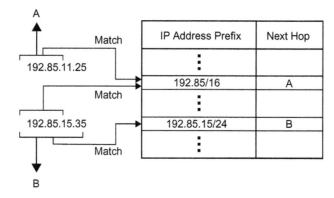

Figure 15.31
Addresses being matched against prefixes; when more than one prefix matches, the longest one wins.

second entry. Therefore the first entry is used to determine the next hop, which will be A.

If a packet arrives with address 192.85.15.35, then the first 16 bits match the prefix of the first entry, and the first 24 bits also match the prefix of the second entry, so the second entry, having the longer, more specific prefix, is used to determine the next hop, which will be B. This is illustrated in Figure 15.31.

The specifics of how these tables are built and searched can vary widely. As shown in Figure 15.30, a content-addressable memory is indicated, where you look up the prefix and get the next hop. But this approach has drawbacks, and is shown only to illustrate the concept.

The DIR-24-8-BASIC algorithm

The specific algorithm we're going to follow for the next-hop lookup is referred to as the DIR-24-8-BASIC scheme by its Stanford-based authors [1]. In the interest of quicker table lookups, this approach "wastes" memory under the assumption we mentioned already that memory is cheap. It is also predicated on the empirical fact that, in the backbone, the vast majority of addresses have 24-bit prefixes or smaller. So the approach is optimized for addresses having 24 or fewer bits in the prefix, with one extra lookup for addresses having longer prefixes.

Instead of a table as shown in Figure 15.30, two tables are created − one for routes with prefixes 24 bits and below, and one for larger tables. The first table, TBL24, can return the next hop for any address that matches a prefix of up to 24 bits in a single lookup; for longer prefixes, the second table, TBLlong, is needed. Figure 15.32 provides a schematic illustration; the details are described below.

Rather than storing the IP address itself in a table, the first 24 bits are used as the address into TBL24. In other words, the first part of the IP address is literally used as the memory address. This means that TBL24 will have 2^{24} entries. What is stored at that location can take one of two forms. Let's start by focusing on routes for prefixes up to 24 bits. In this case, the addressed location contains the next hop information directly. For our 16-bit prefix example above, the first 24 bits of which are 192.85.xxx , this means that the locations addressed by 192.85.0 to 192.85.255 will all contain the next hop for any destination that matches those first 16 bits.

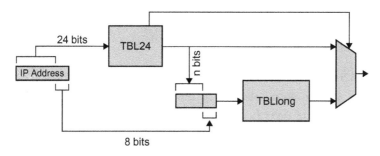

Figure 15.32
DIR-24-8-BASIC block diagram.

TBL24	
Table Address	Next Hop
⋮	
192.85.0	A
192.85.1	A
⋮	A
192.85.254	A
192.85.255	A
⋮	

Figure 15.33
TBL24 entries for prefix 192.85/16.

You might consider this wasteful, since 256 entries all have the same value, but this is where the "memory is cheap" assumption plays out: by using the memory inefficiently, then any IP address starting with 192.85 can be used as an address and yield the next hop in one lookup. Figure 15.33 illustrates the TBL24 contents resulting from this entry. Note that the table address is shown formatted like an IP address in the figure in order to reinforce the fact that the IP address (or a portion of it) becomes the memory address. But, unlike Figure 15.30, those addresses don't reflect IP addresses stored in the table; only the next hop information is stored at that address.

Note that the values for the next-hop routers A and B (and other letters coming up) will depend on the values assigned to various outgoing routes from this router; the specific numeric value doesn't matter for our purposes.

Now let's consider the second routing table entry, 192.85.15. This is actually contained within the space of the previous entry. So, in this case, we'd modify TBL24 from Figure 15.33 to have a different next hop result for those IP addresses starting with 192.85.15. Figure 15.34 shows the result.

Now let's consider what happens for routing entries with prefixes longer than 24 bits. In this case, instead of the TBL24 contents holding the next hop information, the next hop information is put in TBLlong, and TBL24 contains a pointer to TBLlong.

We'll show how that works, but first we have an immediate issue in that the contents of TBL24 sometimes mean a next hop and sometimes mean

TBL24	
Table Address	Next Hop
⋮	
192.85.0	A
192.85.1	A
⋮	A
192.85.14	A
192.85.15	B
192.85.16	A
⋮	A
192.85.254	A
192.85.255	A
⋮	

Figure 15.34
TBL24 modified to include a different next hop for 192.85.15.

TBL24		
Table Address	Next Hop/ Pointer	Extra Bit
⋮		
192.85.0	A	0
192.85.1	A	0
⋮	A	0
192.85.14	A	0
192.85.15	B	0
192.85.16	A	0
⋮	A	0
192.85.254	A	0
192.85.255	A	0
⋮		

Figure 15.35
Extra bit added to TBL24 to indicate whether content is a next hop or pointer.

a pointer to the next hop. So we need a bit in TBL24 that indicates how to interpret the result. If the extra bit is 0, then a next hop is indicated; if the extra bit is 1, then the algorithm must look to TBLlong for the next hop. Modifying Figure 15.34 to add that column yields Figure 15.35.

Let's assume we get a new routing table entry for 192.86.0.0/25, for which the next hop should be router C. We now put a pointer in TBL24 at location 192.86.0 (which, conveniently, is the next location) that contains a pointer to TBLlong (we won't get into how that pointer is determined since we're not going to focus on the details of table updating); assume that pointer is 100. We also set the extra bit to 1 to indicate that this is a pointer (Figure 15.36).

In TBLlong, we have to account for up to 8 more bits in an address, so each pointer from TBL24 commands 2^8, or 256, entries in TBLlong (Figure 15.37). Some number (shown as n in Figure 15.37) of such 256-line chunks is selected to size the memory — it can be sparse since,

TBL24		
Table Address	Next Hop/ Pointer	Extra Bit
⋮		
192.85.0	A	0
192.85.1	A	0
⋮	A	0
192.85.14	A	0
192.85.15	B	0
192.85.16	A	0
⋮	A	0
192.85.254	A	0
192.85.255	A	0
192.86.0	100	1
⋮		

Figure 15.36
TBL24 with a pointer to TBLlong added.

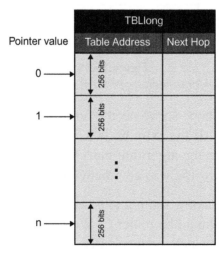

Figure 15.37
Each pointer corresponds to 256 entries in TBLlong.

statistically, there are so few such entries. So pointer value 0 in TBL24 will correspond to locations 0 to 255; pointer 1 corresponds to locations 256 to 511, and, in general, pointer x corresponds to locations $x*256 + 0$ to $x*256 + 255$.

So, given pointer 100, then addresses 100*256 to 100*256 + 255 will be dedicated to all addresses starting with 192.86.0. The routing entry covers half of those, since the first bit of the last octet belongs to the prefix, and is 0. So all addresses in the range 100*256 to 100*256 + 127 will contain next hop C. If this is the only routing entry we have for IP addresses starting with 192.86.0, then we really don't know what the next hop should be for IP addresses above 192.86.0.127, so those entries will be given the default route (Figure 15.38).

If a packet destined for 192.86.0.12 comes in, then the first 24 bits will be used to address TBL24, which will yield 100. That 100 will then be multiplied by 256, and the last 8 bits of the IP address − 12 − is added to that, pointing to address 100*256 + 12 in TBLlong. This yields next hop C (since all of the addresses in the first 127 locations pointed to by 100 all contain C). If a packet headed for 192.86.0.200 comes in, this will still yield pointer 100 from TBL24, but because that's outside the range of the 192.86.0.0/25 entry, it will get the default route (Figure 15.39).

Finally, let's consider what happens if we get a new table entry for route 192.85.15.219/26 that corresponds to next hop router D. We now have two very similar entries, one with a 24-bit prefix − 192.85.15/24 − and

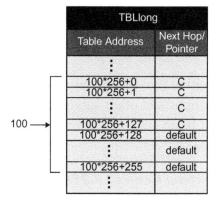

Figure 15.38
TBLlong entries resulting from 192.86.0.0/25 routing table entry.

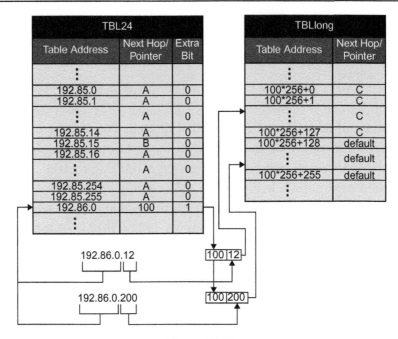

Figure 15.39
Calculating the next hops for 192.86.0.12 and 192.86.0.200.

one with a 26-bit prefix that fits within that same address range. TBL24 at present contains next hop B at address 192.85.15. Because at least one route in that range has a longer prefix, we have to bump this entry into TBLlong using, say, pointer 150. In TBLlong, all of the 256 entries will contain B except for the 64 entries covered by the longer prefix, ranging from 192.85.15.192 to 192.85.15.255 (Figure 15.40).

Now if a packet destined for 192.85.15.20 comes in, it will still go to next hop B as it did before; it just takes an extra lookup to get there. But a packet going to 192.85.15.228 will go to next hop D.

This, then, is the algorithm that is implemented in the specific code below. There are variations and other subtleties that can be included, but we've opted to keep it basic and simple for illustrative purposes.

Example target architecture: Cavium OCTEON CN3020

The device we're going to target with this example is specifically intended for packet processing: the Cavium OCTEON CN3020, one of

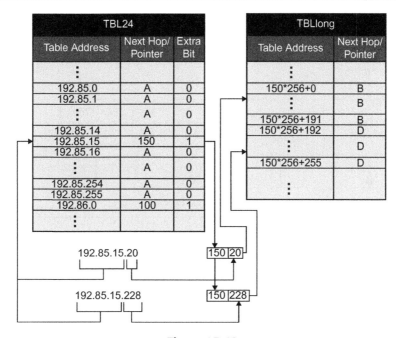

Figure 15.40
Table changes due to addition of long prefix within the 192.85.15 range.

their smallest devices, with two cores. We briefly describe the device here, detailing only those features that are actually used in the code.

One of the advantages of using a purpose-built device like this is the amount of hardware acceleration provided. While we have written full code for the application on this device, an alternative example that assumed no accelerators would have required an enormous amount of additional code to do the things that the accelerators do. We therefore included only a version of the example that relies on the accelerators.

A simplified diagram of the CN3020 is shown in Figure 15.41. It includes only those elements of the architecture that the example uses.

Packets arrive on the Interface RX Port, which, on this device, supports two interfaces with one port each. From here the packets are sent to the PKI, which allocates buffers (using the FPA, which provides 2-Kbyte buffers along with "work queue" buffers, which are essentially packet descriptors). The PKI then DMAs the packet through the I/O Bridge into

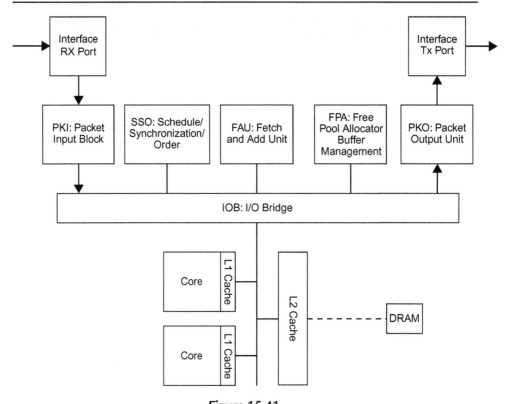

Figure 15.41
Simplified illustration of Cavium OCTEON CN3020, including only those blocks used in the example.

memory. It also validates the packet to make sure that it hasn't been corrupted somehow. All of this is done via hardware acceleration.

The PKI also parses the packet header, grabbing such essential fields as the source and destination MAC addresses, source and destination ports, and the destination IP address. It also tags the packet for ordering, since this processor is capable of maintaining packet ordering. There are three possible ordering tags:

* Unordered, meaning the packet can be sent out at any time without regard to any other packet;
* Ordered, meaning that the packet is part of a flow, and must be sent out after the prior packet and before the subsequent packet; and
* Atomic, meaning that only one core can work on the packet at a given time.

In the example, the default ordering tag is "ordered".

The SSO is the heart of the operation; it schedules work and maintains packet order (if needed). While pipeline examples above dedicate certain cores to certain operations, this device lets the scheduler assign whatever needs doing to whichever core is available. This is done according to the tags, which are managed by the software. Unless the tag is "Atomic", the scheduler will allow multiple packets in a flow to be scheduled at the same time. Transitions from one stage of processing to the next will only happen in packet order if the tag is "Ordered"; they can happen in any order if the tag is "Unordered".

The FAU, for our purposes, is an efficient way to maintain asynchronous counters that can be atomically incremented.

Once processing is complete, the packets go to the PKO, which DMAs the packets out of working memory and recycles the buffers. The packet ordering tag must also be set to "atomic" at this point to ensure in-order send-off. The packets go from there to the Interface TX port, which, on this device, again supports two interfaces with one port each. Each output port has eight queues for use in maintaining any needed QoS traffic management.

Ordinarily, there would be a host processor in the picture as well for implementing the slow path. We haven't done that in the example, since that's not the focus. If there are any issues with a packet or if there is no next hop available in the routing table, those packets would be sent onto the slow path either for dispensation (problem or management packets) or in order to do a query for the next hop. In order to simplify this in the example, where there is no slow path, such packets are simply dropped (which is not recommended for a real implementation).

This Cavium device is operated with the Cavium Simple Executive as a thin services layer; there is no full-up OS running, and programs are written using the run-to-completion model.

Select code examples

In this section, we examine the critical parts of the code to show how it works. We'll start with the main() routine, reviewing other routines as necessary to elucidate the bare-metal execution of this example. Bear in mind that this represents a small portion of all of the code necessary to run the application. In particular, the accelerators require lots of very specific code, much of which is implemented as macros, and it's relatively opaque. Given that this is not the focus of the example, we have tried as much as possible to bury those details.

The complete code is available at http://booksite.elsevier.com/ 9780124160187

Note that calls to the Simple Executive are prefixed by "cvmx_". We will not go into detail on those calls other than to note what is happening.

Main()

This routine handles the initialization and starts the main loop. In fact, the end of the routine is never reached because there is no return from the main loop.

It's critical to remember that all cores run this routine. Yet some of the tasks are only performed by one of the cores, the designated initialization core. So the routine has to include `if` statements that restrict some of the execution only to the initialization core (the "IS_INIT_CORE" test).

There are global items that need initializing, done by the initialization core, and then each core has to do some packet I/O initialization. So the overall routine consists of:

- Initializing the application via the Simple Executive;
- Interrogating the Boot ROM to figure out how many cores to use and which is the initialization core;
- Having the initialization core allocate buffers; there's a barrier after this so that the other cores don't go charging ahead until the buffers are set;
- Initialize packet I/O (all cores do this);
- Initializing the routing table; there's a barrier after this as well;
- Launching the main loop.

The code is shown in Figure 15.42.

```
/**
 * Main entry point
 */
int
main(int argc, char *argv[])
{
    cvmx_sysinfo_t *sysinfo;
    unsigned int coremask_l3fwd;
    int result = 0;

    cvmx_user_app_init();

    /* compute coremask_l3fwd on all cores for the first barrier sync below */
    sysinfo = cvmx_sysinfo_get();
    coremask_l3fwd = sysinfo->core_mask;

    /*
     * elect a core to perform boot initializations, as only one core needs to
     * perform this function.
     *
     */

    if (IS_INIT_CORE) {
        printf("Version: %s\n", cvmx_helper_get_version());
        /* 64 is the minimum number of buffers that are allocated to receive
           packets, but the real hardware, allocate above this minimal number. */
        if ((result = l3fwd_init_simple_exec(1024 + 80)) != 0) {
            printf("Simple Executive initialization failed.\n");
            printf("TEST FAILED\n");
            return result;
        }
    }
    CORE_MASK_BARRIER_SYNC;

    cvmx_helper_initialize_packet_io_local();

    if (IS_INIT_CORE) {
        if ((result = l3fwd_init_routing_table())) {
            return result;
        }
    }

    CORE_MASK_BARRIER_SYNC;

    /* Lets get to work!!! */
    l3fwd_main_loop();

    /* Should not return */
    return result;
}
```

Figure 15.42
The main() routine.

```
void
l3fwd_main_loop(void)
{
    cvmx_wqe_t *work;
    cvmx_buf_ptr_t packet_ptr;
    int32_t status;
    uint8_t next_hop_return = 0;

    while (1) {

        /* get the next packet/work to process from the POW/SSO, the core will block waiting for WORK */
        work = cvmx_pow_work_request_async(CVMX_POW_WAIT);
        if (work == NULL) {
            continue;
        }

        /* Use the OCTEON's L2 & L3 parsing/verification engines to verify the packet */
        if (l3fwd_verify_packet(work)) {
            /* Work has error, so drop */
            cvmx_helper_free_packet_data(work);
            cvmx_fpa_free(work, CVMX_FPA_WQE_POOL, 0);
            continue;
        }

        /* OK now we know with have a proper IPv4 packet, figure out offset into IPv4 header and get
         * the dst IPv4 address and perform an LPM lookup
         */

        /* Get the address of the start of the packet */
        packet_ptr = l3fwd_get_packet_ptr(work);

        /* Get IPv4 header */
        struct ip *v4hdr = l3fwd_get_ipv4_hdr(packet_ptr);
        status = l3fwd_lpm_lookup(lpm, v4hdr->ip_dst, &next_hop_return);

        if (status) {
            /* Route not found, drop packet*/
            cvmx_helper_free_packet_data(work);
            cvmx_fpa_free(work, CVMX_FPA_WQE_POOL, 0);
            continue;
        }

        printf("Next Hop for IP dst: 0x%08x : %d\n", v4hdr->ip_dst,
            next_hop_return);

        /* Reduce TTL */
        v4hdr->ip_ttl--;

        /* Rewrite L2 header, get the dst mac address from the next hop.
         *  H/W will recalculate IP checksum
         */

        struct ether_header *l2hdr = cvmx_phys_to_ptr(packet_ptr.s.addr);
        uint32_t out_port = next_hops[next_hop_return].port;

        /* Write DMAC */
        memcpy(l2hdr->dmac, next_hops[next_hop_return].macaddr, 6);

        /* Write SMAC */
        memcpy(l2hdr->smac, smacs[out_port].macaddr, 6);

        /* Send it to the PKO, output port is from nex hop */
        l3fwd_send_packet_to_pko(work, out_port);
    }
}
```

Figure 15.43

The main work loop.

Next we go into the main work loop, `l3fwd_main_loop()` (Figure 15.43).
You can see the iconic `while(1)` loop that follows the declarations.
Within the loop, the following happens:

- A Cavium executive function is called to see whether there is work
  in the queue; the rest only happens if there is.
- A routine is called to use the hardware accelerators to parse and then
  verify the lower-level (L2 and L3) packet headers. If there's an issue,
  the packet is simply dropped for the purposes of this example.
- Next the L4 (IP) header is extracted. If there is no route in the table,
  then the packet is dropped. Ordinarily, it would be sent into the slow
  path at this point. We'll look more at the lookup routine below.
- The time-to-live (TTL, the number of hops a packet can take before
  the system decides it's hopelessly lost) is decremented to account
  for the current hop it will take. The code is shown in Figure 15.43.

```
static inline int
l3fwd_lpm_lookup(struct l3fwd_lpm *lpm, uint32_t ip, uint8_t *next_hop)
{
    uint32_t tbl24_index, tbl8_group_index, tbl8_index;

    /* Calculate index into tbl24. */
    tbl24_index = (ip >> 8);

    /* DEBUG: Check user input arguments. */
    L3FWD_LPM_RETURN_IF_TRUE(((lpm == NULL) || (next_hop == NULL)), -EINVAL);

    /*
     * Use the tbl24_index to access the required tbl24 entry then check if
     * the tbl24 entry is INVALID, if so return -ENOENT.
     */
    if (!lpm->tbl24[tbl24_index].valid){
        return -ENOENT; /* Lookup miss. */
    }
    /*
     * If tbl24 entry is valid check if it is NOT extended (i.e. it does
     * not use a tbl8 extension) if so return the next hop.
     */
    if (cvmx_likely(lpm->tbl24[tbl24_index].ext_entry == 0)) {
        *next_hop = lpm->tbl24[tbl24_index].next_hop;
        return 0; /* Lookup hit. */
    }

    /*
     * If tbl24 entry is valid and extended calculate the index into the
     * tbl8 entry.
     */
    tbl8_group_index = lpm->tbl24[tbl24_index].tbl8_gindex;
    tbl8_index = (tbl8_group_index * L3FWD_LPM_TBL8_GROUP_NUM_ENTRIES) +
                 (ip & 0xFF);

    /* Check if the tbl8 entry is invalid and if so return -ENOENT. */
    if (!lpm->tbl8[tbl8_index].valid)
        return -ENOENT;/* Lookup miss. */

    /* If the tbl8 entry is valid return return the next_hop. */
    *next_hop = lpm->tbl8[tbl8_index].next_hop;
    return 0; /* Lookup hit. */
}
```

Figure 15.44
The next-hop lookup routine; this implements the DIR-24-8-BASIC algorithm.

- The packet is then modified by putting the new route in the L2 header, setting the source and destination MAC addresses for this hop, and sending it on to the output queue in the PKO.

Finally, we look at the `l3fwd_lpm_lookup()` routine, which implements the the DIR-24-8-BASIC algorithm described above (Figure 15.44). It's actually implemented as an inline function to improve performance. This routine:

- Makes sure there's a legitimate next hop;
- Gets the tbl24 entry to see if it has the next route or needs an index into tbl8 (which is the same as TBLlong in the description above). If complete, the next hop is returned;
- If not complete, then tbl8 is indexed and the result validated and returned.

The code is shown in Figure 15.44.

We won't delve into the remaining routines, but they will be available at http://booksite.elsevier.com/9780124160187

Conclusion

High-performance code can be implemented in multicore systems having little to no OS support. Run-to-completion programs can be executed in pipelines where individual cores have dedicated tasks to implement or where hardware scheduling can maximize core utilization.

While trickier to program and implement, such systems have been crucial for meeting the demand of the communications infrastructure that moves data in enormous quantities all around the world at every second of every day.

Reference

[1] P. Gupta, S. Lin, N. McKeown, Routing lookups in hardware at memory access speeds, Proceedings of the Conference on Computer Communications, vol. 3, IEEE Infocomm, San Francisco, California, March/April 1998.

CHAPTER 16

Multicore Debug

Neal Stollon

Principal Engineer with HDL Dynamics, Dallas, TX, USA

Chapter Outline

Introduction — why debug instrumentation

The ability to analyze and debug the operations and applications of multiple embedded cores and processors and their interactions is critical to system-on-chip (SoC) design. While debug of a single embedded core is challenging, it is a relatively well-understood problem, whereas multicore architectures add new considerations and complexity that factor into the debug solution. As core implementations become more complex, they expand both in subsystem functionality (as an example,

Real World Multicore Embedded Systems.
DOI: http://dx.doi.org/10.1016/B978-0-12-416018-7.00016-X
© 2013 Elsevier Inc. All rights reserved.

moving from "only" a processor core to processor + caches + bus interfaces + dedicated peripherals as a pre-integrated IP block) and in interface complexity (connecting multiple bus interfaces to shared resources). The debug questions for embedded systems design move beyond those of "Is this core and its code working correctly?" to "How do I get my application code operating more efficiently over multiple cores ?" and "How well is this part of the system interacting with other subsystems?" These are problems which require a system debug focus rather than just analysis of a single core.

Throughout the evolution of processor design, analysis capabilities have had to increase with each generation to keep up with designer's needs to validate their designs and applications. In formal terms, multicore embedded systems present an asymmetrical functional debug problem. Borrowing from test terminology, the design controllability is high, since the processor cores can be placed into known states. Their observability is low, however, both in terms of critical signals that are directly available for observation and in terms of the amount of embedded logic and the ratio of internal signals to the available I/O from which to observe the embedded logic.

Dedicated resources and structures to support functional visibility and analysis are often added to increase system observability. For multicore designs, this requires a hierarchical approach, starting with debug instrumentation and resources at the individual core level and rising up to a system communication diagnostic capability that facilitates increasing the observability of core interactions. While a variety of embedded debug instrumentation approaches are common at the core level, system-level diagnostics and multicore-level analysis instruments, tools, and methods are still evolving to address more system-oriented debug issues (Figure 16.1).

As more processing elements, features, and functions are simultaneously included in the silicon, the complexity of debug outstrips the capability of the traditional standalone logic analyzer, debugger, and emulator-based diagnostic tools. This is in large part a result of the rapid growth in gates fueling multicore architecture, compared to the more linear growth in package pins that provide visibility into the SoC. As more of the overall system becomes "deeply embedded", without a direct access

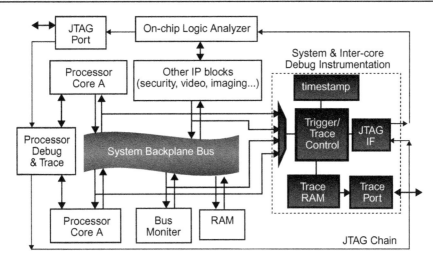

Figure 16.1
A multicore debug implementation.

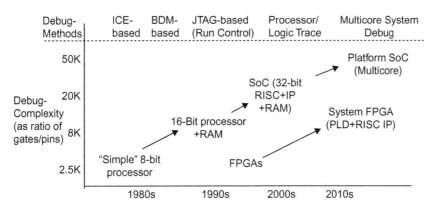

Figure 16.2
SoC and debug complexity over time.

path to the chip I/O, transfer of debug information to the I/O becomes more complex and indirect.

As seen in Figure 16.2, if we consider debug complexity as the ratio of processing gates to I/O in mainstream processing architectures, over time there is an increasing gap in how effectively "deeply embedded" information can be accessed and, as a result, limits on visibility of the internal operations of a multicore system. This is an architectural issue; programmable platforms such as FPGAs have the same issue as SoC

ASICs. Emerging system packaging methods such as 3D die stacking, which further increase the gates-to-pins ratio, will accelerate this issue even further.

This forces the debug and instrumentation element of the SoC world into a constant process of catching up to the ability to add cores and integrate new resources on-chip. With an ever-shortening development cycle, standardized embedded tools and capabilities that enable quick analysis and debug of the embedded IP are being recognized as critical to keeping SoC verification a manageable part of the process.

Like the terms "modeling", "simulation", and even "system", "instrumentation" and "debug" mean different things to different people. We define on-chip instrumentation as an embedded block that provides external visibility into and access to the inner workings of the architecture. When properly implemented, it provides a real-time "peephole" into the operations of key internal blocks that cannot otherwise be accessed with sufficient granularity on a real-time basis.

This real-time visibility and monitoring of key interfaces and buses are increasingly crucial to understanding the dynamics of the operation of system architectures. As a general rule, debug visibility becomes increasingly problematic for highly integrated chips, which have extensive on-chip memory, caches, peripherals, and a range of on-chip buses. The key control and bus signals of interest in a deeply embedded system differ from older and simpler digital systems, and they are often not externally addressable by the physical pins of the device and are therefore inaccessible to traditional instrumentation.

Most modern instrumentation requires an investment in on-chip logic and memory to enable visibility. For cost-sensitive chips, this addition of instrumentation resources, which do not directly contribute to the primary chip functions, has traditionally been a complex trade-off of silicon resources vs. on-chip debug capabilities. These trade-offs have shifted over recent years by dramatic increases in complexity, gate availability, and overall speed of system operation. They have changed the norm in chip design from a "core- and gate-limited" focus to a "pin- and I/O-limited" focus, which makes a small increase in gates for instrumentation much less of a concern. The debug implementation question is migrating from "how many gates can be spent on debug" to "how many of the system resources and I/O are needed to debug the system successfully".

System-on-chip debug, in common with most verification philosophies, seeks to maximize test functionality and ease of verification while reducing the overall cost to the end user. There is a constant trade-off that must be made between system analysis and debug resources and the cost of including these features. The value of debug is mainly perceived during the development cycle (be it hardware or software) to resolve operational questions and integration issues for the key processing blocks. After the system is "fully debugged", the debug capabilities become much more application focused, serving as a tool for software testing and optimization.

While we focus in this chapter on instrumentation for debug and verification, engineers will recognize that instrumentation is complementary to simulation. Simulation is critical for embedded SoC verification, but it is not a total verification solution, as even detailed simulations cannot address all the facets and nuances of physical hardware. Simulation is compute-intensive, so, in many cases, it is not realistic to simulate large, multiprocessor architectures for the extremely large numbers of cycles required to evaluate the software-specific aspects of system operation and real-world system performance. While accelerated simulation, co-simulation, and emulation environments improve simulation performance, these approaches often have costs and complexities beyond the resources of many projects. On-chip instrumentation and debug approaches have therefore needed to evolve as a low-cost, efficient addition to simulation for increasing visibility for analysis of the total hardware and software system.

The integration of deeply embedded memory and buses — along with limited I/O for such embedded subsystems available for test purposes — limits the visibility of the embedded processors and data flow in SoC operation. To support "deep encapsulations" of multicore system functions, along with higher internal bus speeds, new logic analysis and debug approaches such as on-chip instrumentation are used. Dedicated on-chip instrumentation is typically implemented as a set of shadow subsystems that provide a "debug backplane" to capture, extract, and analyze debug signals and operations. Some approaches are discussed in this chapter. But we will first look at some basics of debug instrumentation as a point of reference.

How does multicore differ from single-core debug?

Multiprocessor debugging covers a range of architectures and approaches, and, depending on the level of detail required, can range from incrementally to exponentially more complex with the addition of each processor.

In general, the biggest conceptual leap in multicore debug occurs in moving from one to more than one processor, since now, instead of the main debug concern being whether an action or an operation has occurred, additional information needs to be captured on what processor it was associated with. Consider an on-chip instrument that monitors on-chip bus activity. For a simple single processor, it is sufficient to see what was written to a memory location; it is a given that the processor is the master of the bus transaction. With two processors, the instrumentation needs to identify which processor made the write, along with the status of both processors, in order to determine whether there are shared memory contention issues, track the source(s) of previous memory accesses, and provide context for the memory debug. This requires additional instrumentation hardware to maintain and manage separate debug resources for each core or to interleave the trace buffers between processors. Additionally, bus monitoring requires global breakpoints to halt or otherwise control both processors on the same cycle. These global breakpoints come in addition to, not instead of, the local breakpoints for each processor, which are still required for single-core debug. All of this is required just to process a fairly simple memory access debug request.

In a more comprehensive debug scenario, the amount of debug information sources that must be potentially considered, including different cores and their common infrastructure (embedded buses, shared peripherals, etc.), is much larger than that required for single-core debug. Different cores from different IP vendors have different types of debug features, with varying levels of inherent debug capability. Since subsystems interact on chip, implementing debug controls for a single core, without looking at it in the context of how other processors are operating, can be ineffective and misleading. Having debug information available for several interacting cores can make debug much more straightforward.

However, extracting debug information from several cores in their correct ordering and context can be complex. Synchronizing multicore debug accesses, especially for different types of cores, presents the additional challenge of synchronizing the debug information needed for coherent debug interface integration. This may require additional instrumentation resources for "embedded intelligent" debug operations and to allow analysis features such as system-wide error recognition and filtering and cross-triggering and performance analysis between different subsystems of a complex architecture. At a minimum, a multicore debug environment must address three major facets of the problem:

- Multicore debug concurrency: the need to concurrently access debug and JTAG ports for all the cores in a system. To analyze problems and optimize performance in multicore operations, the designer should be able to exercise any and all core debug features and interfaces through them. This capability is not supported in current JTAG debug architectures. New debug schemes need to address concurrent JTAG or debug port communications with debug tools that support them.
- System debug integration: triggering and trace need to be addressed at a system rather than core level. System-level instrumentation for a debug, trace, and triggering environment needs to support multiple on-core and inter-core conditions (breakpoints, tracepoints, other specific control or status conditions) and send global actions to all or a subset of the cores (halt being the most obvious example). Most systems communicate over a range of buses. The debug architecture must be able to monitor signals on the bus, support the specifics of widely used on-chip bus schemes, and trigger and trace bus operations based on specific conditions.
- Debug synchronization: systems debug must address blocks running over several time and clock domains. Robust approaches to global

timestamp synchronization are needed to support coherency between multiple debug environments that may be involved for different cores on a single chip.

Background — silicon debug and capabilities

To discuss on-chip instrumentation in the proper context, it is useful to examine traditional embedded systems debug, which is the basis for many of the modern approaches in embedded system development that incorporate diverse processing resources. On-chip instrumentation builds on the integration of older ICE (in-circuit emulator) and other probing-based debugger implementations, integrated BDM (background debug mode) and JTAG interfaces, and logic analyzer capabilities. A key difference is that much of the logic analysis logic that was previously in external instruments is now included on chip. Where JTAG or BDM has traditionally provided a processor operation snapshot that is sufficient for simpler systems, the dynamic interaction of multiple processors requires a more dynamic and robust means of providing diagnostic information.

The embedded systems market has seen a proliferation of new processor architectures (composed of RISCs, DSPs, controllers, and application-specific co-processors) focused on SoC integration. Many currently available cores provide some form of JTAG interface for run control and debug functions. In addition, instruction and data trace capabilities, usually in the form of a trace port such as the Nexus (IEEE 5001), ARM ETM (embedded trace) or the MIPS EJTAG + TCB (trace control block), each of which is discussed later in this chapter, provide more productive system debug.

Debug tools that address processor analysis may not be the best way to facilitate the debug of a multicore system application. For processor in-circuit emulators and traditional JTAG-based debugging, the system must be placed in special debug modes or halted before being able to probe processor registers or reading/writing to the embedded memories. For deeply embedded and integrated multicore processors, traditional development approaches for system debug applications (such as halting or single-stepping a processor) may be very intrusive. In many cases, this interruption of the steady-state performance of the system introduces new debug factors and elements into the system operation that can complicate or invalidate the data or operations being debugged. This problem grows proportionally with the operating frequency and complexity of high-performance embedded processors. More general-purpose trace control blocks for on-chip instrumentation (such as Nexus 5001) provide a vendor-independent means of enabling trace, triggering, and run control capabilities.

The logic and memory investment for even a comprehensive debug and trace system is small for leading-edge, deep-submicron, high-performance multicore SoCs. A typical 25 K−50 K-gate instrumentation block per core to enable debug, trigger, and trace systems has a small impact on a multicore chip, especially when weighed against the potential of loss of additional weeks or months debugging the systems in the lab.

Trace methods for multicore debug analysis

Run-control debug, monitoring, and trace are the basis for debugging systems that react with another system (e.g., the real world). Trace solutions can provide for almost any debug requirement from high-level application-generated trace and software debug and profiling of multicore code to hardware-intensive analysis, where the instruction-by-instruction code and data are recorded over a period of thousands of cycles. By recording system execution while the system is operating, trace allows data-dependent or asynchronous observations in situations where halting the system is not a viable option.

Two widely used low-overhead trace methods are:

1. Software trace, where software executing on the cores generates the trace data by writing to an area of system memory. At some convenient point, the memory is then emptied and the trace data is sent back to debug tools via an available communication channel, most commonly over a JTAG test access port (TAP). Performance monitoring is widely incorporated in many processors as a feature in software trace. In this case, code in the processor updates performance monitor registers providing system performance

profiling data such as the number of cycles executed, branches taken, mis-predictions, and cache hits and misses.

2. Instrumentation-based trace, which allows for the capture of both data and instructions without changing the flow of processor operations, provides the most flexible level of multicore processor debug support. This capture of sequential information and processing interactions allows both performance optimization of an application and identification of the causes of errors. Instrumentation-based trace uses both logic and memory, so choosing an appropriate level of trace has an impact on the costs of implementing the on-chip debug system. Instrumentation-based trace uses dedicated logic that monitors address, data, and control signals within the SoC, triggers on specific events, captures and compresses the event-related information, and stores it to a trace buffer. It does not rely on processor software, which makes the trace operations transparent to other parts of the chip. Instrumentation-based trace can also operate with higher-bandwidth debug channels that allow for more instrumentation points to be captured and enable real-time delivery of trace data to very deep off-chip buffers. For multi-processing systems, instrumentation trace will include signals for understanding context, e.g., which thread or core a trace stream is coming from. It can also add useful higher-level context information such as timestamps or external triggering condition information.

Trace can be subdivided into program/instruction trace, data trace, and bus (or interconnect fabric) trace. Each of these functions has different usage models and costs.

Program trace is the most widely implemented trace method, for both hardware and software debugging as well as profiling of operations. Program trace data can be compressed to be as small as 1 bit per core instruction, so its implementation cost is relatively small; in some cases, a trace block may be less than 10 K gates. Adding cycle-accurate instruction trace, useful for correlation of multiple processor interactions, increases the needed bandwidth to about 4 bits/instruction; this can substantially increase the required bandwidth of a trace port. Adding data and bus (addresses and/or data values) trace, which is more difficult to compress, also significantly adds to the trace overhead.

Where on-chip trace buffers are implemented, higher-end debuggers can continuously process program trace profiling data in real time for runs of several hours. More powerful debug tools contain a rollback feature where all debugger windows, including processor register values, may be subsequently recreated from the data trace. As a result, a programmer can step forward (or backward) through code actually executed in real time in the real environment, showing behavior that may not be captured by the trace operation.

The size of the trace subsystem, driven in large part by port I/O and trace memory, depends on the required trace functionality. Multicore SoCs may use a combination of solutions. For example, a system with three parallel trace systems may use a variety of ports for different subsets of trace data: a very high-bandwidth interface to on-chip trace buffers, a medium-bandwidth trace port to a very deep off-chip buffer, or a very narrow (even single-pin) interface for continuous monitoring of simpler on chip information.

To address the I/O limitations of data trace, trace ports using gigabit physical layers (PHYs) can support multiple gigabits of trace data. As an example, a complete cycle-accurate simultaneous program and data trace of three cores running at about 600 MHz requires approximately 6 Gbits/s, which is achievable in 2–4 gigabit lanes.

Types of instrumentation logic blocks

Debug can be thought of as three interrelated processes, each of which requires a specific set of instrumentation logic blocks. These processes are monitoring and triggering, trace capture and formatting, and chip-level interface.

The types of monitoring and triggering required for debug can vary significantly depending on whether you're monitoring a processor, bus, or specialized subsystem. These will all look for different events. The events may be sequential, where a second event occurs some time after a first event; they may be asynchronous; and they have combinations of analog and/or digital events. The triggers themselves may vary as well: they could be pulses, level changes, register updates, etc. Defining and developing the most appropriate monitoring and triggering logic is, in general, the most complex and labor-intensive part of developing a debug environment.

Trace capture and formatting can vary widely as well. One will generally trace some combination of control and data signals; these signals are captured and are usually placed into a trace buffer. The tricky part is that, in general, there is not enough buffer space to store all the required trace information without some sort of data reduction. The processing may consist of elimination of unnecessary data (unconditional instructions from a processor, for example), consolidation of trace information (concatenation of request and response transaction cycles in a bus operation), or compression (removal of unnecessary formatting information or leading zeros, truncation, etc.). These operations are in many cases followed by reformatting of the trace data to make it compatible with the memory configuration.

The chip-level interfaces take either stored or real-time data and make it available to the external tools. In principle, any chip-level interface may transfer debug data, but, over time, the industry has, for diverse historical reasons, standardized on the JTAG TAP. Since JTAG is relatively small (single data in and out wires) and slow (practically limited to less than 100 Mb/s), the JTAG interface is usually augmented with additional data transfer pins for significant trace operations. These debug extension signals may be associated with the JTAG interface, or they may form a separate (usually parallel) port that operates asynchronously and/or with separate control signals from the JTAG interface (as shown in Figure 16.3).

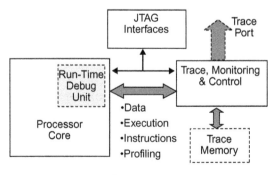

Figure 16.3
Processor OCI example.

Processor debug instrumentation

On-chip instrumentation integrated into processors is usually tailored to the in-silicon debug needs of embedded processor subsystems and system application code. This includes real-time debug and tracing of code execution and interfaces of embedded processors and buses that are unobservable since they are "buried in the silicon". Processor instrumentation is primarily concerned with access to a limited and, for different types of processor, well-defined set of signals. Required debug signals may include instructions and data along with their respective address buses, internal registers (such as program counters and status registers), and execution history.

Most processor cores have some JTAG feature to allow an external controller to implement run control: monitoring and halting or single-stepping the processor and reading a range of register values. JTAG-based processor run-control instrumentation can interface to a variety of other debug blocks for specific processor cores, allowing JTAG interface control of start/stop execution, single-step, breakpoint monitoring, and register/memory access in the processor.

Processor run control as a debug method is, however, limited; instructions typically depend on prior instructions and register information. The halt-and-read approach of JTAG provides an even more limited snapshot of sequential information, which does not allow examination of processor operations in a real timing context. In a multicore environment, halting a processor to extract internal information introduces many issues in its communication with another processor, such as the synchronized halting of multiple processors. To address this limitation, many processors integrate systems debug trigger and trace features into and adjacent to the cores. We will discuss some of these later.

Different processors have different levels of integrated debug support for more complex features such as trace. As examples, Nexus 5001 provides a standard for trace instructions and transfer operations; ARM cores have a proprietary ETM interface that supports run control and trace; and MIPS integrates similar features with their proprietary EJTAG and PDTrace debug blocks.

Trace may focus on different parts of the processor operation; typical trace information includes instructions being executed at stages in the processor pipeline, data that is being manipulated by an instruction, and profiling information being captured for performance analysis. The processor trace operations are usually enabled and disabled by hardware breakpoints (called tracepoints) or by external event signals set to generate trace actions.

As an example, consider MIPS PDtrace and ARM ETM, which are debug blocks provided by the two processor IP companies. While their specifics differ, both include a trace instrumentation block integrated with a processor core that selects, collects, compresses, and transfers real-time trace data, as shown in Figure 16.3. Processor trace instrumentation allows the trace history to be captured in several modes, depending on the available information and bandwidth. It combines trace messages of various lengths into trace words of fixed width suitable for writing into memory. The trace operation may be internal (stored using on-chip trace buffering and transferred over a JTAG interface), or streaming (exported in real time via a trace port to off-chip trace storage). Internal trace requires more on-chip trace memory and logic; streaming uses less memory resource, but requires more pins. The trace collection and hardware breakpoint registers set to generate trace actions operate the same in either case.

Figure 16.4 shows an example of typical sub-blocks for processor trace. More specialized trace processing blocks allow capture of processor bus

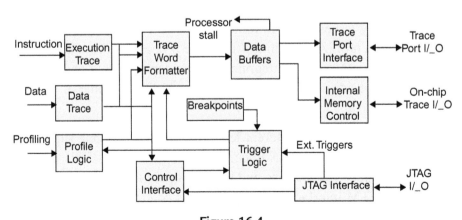

Figure 16.4
Typical processor trace.

trace, execution history, profiling data, and other real-time information from the core as well as managing trace storage.

Because access to I/O may be delayed or multiple clock cycles may be required to output one trace word, trace data can overflow the buffers. To avoid loss of debug data or synchronization, the control logic can either request stalls to the processor so that no trace information is lost or set a flag in the trace buffer indicating that trace information has been missed. As discussed, stalling a processor for debug purposes can be very complex operation for multicore architectures.

Reducing the required trace bandwidth from each core is critical to supporting multicore debug. Many debug systems reduce the information sent over the debug port and have inferred information derived by the development tools. A common method of reducing execution instruction trace is known as branch trace messaging. In this method, only trace instruction sequences due to conditional operation (i.e., a conditional jump or branch) or other discontinuity (i.e., an interrupt) are required; periodic checkpoints ensure that the debug tools are synchronized to processor operations and between concurrent multiprocessor traces.

Bus-level trace

Bus-level analysis and trace take on much greater importance in multicore debug subsystems, since multicore systems require much greater bus complexity for efficient operation. In a single processor system, buses are relatively well-controlled components. Even with more complex multi-master bus architectures, the various controllers that are bus masters are ultimately configured and controlled by a single processing element. This makes both control and operation deterministic based on processor requirements. Having a multicore bus architecture introduces new issues: operation priority, cache ordering, and synchronization of bus transactions.

With respect to bus architecture, there are numerous types; for brevity, we will look at two basic classes of interconnect structure, crossbar exchanges and shared links; each has its own analysis requirements. Crossbar exchanges allow the fastest connectivity between cores, while shared links require fewer gates but allow more dedicated transfer of data between cores that are in different subsystems. Other bus interfaces are optimized to connect slower peripherals or optimize DRAM controller operations and the effective bandwidth utilization of external DRAM.

Bus-level instruments can vary from basic performance monitors to full trace systems that support multi-threading and complex request-response interactions. As an example, a bus transaction monitor (or "transactor", not to be confused with a TLM transactor) is an on-chip instrument used to configure cores, load or read information in peripherals, or gather information to analyze bus timing (latency, throughput, saturation, deadlocks, etc.) without involving the processor or other on-chip system resources. The bus transactor allows the coordinated set of JTAG-controlled registers and triggering resources to implement a simple bus master for controlling and debugging bus traffic operations. An example developed for an OCP bus is shown in Figure 16.5.

A basic bus transactor can initiate and monitor read and write data operations and the synchronization of signaling between data and control. More advanced bus operations such as bursting may require additional transactor logic to support burst synchronization and alignment.

A transactor can help debug a bus by operating as a "debug" master for initiating bus transfers or as a slave reacting to specific bus monitoring operations. Address and data for individual bus transactions can also be

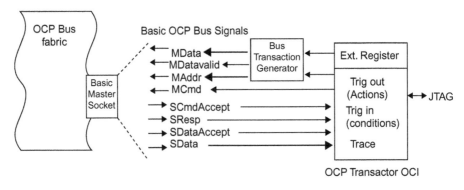

Figure 16.5
An OCP transactor for bus master operations.

written over multiple cycles. Bus data and control signal data traces may be exported during a JTAG data download phase. Additional trigger or state signals may be added for debugging and controlling basic memory maps or domains transfers.

Bus instruments allow several methods of accessing performance and analysis information from the interconnect structure. Debug ports associated with different segments of the interconnect fabric allow access to data flow with minimal impact on the interconnect structure. Bus trace may be captured in several formats:

- Raw bus state: the "raw" acquisition is a state display that contains all the basic trace fields: transaction type (request or response), master name, slave name, transaction ID or outstanding request count, read/write, and trace timestamp. This is the simplest capture method, but it is not very memory-efficient, as it requires a separate trace capture of the request or response of each transaction.
- Aligned bus state: this view concatenates two frames − a request cycle and its matching response cycle − along with a delta timestamp between the current transaction and the start of next transaction. This reduces the amount of trace information per transaction, but requires monitoring of each transaction until completion, which can become complex for multi-threaded or out-of-order transactions.
- Correlated processor and bus state: this is a combined view of processor trace and bus trace data captured with a common timestamp. It also allows processor and bus operation to be locally correlated based on calculated or user-defined offsets to requests and responses in order to account for differences in bus clock speed, latencies, etc.

Cross-triggering interfaces

The information being computed in a multicore SoC is complex and distributed, so that global event cross-triggering and system-level control for multicore debug and triggering are required to identify and isolate events occurring throughout the system. The ability to stop and start all cores synchronously is extremely valuable for multicore systems that have inter-process communication or shared memory. To ensure that this synchronization is within a few cycles, a cross-trigger matrix consisting of event monitoring and triggering blocks can be used to monitor signals from diverse points in the architecture

Cross-trigger interfaces can be implemented between multiple processors, buses, and other on-chip components and interfaces to provide at-speed and low-latency trigger monitoring and control. Sophisticated on-chip trigger systems can include a range of combinatorial, sequential, and timing-based triggering conditions and actions. Cross-trigger configurations can be created using relatively moderate amounts of on-chip logic, with JTAG-controlled setup of the configuration providing a wide range of programmable flexibility (Figure 16.6).

Examples of types of control operations supported by cross-triggering include:

- synchronized configuration and changing of system registers,
- setting of hardware interrupts and other system conditions,
- forcing processors into special modes,
- driving control inputs to IP blocks,
- setting processor mode signals,
- controlling system halts or stalls,
- controlling or synchronizing off-chip systems or instruments,
- and otherwise allowing real-time on-chip control based on system conditions.

Cross-triggering can also involve triggering signals from and controlling actions to off-chip components of the system, including other types of

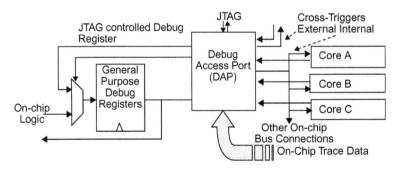

Figure 16.6
Bus analyzer register debug control, trace, cross-triggering.

instrumentation. Since the trigger system is capable of operating in real time, the cross-triggering can be set up to capture and respond to real-time events in the SoC.

The configuration registers of the cross-trigger interface enable the developer to select the required cross-triggering behavior, e.g., which cores are halted on a breakpoint hit by another core. If, on the other hand, the cores have widely separated and non-interfering tasks, it may be sufficient to synchronize core stops and starts with the debug tools. Methods of synchronous starting of cores can be via a cross-triggering mechanism or the TAP controller of each core.

Event recognition and triggering is used to control trace operation and to capture information on events and operations in the SoC. Conditions are monitored and compared to generate real-time triggers in a cross-trigger manager. These triggers can in turn be used to control event actions such as configuration, breakpoints and trace collection. More complex implementations can be programmed to trigger on specific values or sequences such as address regions and data read- or write-cycle types.

The cross-trigger block is a distributed subsystem that connects to some or all cores in a subsystem, as illustrated in Figure 16.7. As an example, if the cross-trigger wiring is in the bus fabric, then some pre-processing or wrappers that have trigger logic (condition/action nodes) at each core interface can be used to simplify the cross-trigger information. Wrappers can be programmed via JTAG debugger (or native to a processor) for various system-level trigger conditions. Generally, triggers would be configured so

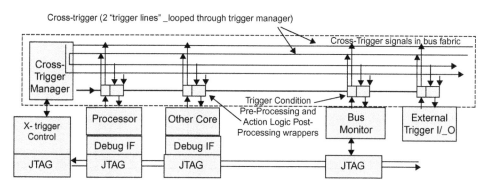

Figure 16.7
A distributed cross-trigger.

that any block can send a trigger and receive a trigger. The debugger or processor architecture may allocate specific edge- or level-sensitive trigger lines on different cores for condition and trigger signals.

As an example, each trigger line can consist of trigger condition and enable signals. To implement a cross-trigger, there would need to be at least two independent trigger lines. The trigger in, out, and enable line values may be a logic combination of several signals for each core. Trigger lines connected to package pins can enable cross-triggering between several chips.

JTAG interfaces for multicore debug

Since the release of the 1149.1 specification in the 1990s, JTAG has been used extensively in processor debug systems and remains the most widely used debug interface. JTAG's test signals and its serial bus operation can double as default interfaces for most basic embedded core debug functions. Assuming a multicore system with debug instrumentation and JTAG interfaces defined for all cores, a JTAG debug interface can be configured in some combination of serial and parallel operations as seen in Figure 16.8. The trade-offs between serial and parallel JTAG implementations are:

1. Have multiple cores on a single JTAG chain, which reduces the number of interfaces but limits JTAG information access to one core at a time (this is shown in cores 1–3 in Figure 16.8), or

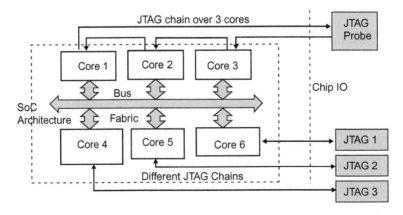

Figure 16.8
Debug alternatives over JTAG.

2. Have separate JTAG interfaces for each core, in which case each core may be debugged simultaneously but independently, with the cost of an additional JTAG port for each core (as shown with cores 4–6 in Figure 16.8).

Neither solution has proved optimal in practice. JTAG's limited bandwidth was never designed to support real-time debug. Separate JTAG TAPs are used in some cases for limited core debug, but they are too pin-intensive for more than two or three cores. Alternative approaches to support multicore architectures have focused on different encapsulating architectures for JTAG, including packetized debug messages, parallelizing of JTAG operations, and hierarchical partitioning of the JTAG flow to provide sufficient bandwidth, flexibility, and integration. Of these, packet-based debug messages are the most mature and widely used in cases where cores can be stalled to match processor operation to JTAG bandwidth.

External interfaces for on-chip instrumentation

One of the major factors limiting the use of instrumentation in SoC and multicore architectures is the ability to quickly export data as it is generated. On-chip instrumentation addresses many of the operations associated with large amounts of on chip debug by supplementing and buffering JTAG in many systems. There is, however, often a need for viewing things like the instruction or data trace from a processor, which means data must be exported off-chip. Trace and probe ports provide additional I/O bandwidth needed for on-chip trace, which is a data-intensive means of providing on-chip visibility. The ability to transmit debug signals, most notably trace, is a hard limited function of two parameters:

1. The number of I/O pins that can be dedicated to export of debug information at any given time, and
2. The speed at which these signals can transmit the data off chip.

These bandwidth problems of exporting debug data are compounded in multicore architectures, since each core has internal address, data, and control signals as well as inter-core monitoring and peripheral bus signals. One approach is to call on on-chip memory to buffer between traced data and the available export bandwidth. Buffers of modest size, however, are easily overloaded given the amount of trace data that can be generated from multiple IP blocks or internal buses.

Even with these additional ports, the amount of debug information required when tracing from multiple cores can easily exceed the allocated debug interface bandwidth of an SoC. To reduce the information sent over the interface, approaches such as data compression and filtering can be used to increase debug interface performance.

Debug flows and subsystems

A common criterion for embedded-system debugging is the ability to debug through reset and power cycles, requiring careful design of power domains and reset signals. Critically, reset of the debug control register should be separated from that of the functional (non-debug) system. Power-down can be handled in different ways when debugging, such as disabling power-down signals or putting the debug logic in different power domains that aren't powered down.

Multicore SoCs that place cores in multiple clock and power domains for energy management should replace a traditional JTAG chain with a system that can maintain debug communications between the debug tool and the target despite any individual core being powered down or in sleep mode.

The debug infrastructure should address at least 4 major concerns:

1. It should reduce the chip I/O requirements to enable concurrent debug of each of the interacting blocks of a design.
2. It should provide synchronization and interactive debug communication for each part of the design to enable recognition of system events and the triggering of debug actions in any other part of the architecture.
3. It should be scalable, within reasonable limits, to allow for minimal increases in debug internal wiring and I/O as the size and architecture complexity increase.

4. It should be extensible to address diverse multicore debug subsystems that operate on different core types, levels of hierarchy, and debug features with a similar set of debug resources.

Simply adding independent debug blocks for debug support for each of the major components, as seen in Figure 16.9, does not result in a desirable solution from a wiring, I/O, or integration point of view. Fitting multiple debug ports, one for each core, has obvious silicon and pin overhead. It also leaves the synchronization and power-down issue to be managed by the tools. This approach has merit in completely different cores with completely different tool chains, where the re-engineering costs of sharing a single debug port with a single JTAG emulator box are substantially higher than the costs of duplicating debug ports and debug tool seats.

While having multiple JTAG chains on a chip is a small cost for the ability to simultaneously debug multiple cores, having separate JTAG I/O for each chain is often unacceptable in a world of pin-limited chips. Ideally the data from each JTAG interface should be concentrated in order to be accessible from a smaller number of external pins. This concern is addressed by a new generation of lower-I/O JTAG-based interfaces, with 1149.7 as the standard in this area.

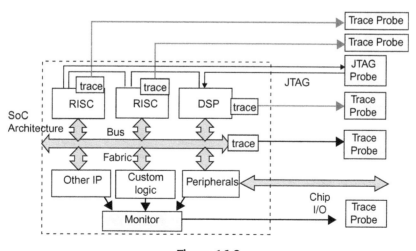

Figure 16.9
Adding instruments for multicore debug.

For reducing the I/O requirement, an on-chip debug collection, concentration, and concatenation block is used to combine the larger number of debug interfaces into a single debug port (Figure 16.10). As a general rule, each on-chip JTAG chain communicates with the JTAG debug interface for one core. So the number of cores being concurrently debugged is proportional to the number of JTAG chains on chip. While some processor debug interfaces allow access to multiple identical cores on one chain (ARM MultiCE as an example), these solutions are not general.

Multicore debug information collection requires some aggregation blocks to support simultaneous acquisition of data for debug of different cores. This may be done by several means:

1. Real-time multiplexing between core instruments; this increases the effective bandwidth for a relatively slow JTAG interface and allows a separation of different debug streams.
2. Using packets or other encapsulation of instrumentation data from different cores; packetizing the instrumentation data along with its control and ID fields allows simpler back-to-back transactions from different cores.
3. Combining debug data from different cores into a common JTAG output; this may include download of an entire buffer of data bits in a single JTAG transfer. Since a significant amount of the time JTAG is in non-data-transfer states for several cycles each transaction,

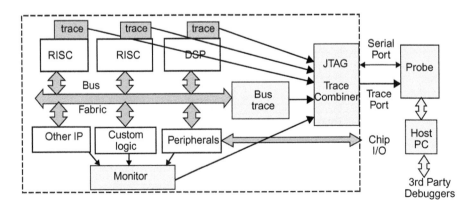

Figure 16.10
Instrumentation funneling.

being able to stream large amounts of data in a single transaction operation allows for significant increases in transfer efficiency.

The second of these solutions requires a distributed monitoring architecture. The instrumentation blocks from different cores may be integrated by providing a layer of synchronizing encapsulation and system-level monitor and trigger integration to combine core-specific instrumentation blocks into system-level debug operations (as shown in Figure 16.11).

As previously mentioned, there are limitations to the number of cores that can realistically be debugged concurrently. In additional to the amount of information involved that would need to be transferred off chip, there is the more fundamental question of how many cores are involved in a typical multicore debug scenario. While there are circumstances where many cores may need to be monitored and used for complex triggers, the need for trace information from large numbers of cores is very limited in most heterogeneous multicore architectures. Since the primary multicore-level information that needs to be debugged often relates to communication between cores, the numbers of cores that would need to be traced is generally limited to how many cores are interacting with each other at any given point in the chip operations. For most RISC architectures, this is limited by their interface architecture to two or three concurrent.

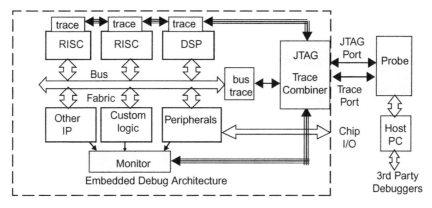

Figure 16.11
Serial integration of on chip instruments.

Commercial approaches

In this section, we look at some standards and commercially developed debug systems.

The OCP debug interface

OCP-IP (Open Core Protocol − International Partnership) is a consortium that develops IP-level multicore solutions based on standard sockets that plug into an OCP-based bus fabric. The OCP-IP Debug Specification [1] defines a synchronous debug socket interface that provides "multicore debugging-aware" functionality for four basic groups of signals: debug control, JTAG, debugger interface, cross-triggering interface. It also provides for optional additional groups that address application-specific debug and analysis requirements such as timestamps, performance analysis, multicore security, and power management domains.

The synchronous debug interface, designated as a Debug Interface Port, can be added to cores and other IP for debug access. The Debug Interface Port contains debug control and data signals that may be part of a core-level data socket or an independent OCP port configuration. Figure 16.12 shows a simple system where debug signals are integrated into a modular OCP system interconnect. In this example, OCP debug port operation is separate from the data port, which allows for more independent control and dedicated registers. Debug wiring goes through

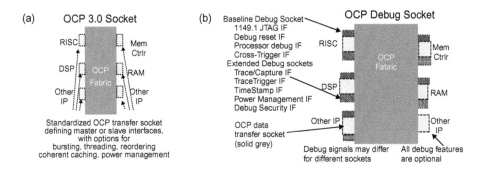

Figure 16.12
(a) OCP sockets for various cores. (b) OCP debug interface sockets.

the OCP bus and is accessed through a debug I/O port. This may be JTAG or a dedicated trace interface.

An advantage of the OCP debug architecture is that, since the debug blocks are connected to the bus fabric, configuration and transfer of information from the debug blocks may be JTAG-mapped (i.e., communicating via the OCP-defined JTAG wires) or memory-mapped (transferred over available data bus channels).

In the memory-mapped case, the master port of the main debug core provides the debug instructions and configuration based on an address in the main memory space. This allows one core to be the main debug agent, receiving instructions over a JTAG slave interface. The core then accomplishes all debug-related actions, including configuration of debug and trace conditions and accessing debug data from other cores in the system through the main system bus. The core may need access to processor privileged states for debug. The OCP interface defines "Abort" and "Force" type signals for this purpose as part of the debug control interface.

The more general, system-independent, JTAG-mapped architecture uses the OCP Debug Interface Ports to communicate with all debug blocks. Instructions are sent over JTAG interfaces to the core-level Debug Interface Port using debug registers that are part of a JTAG TAP controller to configure and control debug operations (including "Abort" and "Force") as well as to configure trace compression and time stamps for trace of each core.

For simplicity in OCP debug examples such as Figure 16.13, debug ports by default are connected to JTAG interfaces, which are included in OCP specification. The implementation of a OCP Debug Interface Port is not limited to JTAG interfaces; a port can employ other debug interfaces as well. The OCP-IP Debug Interface specification includes discussion of integration with other debug and trace instrumentation, including Nexus 5001 and Infineon MCDS (Multicore Debug System).

Nexus/IEEE 5001

The Nexus 5001 activity was initiated to define a specification to address a standardized interface for on-silicon instrumentation and debug tools,

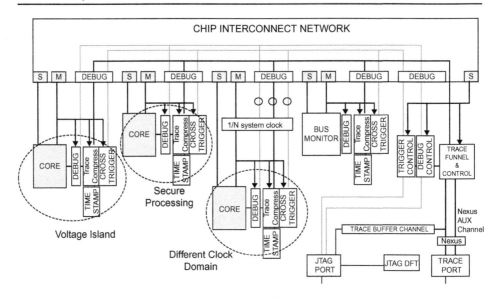

Figure 16.13
Multicore synchronous debug implementation.

providing expanded features and higher performance required for high-performance architectures (Figure 16.14). The Nexus 5001 infrastructure supports multicore development and multi-featured trace and configuration/control. Nexus, at its simplest (Class 1) level, is compatible with JTAG, but recognizes, as a starting point, that JTAG bandwidth limitations are not realistic for the debug requirements of complex or multicore environments.

Nexus 5001 is supported as an IEEE_ISTO-5001 specification [2]. Versions of Nexus architectures have been used extensively in US automotive applications as well as several other processor application areas. The Nexus architecture defines a high-performance data interface, protocol, and register that can be used to implement a variety of trace and control instrumentation options.

Nexus 5001 communication uses packet-based messages, which transfer control and data over standard and user-defined instructions. Standard instructions include industry-proven and best-in-class features for instruction and data trace, data transfer, monitoring, breakpoints and run control for system debug, as well as the importing of data for calibration and port replacement operations.

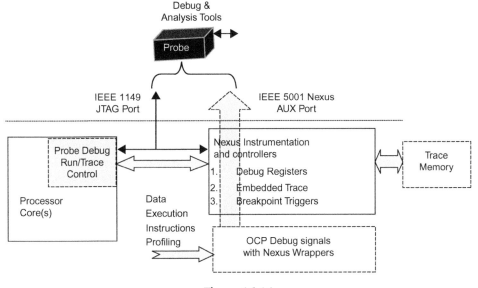

Figure 16.14
Nexus interfaces.

Since debug-related concerns and trade-offs vary, ranging from the increased logic budget required for effective system debug to adequate bandwidth and resources for debugging complex architectures and systems, the IEEE 5001 standard groups these features in four classes of increasing complexity, as summarized in Table 16.1. These classes are guidelines, and not all system debug features will, or need to, fit into a given class. A comprehensive definition of the various services and features, along with other topics in this section, is discussed in the IEEE 5001 Nexus specification.

The Nexus specification defines a vender-neutral I/O signal interface and communications protocol that supports high-bandwidth parallel debug and instrumentation support for data-intensive operations such as trace. The Nexus interface defines a small set of state machine control signals for auxiliary parallel (AUX) data ports that may be used in conjunction with JTAG or as a self-contained port. The additional data pins provided by the AUX interfaces are scalable to allow higher read/write throughput between a multicore system and debug and analysis tools.

The AUX interfaces are uni-directional (either Data In or Data Out) and may be asynchronous to each other, with each AUX port having its own

Table 16.1: NEXUS 5001 Implementation Classes

Nexus class	Services	Features
Class 1	Static debugging	Single step
Basic run control	Breakpoints	Set two (min.) breakpoints/ watchpoints
		Device identification
		Static memory and I/O access
Class 2	Watchpoints	All Class 1 features
Instruction trace watchpoints	Ownership trace	Monitor process ownership in real time
	Program trace	Real-time program tracing
Class 3	Data trace	All Class 2 features
Data trace read/ write access	Real-time read/write	Access memory and I/O in real time
	Transfers	Real-time data tracing
Class 4	Memory substitution	All Class 3 features
Memory and port substitution	Port replacement	Start traces on watchpoint occurrence program execution from Nexus port

Source: Nexus 5001 specification.

clock. The Data Out pins of an AUX interface are typically used for trace, and the Data In mode is typically used for configuration or calibration of an IC. AUX Data In and Out ports may be operated concurrently. Nexus also specifies how a JTAG interface can be used in conjunction with the AUX ports. JTAG interface operations in Nexus may be used both for configuration and control of the on-silicon instrumentation and for embedding the Nexus protocol and data into a JTAG message. Both AUX and JTAG interfaces are controlled by FSM-based controllers allowing a variety of transfer operations.

Figure 16.15 shows a Nexus multicore debug example that supports the concurrent debugging of both processor and bus operations. While each processor or logic/bus element in a design may have a native debug environment, debug information can be reformatted using Nexus interface wrappers that package debug information into Nexus messages. Nexus messages can be merged at a Nexus port control level to allow packets from many debug sources to be communicated over a common Nexus port. This may be as simple as a translation layer to standardize

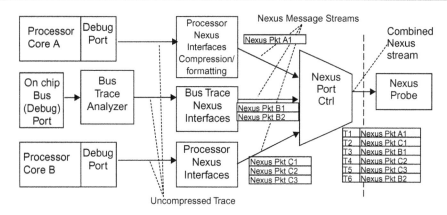

Figure 16.15
Basic Nexus multicore debug flow.

trace compression formats and convert information in debug registers into Nexus-compliant messages.

A Nexus Port interface or multiplexer manages the multiple Nexus streams by selecting messages and funneling them through a single combined Nexus stream at the port interface. Since each debug block can be assigned an independent identification (DID) value, debug information can be redirected once off-chip at the probe interface or as a software operation.

One multicore system debug issue is that, when debug information from different blocks combines into a single Nexus stream, the trigger control and synchronization of different cores or subsystems is handled by different logic and different trace streams that may start and stop independently of each other. Coordinating control and synchronization of different debug resources is desirable for the chip to present the debugger with a more comprehensive view of transactions under observation. Synchronizing the trace interfaces may be done using on-chip debug resources, such as cross-triggering between the Nexus blocks or a Port Control instrument (Figure 16.16), for synchronizing the starting and stopping of execution, and by system timestamp resources, for cross-referencing the timing of concurrent debug operations in different parts of the architecture.

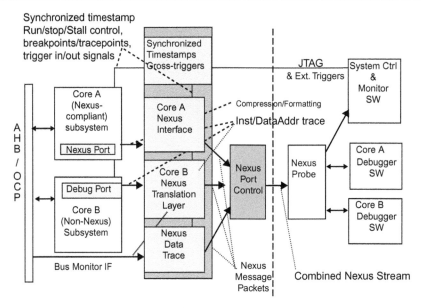

Figure 16.16
A Nexus-compliant OCP multicore embedded debug environment.

In the 2012 Nexus 5001 specification release, IEEE 1149.7 and the Xilinx-developed Aurora (a SERDES standard) interfaces are supported, providing significant debug instrumentation advantages. IEEE 1149.7 is an extension of the 1149.1 JTAG standard that allows for both 2-wire JTAG interfaces as well as parallel chip-level data interfaces (as opposed to 1149.1, which only supports daisy-chained serial data interfaces between chips). IEEE 1149.7 also adds advanced test and debug features such as low-power modes and Custom Data Transport and Background Data Transport modes that improve JTAG data transfer flexibility and throughput.

Aurora is a scalable low-latency protocol that was originally developed for logic-constrained FPGA implementations. It provides a transparent and flexible framing interface for high-speed serial links. The instrumentation interface, layered over the Aurora physical layer, allows multiple channels to move trace and other bandwidth-intensive data across point-to-point serial links.

These standard interfaces allow for the introduction of Nexus 5001 instrumentation into pin- and wire-limited systems such as mobile devices as well as complex networking and computing systems that

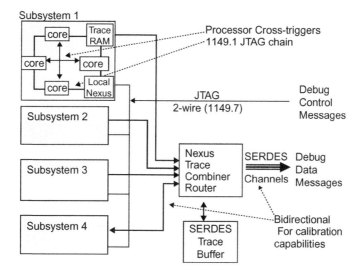

Figure 16.17
Nexus system implementation with 1149.7 and SERDES interfaces.

require large amounts of debug data to provide a useful view of system operations. An example using 1149.7 and Aurora Gigabit SERDES in a system-level configuration is shown in Figure 16.17.

Additional instrumentation interfaces may also be custom-implemented based on Nexus commands. Nexus 5001 supports memory accesses to both on- and off-chip memory elements, transferring information to local memory for transfer and capture using other peripheral interfaces in a system. It also allows the import of specific data from memory for use in debug configurations. As an example, blocks of debug-specific instructions may be imported via Nexus 5001 messages and applied to the processor instruction cache to substitute for normal instruction blocks. This allows the export of trace and other debug operations under processor control.

ARM CoreSight

CoreSight is an ARM debug architecture for providing on-chip real-time debug visibility to multiple cores. It supports real-time processor and bus trace, multicore debug, and cross-trigger features. While it is extensible to non-ARM cores, it is designed to operate on ARM Debug Interface (ADI) architectures, in particular ARM Embedded Trace Macrocell and Program Flow Trace.

CoreSight is a modular set of debug IP blocks provided by ARM communicating through a Debug Access Port (DAP) — a gateway (typically a JTAG TAP) for transmitting debug control interface information between external debugger and on chip subsystems — and various blocks and instruments. The DAP is a modular interface that supports system-wide debug. Internally, it is extendable to support access to multiple systems, including AHB system interfaces and CoreSight logic. The DAP may also include optional external interfaces such as the Serial-Wire Debug Port, which supports bi-directional two-wire operation via an ARM-developed serial debug interface.

The Advanced Trace Bus (ATB) is used by trace devices to allow concurrent trace of multiple cores and to share CoreSight capture resources through a single debug interface. The ATB dispatches trace data (via the Replicator and Trace Funnel blocks) to the ETM and/or trace port interface unit (TPIU). Other parts of the CoreSight environment, as shown in Figure 16.18, are

- ETM, the core-level trace module that connects to the processors to capture, compress, and transfer trace data to the ATB
- AMBA AHB Trace Macrocell (HTM) for bus transfer trace support
- Instrumentation Trace Macrocell (ITM) for configuring non-ETM trace information
- TPIU, which collects trace data to communicate with an external debugger ETB (embedded trace buffer) for storing trace information on chip for subsequent export
- CTM (cross-trigger matrix) and CTI (cross-trigger interface) to control cross-triggering and to transmit trigger conditions, events, and actions from one core to another.

The DAP bridges between the external debug interfaces and the debug interfaces of the cores as shown in Figure 16.19. It maintains debug communications with any core at the highest frequency supported, rather than the slowest frequency of all cores on a JTAG daisy chain.

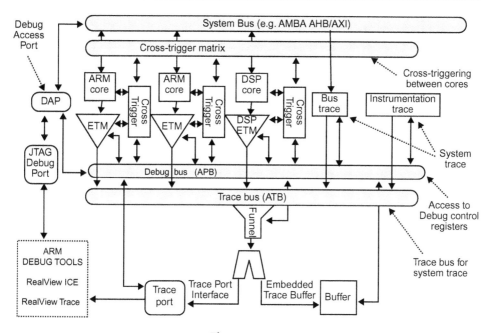

Figure 16.18
The ARM CoreSight architecture.

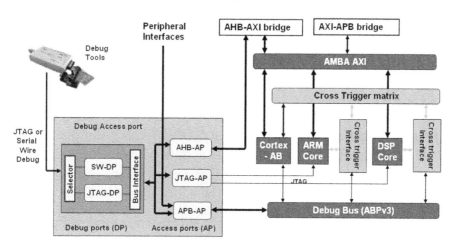

Figure 16.19
CoreSight multicore debug connectivity.

A CoreSight trace funnel combines multiple asynchronous, heterogeneous trace streams into one stream for output via a single trace port or trace buffer (Figure 16.20). This reduces the implementation overhead while allowing better utilization of deeper

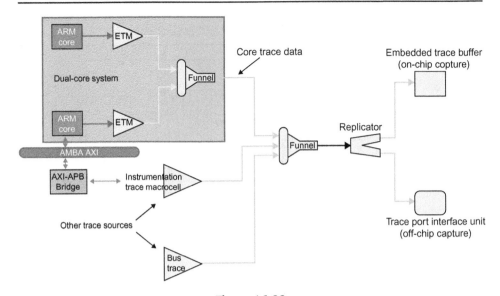

Figure 16.20
CoreSight multiple trace funnel example.

buffers or higher-bandwidth ports needed for simultaneous trace of multiple cores.

Example: MIPS PDTrace and RRT analysis

The following is an example MIPS multicore debug environment that supports a system of multiple bus masters (MIPS cores, third-party cores, and DMA) with different data rates and system core clock rates and slave peripherals (including off-chip DDR memory and on-chip SRAM) over several buses of varying widths The subsystems consist of:

1. Two MIPS 34K cores operating in multi-threaded modes. Each has instrumentation for capturing 34K core operation information. The 34K typically operates at twice the system clock frequency.
2. An imaging subsystem, made up of an array of vector cores connected via a complex crossbar exchange.
3. A 16-channel DMA capable of burst modes.

The goal was to tune and debug a system of interacting cores in a vision analysis and response system. The architecture and debug environment is

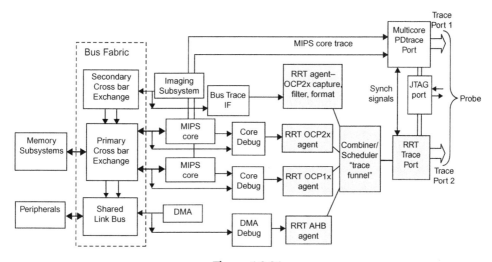

Figure 16.21
MIPS-based multiple core trace example.

shown in Figure 16.21. The system analysis environment consists of two major subsystems:

- Processor Debug: a PDtrace block for each MIPS core. The PDtrace interfaces to an aggregated trace port handling the trace outputs of both processor cores when running at a reduced speed to allow the trace port to handle the combined trace traffic. The PDtrace interface includes a common JTAG connection and PDtrace trigger pins for trigger and trigger acknowledge.
- Bus Debug: a Request-Response Trace (RRT) block, which is configurable to trace one single bus, monitor all bus masters simultaneously, or provide selective trace information from several cores. The RRT requires buffering on each request-response output and a trace funnel to route the buffered outputs to the off-chip trace port. Like the PDtrace port, the system clock speed can be reduced so that the RRT port can transfer a larger amount of trace information.

Post-trace software provides post-processing and views of transactions and delay times over varying periods of time for both single and multiple cores. RRT data is correlated to and operates in conjunction with PDtrace data to provide a picture of system operation. Post-trace software can display per-thread bus transaction information, providing valuable insight into the density of transactions over time and the delays associated with those memory accesses, generated for each hardware thread.

The RRT and PDtrace data are transmitted off-chip over a dedicated 16-channel trace port, with each interface having its own independent clock source. A common EJTAG interface configures and controls the RRT and PDtrace operations. An external probe with a very large cache combines the trace inputs from the two sources and records them in a common memory.

Trace collection allows the data produced and transferred between cores by the application (for example, a video frame) to be captured over an extended period using the memory buffer in the probe. Using probe memory in general is a good cost vs. efficiency trade-off to buffering the data on chip. Concatenation of multiple frames for more extensive analysis is a post-processing step done by software on the RRT trace files.

RRT implementation

The on-chip component of RRT as shown in Figure 16.21 consists of three primary On-Chip Instrumentation (OCI) blocks:

- RRT agents, specific to the processor or core interface, for capturing and buffering relevant trace information based on system operations and trace configuration.
- The RRT trace funnel, which provides the aggregation of trace information from all RRT agents and combines and schedules the trace information for export, and
- The RRT Trace Port, which handles communications with the off-chip probe.

Setup or programming of each block is performed by a common JTAG port. The RRT provides for on-chip capture and collection of basic bus operation metrics by the RRT agents; these are then periodically exported via the RRT port. Some of the metrics captured are:

- Master-slave bus transactions and the number of clocks of delay between each request and response.

- The timing and latency of read cycles. Burst reads are reported on arrival of the first requested word or on the last burst arrival.
- Bus transactions between one master and all the slaves it transacts with, or several masters at the same time. These masters may include the MIPS processors, a video-processing channel selected from the output of the crossbar, and active channels of the DMA.

This allows the RRT to capture the following during normal chip-level operations:

- Measurements of a processing loop and frame time.
- Information for aligning bus trace with core processor execution trace to correlate cause and effect of code execution to bus traffic.
- Information on aligning bus measurements to correlate hardware threads to the data transfers that each processor generates.
- Thread information from bus address bits of each processor.
- Timestamp values stored along with trace, which use bus protocol and ID bits for post-processing to match up requests and responses and calculate the delays between them.

Bus RRT operations are broken down into two categories — Fast and Full. Fast is limited to a single cycle frame and includes bus-level control signals characterizing the bus transfer along with buffer overflow and/or trigger indicators. Full mode includes control signals as well as full address trace, based on a memory map of necessary upper addresses, and is typically transmitted over multiple trace clock cycles. Bus records captured by the RRT include:

- The master ID (if multiple masters are being recorded at one time).
- Addresses or partial address fields to determine code vs. data memory-mapped regions.
- Slave ID based on address bits that identify slaves and other memory-mapped areas.
- Protocol bits that determine the alignment of a read-response cycles to request cycles.
- Request and/or response cycle type — read/write, single/burst.
- Buffer overflow/underflow indicators to monitor whether the RRT has lost synchronization.
- Trigger signals between external probes and the on-chip subsystems.

For flexibility and reduced on-chip impact, the RRT triggering system is largely managed within the off-chip probe and includes event monitoring of captured control and address signals that start and stop the capture of trace information in the probe. The probe and on-chip logic have a common triggering communication to allow the probe to enable and disable or stall RRT operations in conjunction with PDtrace operations. The triggers may put one or more cores in debug mode and also communicate with the processor and PDtrace subsystems. On-chip trigger output pins indicate the probe status of the processor cores. The triggering scheme also communicates stalling of the trace capture based on processor status.

Post-trace analysis tools for RRT and PDtrace consist of a set of control and display views and utilities to support basic analysis of RRT and PDTrace data, including reading trace data and formatting it into columns of readable data. Data can also be exported to third-party tools for more complex visualization. Control setup includes the setting of trace priorities, selecting which masters are included in the trace run, and trigger configuration based on different address values, common on-chip triggers, markers, or instruction types (read/write/burst) that are captured in both the PDtrace and RRT. The purpose is to correlate bus traffic with instructions and determine which thread caused the read or write bus cycle.

The future of multicore debug

Many of the current approaches to multicore debugging are rooted in the extension and integration of single-core debug approaches. Relying on instruments and data from single-core debug and trace subsystems can fall short when used to address the complexity of interactions in multiple cores and processors interfacing with equally complex application-specific IP.

Multicore debug introduces a new layer of challenges in the debug process. SoCs integrate multiple types of cores, such as DSP processors or other complex application-specific IP, for a myriad of functions that may be running asynchronously or with variable or indirect communications with each other. Complicating multicore issues further is the fact that IP and cores coming from a variety of vendors have different compile and debug environments or levels of debug features that may be complex to integrate into a consistent debug data format.

So combining these subsystems into a cohesive debug view can be like building a Tower of Babel. Tasks such as processor interfacing, inter-processor communications, run-time execution and coordination, and data presentation place significant overhead on the debug requirements for heterogeneous and multicore chips. Instrumentation blocks must be customized and extended to support the specific verification and debug requirements of processors both on a standalone basis and in a multi-processing configuration. System-level instruments are needed to correlate debug information timing between cores. Instrumentation blocks must be diverse enough to effectively communicate debug data with their respective cores and interface to each other in order to coordinate all of their activities.

Designing for debugging is the art of planning for what might happen. As the number of processor cores continues to increase, the challenges of multicore debugging will also increase, so there is a lot of work left to do in this area. Perhaps the greatest near-term problems are in providing visibility into the internal operation of multiple processors at full clock rates and efficient methods that not only store and transfer the resulting trace, but synchronize and coordinate this information for different processors. This requires better methods to allow debug and trace of synchronous-synchronous signal boundary crossings and of multiple on-chip clock domains, which are required for many types of multicore architecture.

In addition to processor-level debug, instrumentation for on-chip buses and other multicore infrastructure provides visibility and debug in multi-processor designs. For systems that have multiple processors communicating over a standard bus, access to common information, such as which processor owns which parts of an on-chip bus, is a necessary context for understanding and debugging communication and multi-processor execution. As the problems are compounded with increasingly complex bus architectures such as on-chip networks, multiprocessor debug will rely on more extensive tracing and triggering of bus operations to address inter-processor communication issues in conjunction with more specific point solutions for processor-specific analysis.

Looking ahead to more complex systems, instrumentation will need to have sufficient embedded debug intelligence to interpret information passing between cores, figure out whether there may be problems, determine what information is needed to be extracted for debug, and perform other task-aware on-chip debug or network protocol analysis. Equally challenging is presenting and visualizing all the diverse debug information in a coherent, understandable way. As in many areas of multicore SoC design, extending the way things are done for single-core architectures is only a first step, and new thinking and classes of solutions are needed to address the diverse debug and analysis requirements of these diverse and complex architectures.

References

[1]   OCP-IP Debug Specification: <www.ocpip.org/socket/ocpspec>.
[2]   Nexus 5001 (IEEE_ISTO-5001) Specification: <www.nexus5001.org>.

Further reading

On Chip Instrumentation: Design and Debug for Systems on Chip. Springer, 2011. ISBN 978-1-4419-7562-1.

Multi-core analysis made easy with Nexus 5001 debug specification. Embedded Systems Design, March 2008.

Bus request-response trace for SMART interconnect systems. GSPx Conference Proceedings, Fall 2006.

Nexus based multi-core debug. DesignCon 2006 Proceedings, Jan 2006.

Embedded multi-core debug techniques. ARM Developer's Conference, Oct. 2004.

Multi-core embedded debug for structured ASIC systems. DesignCon 2004 Proceedings, Feb. 2004.

Index

Note: Page numbers followed by "*f*" and "*t*" refer to figures and tables, respectively.

Printed and bound by CPI Group (UK) Ltd, Croydon, CR0 4YY

03/10/2024

01040329-0003